PRENTICE-HALL SERIES IN ELECTRONIC TECHNOLOGY

Dr. Irving L. Kosow, editor
Charles M. Thomson, Joseph J. Gershon, and Joseph A. Labok, consulting editors

PRENTICE-HALL, INTERNATIONAL, INC., *London*
PRENTICE-HALL OF AUSTRALIA, PTY., LTD., *Sydney*
PRENTICE-HALL OF CANADA, LTD., *Toronto*
PRENTICE-HALL OF INDIA PRIVATE, LIMITED, *New Delhi*
PRENTICE-HALL OF JAPAN, INC., *Tokyo*

To My Wife, Elsa

ELECTRICAL WIRING

Design and Construction

ROBERT C. JOHNSON

Wentworth Institute
Boston, Massachusetts

Prentice-Hall Inc., Englewood Cliffs, New Jersey

The publisher would like to thank Dean Charles M. Thomson of Wentworth Institute, who reviewed the manuscript and proofs and lent his technical assistance after the death of the author.

Current printing (last digit):

10 9 8 7

13-247635-5

Library of Congress Catalog Card Number: 78-147807
Printed in the United States of America

Preface

This text has several objectives. The first five chapters present the fundamentals of electrical theory necessary to those beginning careers as electrical technicians or electricians. The remainder of the text concentrates on the theory and practical applications of design problems associated with wiring systems in buildings.

Some readers may wish to begin with Chapter 5. Here three-phase power supplies, delta- and wye-connected loads, voltage current, and power measurements are studied in detail.

Chapters 6 and 7 present an operational study of motors and transformers. Theory of operation is emphasized rather than design, with particular attention to voltage, current, power, and speed variations with respect to mechanical or electrical loading.

Chapter 8 consists of information and definitions that are intended to provide the road map for the last six chapters. Chapter 9 introduces the details of electrical wiring systems in buildings.

Chapter 10 on grounding is a part of the entire sequence and has the further purpose of assisting those in the electrical industry who do not fully understand this important requirement. It also enables qualified technicians to study in detail the design procedures of lighting and space heating, as exemplified in Chapters 12 and 13.

Chapter 14 will help architects and others to understand the problems associated with the distribution of electrical power in buildings.

The National Electrical Code is referred to in a general manner to assist the reader in design problems. A further and more detailed study of the NEC is recommended.

The mathematics of the entire book is limited to elementary algebra and right angle trigonometry. The problems at the end of each chapter relate closely to their particular and preceeding chapters. They range from quite simple to difficult. Many problems become more difficult with each succeeding part. Thus part (d) of a problem is usually more of a challenge than part (a).

Since this book makes limited references to electrical hardware and installation problems, a concurrent course in electrical shop practice would supplement the text material.

The author is deeply indebted to many individuals who assisted in making this text a reality: to Mr. Edwin Francis and Mr. Matthew Fox of Prentice-Hall for their interest and much needed motivation; to Dr. Irving Kosow for his careful review and most welcome suggestions; to my wife, Elsa, who typed faster than I could write and patiently kept up with the changing changes.

I wish also to pay tribute to two other individuals who, although remote from the scene, made a very important contribution: to Mr. William P. Lawlor, formerly of Medford, Massachusetts, now of Inglewood, California, my first employer, who taught me that "a job worth doing is worth doing well"; to Mr. L. C. King, my department head at Wentworth Institute for seven years, who by his own example taught me to live and work by good ethical principles.

ROBERT C. JOHNSON

Boston, Massachusetts

Contents

ELECTRICAL WIRING

Design and Construction

Chapter One

The Electric Current

1-1 ***Introduction.*** In all of history no discovery has had such an effect on human existence as electricity. We have only to look around us and pause for a few moments to contemplate how dependent we are on its versatility.

Our homes are lighted by electric lamps of various types. Electricity plays an important part in heating our homes. Household appliances by the dozen, from electric ranges to toothbrushes, make our daily lives less burdensome. Our automobiles and trucks though propelled by gasoline, must have their fuel ignited by an electric spark.

The field of communication, including telephone, radio, and telemetry from outer space, is only possible through the use of electricity. The industrial might of the world is powered by electric motors of all possible sizes and shapes, from a doorbell to a motor of thousands of horsepower that drives our largest machines. Even our clocks keep exact time because of the particular design of a small electric motor.

The electric utility industry in the United States is growing by leaps and bounds as the demand for electric power doubles every 10 years. It is of great concern to the people who are responsible that when we turn on a light the electricity will be available without prior notice. Also, the response to this demand must be instantaneous.

To the layman or the casual observer, electricity has an air of mystery or magic. How can this unseen "servant" perform so may different tasks, do them so well, and require so little attention? The fundamental principles of electricity are extremely simple in nature and few in number. We can simplify the concepts of this great servant by first describing it as a means of converting energy from one form to another. We must be able to visualize that the heat produced by coal or oil, burned at an electrical generating station, is the *same* heat that appears when we switch on an electric stove or is the heat that causes the tungsten filament of an incandescent lamp to glow and give off light. Likewise, the mechanical rotation created by the steam turbine powered by combustion of oil or coal, will be re-created in electric motors many miles from the power plant. (It's the *electricity*, which re-creates the mechanical rotation.)

Through the following pages of this book we shall trace very carefully this "flow" of electricity and analyze the equipment required so we can design the electrical installation from the smallest to the largest buildings. However, we shall limit the scope of this book to the building. We shall *not* investigate the power company's problems of generating, distributing, and delivering the electric power, since this is a complete study in itself. One point must be made absolutely clear, however: The fundamentals covered in this book can be applied to any electric system. Electricity behaves the same whether it travels 3 ft from the battery of your car to the starter motor or 300 mi from the Hoover Dam generators near Las Vegas to Los Angeles. Obviously, the equipment used to deliver the electricity in these two cases have different design problems.

1-2 ***The electric current.*** "Electricity" is the only word that has been used so far to describe the phenomenon of energy conversion. *Webster* defines electricity as "one of the fundamental quantities in nature," which helps us to understand that electricity was not invented but *discovered*. *Webster* continues, "electricity is characterized especially by the fact that it gives rise to a field of force possessing potential energy and that, when moving in a stream (an electric current), it gives rise to a magnetic field of force with which kinetic energy is associated."

It is with apology that the first time we see the word "current" in this book it is somewhat obscured by parentheses.

The unseen current is the medium by which the more general term "electricity," carries out its mission. An electric current flows in a material if its atoms have *free* charged particles (positive ions and negative ions) or electrons that actually can move from one atom to another. All metals have this characteristic. Some (such as silver, copper, or aluminum) have a large number of free electrons (negatively charged particles) in their atomic structure and, therefore, are *good* conductors of electricity. In the presence of an electric field these free electrons stream from atom to atom through a conductor. We shall call this part of the electrical phenomenon *current flow*.

Other metals such as iron, lead, and zinc have *some* free electrons but fewer in number and are *not* considered good conductors. Nonmetallic materials such as porcelain, rubber, and wood have *very few* free electrons. They will *not* easily conduct an electric current and hence are used as *insulators*.

Since we are introducing new terms, we shall give current a unit so we can evaluate it in terms of its magnitude. The unit of current is called the ***ampere*** (abbreviated A) named after the pioneer physicist André M. Ampère. The symbol for current is *I*. (At the outset of electrical experimentation more than 100 years ago, current was called the "intensity" of electricity, and this symbol has not been changed.) Charge movement that causes a current flow in a conductor is often considered analogous to the fluid used in a hydraulic system as a medium to transmit power a moderate distance to perform some work function. There are obviously mechanical limitations to the distance that a hydraulic system can operate effectively, but we shall see in the ensuing chapters how proper design can overcome most problems when transmitting power over long distances by means of an electric current.

Obviously, a hydraulic system must have a pump so that the fluid can (overcome) the friction of the load. Similiarly, the electrons of the electrical system will not move unless required to do so by some force, or *electric field.* This force is created in a most ingenious way, which introduces one of the basic concepts of electricity: When a conductor is moved through a magnetic field an *electromotive force* is generated in that conductor. This electromotive force is called a *voltage* and is measured in *volts.* This common electrical unit was named in honor of Allesandro Volta, an Italian physicist. How magnetic fields are produced, maintained, and controlled is discussed in a later chapter. We are all familiar with common magnets such as horseshoe magnets and have witnessed the effects of the mysterious magnetism that appears in the air space between the two ends of the horseshoe-shaped magnet.

The electric generator could be called an *electron pump.* It is made up of many conductors forced to rotate through a magnetic field by a mechanical prime mover: a gas engine, a steam turbine, or a water wheel. The generator is the source of energy of the electrical system. The conversion of mechanical to electrical energy takes place within the generator. The generator, for this reason, is an *energy conversion* device.

An hydraulic system can be described as consisting of three parts: the hydraulic pump, the connecting tubing or pipes to direct the fluid, and the load that uses the hydraulic fluid to perform some useful function. Likewise, an electrical system has its *generator* to furnish the electromotive force (voltage) and the *conductors* to permit the current to flow to some electrical device, which we shall call a useful *load.* These three items—source, conductors, and load—comprise what is known as an *electric circuit.*

1-3 ***Use of proper terminology.*** As in the study of any technology, the study of electricity introduces new words, definitions, and rules. To learn this material a firm commitment is required from the student: He must use the proper word or unit at the proper time and in the proper way. The technique does not develop haphazardly but by continual practice and concentrated study in a step-by-step process, without compromise or procrastination along the way.

There are few electrical rules or laws. The first objective is always to understand how the electrical device or system functions; mathematical analysis can follow. This simply means that electrical or physical quantities and their relationships with each other must be understood.

Chapter Two

Resistance

2-1 ***Definition of resistance.*** Various metals have different atomic structures that affect the number of free electrons in the material. The greater the number of free electrons, the easier it is for an electric current to flow through that material. *Resistance*, as the word implies, is the *opposition* to current flow. Since all materials that conduct current have definite values of medium, high, or low resistance, it is necessary to establish a method of measuring these values of resistance. Such a method would also measure the resistance that restricts current flow.

It is very important to note here that resistance is a physical quantity in that it is a property of one or more of the natural elements. This property varies with the *geometry* of the material (length and cross-sectional area) as well as with the *temperature* and the nature of the *material* itself.

2-2 ***Function of resistance.*** We have described resistance as the opposition to current flow, yet we have stated that conductors must conduct the current from the generator to the electrical load. This means that these conductors must possess as low a resistance as possible so as to offer the *least* possible opposition in this part of the electric circuit. Since silver, copper, and aluminum have low resistance, we use these materials for *conductors*. However, it must be understood that even these low-resistance materials have some resistance that must be considered. There is no material that has zero resistance or, conversely, infinite resistance. There is no perfect conductor or perfect insulator.

One of the most important uses of an electric current is to produce heat. This heat energy comes from the original mechanical energy that turned the generator. A resistance material to produce heat must be selected for fewer free electrons or much greater resistance than a conductor. The same current that flows with little resistance opposition through the conductors will produce heat when it flows through a resistance material having much greater opposition to current flow. Such a resistance device is called a *resistor.*

Resistance may perform many functions in electric circuits. One function of a resistor is to produce heat. Another is to limit current flow in a circuit. A combination of resistors also may serve to provide a variety of voltages; such a combination is known as a *voltage divider.*

2-3 ***Unit of resistance.*** Some physical quantities are measured in feet and others by the pound or the mile or the gallon. For evaluation and design purposes we must provide measures for all electrical quantities. Before we investigate the effect of resistance in an electric circuit, we shall evaluate its physical properties. The unit for resistance is called the *ohm,* named for the German physicist George S. Ohm. Its symbol is R, and its unit abbreviation is **Ω**. It is the measure of a resistor's ability to oppose the flow of an electric current. A high resistance is expressed in thousands or millions of ohms.

2-4 ***Unit of area measurement; the circular mil.*** All resistances, whether encountered in the wires of an extension cord or the heating element of a toaster, have two dimensions: *cross-sectional area* and *length.* The usual unit for length is the foot. Trying to measure the cross-sectional area by square inches, however, becomes quite difficult because of the small sizes involved. It would be like measuring the length of a football field in miles. The field would be 0.0568 mile long, so instead of saying "first and ten on the 20 yard line we would have to say "first and 0.00568 on the 0.01136 mile line."

Wires are usually round, as well as small, and the unit used for measuring the cross-sectional area of wires, therefore, is the *circular mil* (abbreviated cmil). A mil is 0.001 inch (in.); a circular mil is a circle 0.001 in. in diameter. To understand this circular measure let us compare it with square measure: A square inch is a square area, 1 in. on each side. A circular inch, then, is a circle 1 in. in diameter. The area of such a circle is 0.7854 sq in. or one circular inch. To find the area of a circle by square measure we must use the relationship πR^2. But *to find the area in circular measure* we need only square the diameter, and then the area obtained is in circular inches or circular mils, depending on whether the diameter was measured in inches or mils.

The following equations and examples show this difference:

$$\text{Area in circular mils} = (\text{diameter in mils})^2$$

$$\text{cmil} = D^2 \tag{2-1}$$

where cmil is the area in circular mils and D is the diameter in mils.

$$\text{Area in square mils} = \frac{\pi}{4}\,(\text{diameter in mils})^2 = \pi\,(\text{radius})^2$$

$$A = \frac{\pi}{4} D^2 = \pi R^2 \qquad (2\text{-}2)$$

where A is the area in square mils, D is the diameter in mils; R is the radius in mils.

Example **2-1**. Given a wire $\frac{1}{8}$ in. in diameter. Find the area (a) in circular mils and (b) in square mils, and (c) the ratio of square mils to circular mils.

Solution

$\frac{1}{8}$ in. = 0.125 in. or 125 mils

(a) $\text{Area in cmil} = D^2 = 125^2 = 15{,}600 \text{ cmil}$ (2-1)

(b) $\text{Area in square mils} = A = \pi R^2 = (3.1416)(62.5)^2 = 12{,}230 \text{ mils}^2$ (2-2)

(c) $\text{Ratio} = \frac{A}{\text{cmil}} = \frac{12{,}230}{15{,}600} = 0.7854 = \frac{\pi}{4}$

Example 2-1 shows that the ratio of circular mil area to square mil area is 0.7854 or $\pi/4$. Expressed as an equation,

$$\text{Circular mil area} = 0.7854\,(\text{square mil area})$$

$$\text{Square mil area} = \frac{\text{circular mil area}}{0.7854}$$

$$A = \frac{\text{cmil}}{0.7854} = \frac{4\text{ cmil}}{\pi} \qquad (2\text{-}3)$$

where A is the area in square mils and cmil is the area in circular mils.

The circular mil area of a wire of any shape other than round may be found by first calculating its area in square mils and then dividing by 0.7840. Since a circular mil is smaller than a square mil, the number of circular mils in a given area will always be greater, as shown by Eq. (2-3) and Ex. 2-2.

Example **2-2**. Calculate the circular mil area of a rectangular conductor $\frac{1}{4}$ in. $\times$ $\frac{1}{8}$ in.

Solution

$$\tfrac{1}{4}\text{ in.} = 0.25\text{ in.} = 250\text{ mils} \qquad \tfrac{1}{8}\text{ in.} = 0.125\text{ in.} = 125\text{ mils}$$

$$A = \text{area in square mils} = (250)(125) = 31{,}250\ (\text{mils})^2$$

$$\text{cmil} = \text{area in circular mils} = \frac{31{,}250}{0.7854} = 39{,}800\text{ cmil} \tag{2-3}$$

Conductors of varied shapes should *always* have their cross-sectional areas expressed in circular mils. This facilitates their resistance calculation by the use of tables or equations that generally use this unit of area.

In Ex. 2-2 the diameter of D, the round conductor that would have the same area as the rectangular one, can be found as follows: Since D in $(\text{mils})^2$ is the area in circular mils, then

$$D \text{ in } (\text{mils})^2 = \text{cmil} \tag{2-4}$$

$$D = \sqrt{39{,}800} = 158\text{ mils} \qquad \text{or } 0.158\text{ in. in diameter.}$$

2-5 *Resistivity.* The *resistivity* of a conducting material is defined as the resistance in ohms of a *specific sample* of given cross-sectional area and given length. These dimensions could be given in any convenient units if the appropriate value of ohms were allocated to this sample. The most common specific dimensions used in such a sample is the *circular-mil-foot.* This sample is a piece of wire having a cross-sectional area of 1 cmil (or 1 mil in diameter) and a length of 1 ft. This is a very fine wire sample since a human hair is about 3 mils in diameter or has an area of 9 cmil.

The resistivities (specific resistances) of some common materials at 20°C is listed in Table 2.1.

Table 2.1. Resistivity of some conducting materials

Material	Resistivity at 20°C (Ω-cmil/ft)
Silver	9.8
Copper	10.4
Aluminum	17
Tungsten	33
Nickel	50
Iron	60
Manganin	290
Nichrome	660

The resistance of *any* conductor of any diameter, cross section, and length can be calculated if we know its resistivity. Since the resistance of a conductor varies directly with its length and inversely with its cross-sectional area, this calculation for resistance may be expressed as

$$R = \frac{\rho l}{D^2} \quad \text{or} \quad R = \frac{\rho l}{\text{cmil}} \tag{2-5}$$

where R is the conductor's resistance in ohms; ρ represents its resistivity in Ω-cmil/ft; l is its length in feet; D is its diameter in mils; and cmil is its cross-sectional area in circular mils.

The relationships in Eq. (2-5) should be obvious. As a conductor gets longer, its resistance increases because there is more opposition to current flow for a greater distance. As its diameter increases, its resistance decreases because there are more paths for current flow and, therefore, less opposition to the electron movement.

Example **2-3.** Calculate the resistance of 350 ft of copper wire having a diameter of $\frac{1}{8}$ in.

Solution

$$D = \tfrac{1}{8}\text{ in.} = 125\text{ mils} \qquad l = 350\text{ ft} \qquad \rho = 10.4\ \Omega\text{-cm/ft}$$

$$R = \frac{\rho l}{D^2} = \frac{(10.4)(350)}{125^2} = 0.232\ \Omega \tag{2-5}$$

2-6 ***Effect of temperature on resistance.*** Resistivities are usually given at 20°C (which is 68°F), the usual ambient temperature for most installations. When the temperature of a metallic conductor increases either because of an increase in its ambient temperature or the current it is carrying, its resistance will usually increase. A metal is heated either by molecular friction or the result of electron movement. A temperature increase results in prior electron movement before any current flows. This means that the current will not flow as easily and, therefore, the heated conductor will present more resistance. The opposite is also evident: As the conductor is cooled its resistance decreases. For conducting materials there is a low extreme temperature at which all molecular activity ceases. This is at −273°C or 0°K. In this region of absolute zero the change in resistance with respect to temperature is somewhat erratic. So that we may use a linear relationship between temperature changes and resistance, an *inferred* absolute zero has been determined for all conducting materials.

Assuming the material has zero resistance at this inferred temperature, graphically we can extend a straight line from this point to any known resistance at a known temperature. With this graph we then can determine the resistance of the material at any temperature.

Table 2-2 lists this inferred temperature for zero resistance for some common materials.

Table 2.2. Inferred zero resistance for some common materials

Material	Inferred Temperature for Zero Resistance (°C)
Copper	−234.5
Aluminum	−236
Tungsten	−202
Nickel	−147
Iron	−180
Nichrome	−6250
Silver	−243

Refer to Fig. 2-1. Point A is the inferred temperature of copper at which it has (zero) resistance. B is its temperature when its resistance is R_1. Then T_2 (point C) would represent the temperature that causes its resistance to increase to R_2.

By the law of similar traingles, $AB/BD = AC/CE$, where the length $AB = 234.5° + T_1$; $R_1 = BD$ is resistance in ohms at T_1; $AC = 234.5 + T_2$; and $R_2 = CE$ is resistance at T_2.

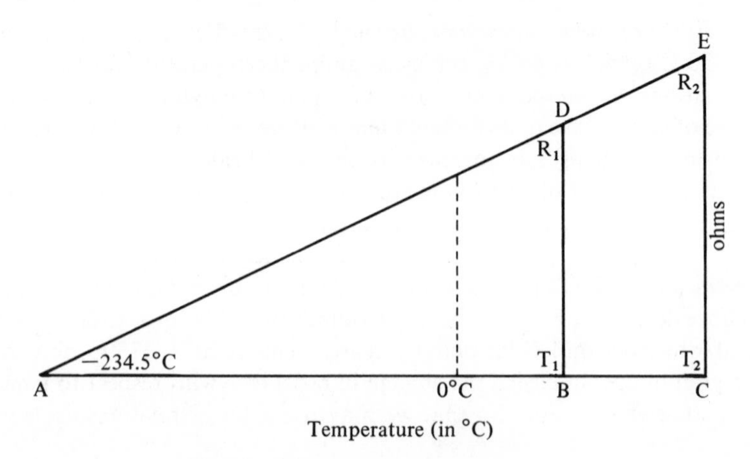

Fig. 2-1. Temperature vs. resistance.

This relationship between resistance and its temperature indicates that a moderate increase of 10 or 20°C will cause only a small increase in resistance. It must be remembered that this increase in heat can result from its ambient temperature or from the current the resistance is carrying.

The above relation may be summarized into an equation to solve for either the higher or lower resistance at any temperature

$$R_2 = R_1 \frac{234.5 + T_2}{(234.5 + T_1)} \tag{2-6a}$$

$$R_1 = R_2 \frac{234.5 + T_1}{(234.5 + T_2)} \tag{2-6b}$$

where all terms have been defined above.

Example **2-4.** The resistance of a copper winding at 20°C is 35 **Ω**. Calculate its resistance at 80°C. Refer to Fig. 2-2.

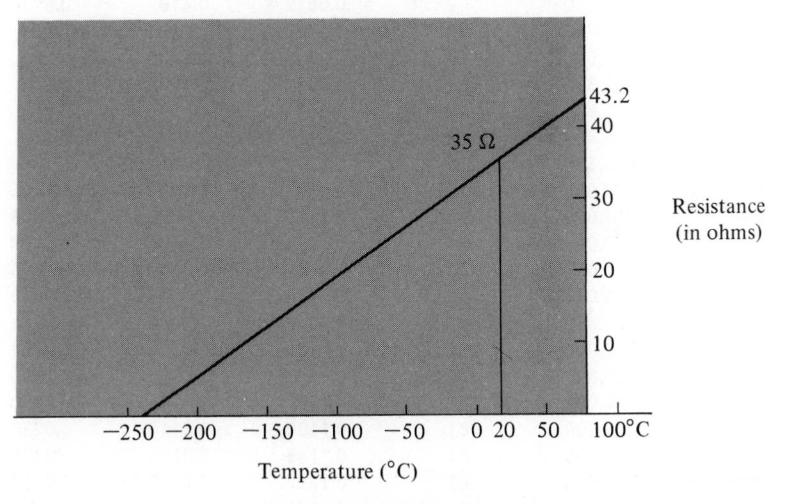

Fig. 2-2. Graphical solution of Ex. 2-4.

Solution

$$R_1 = 35\ \Omega \qquad T_1 = 20°\text{C} \qquad T_2 = 80°\text{C}$$

$$R_2 = R_1\left(\frac{234.5 + T_2}{234.5 + T_1}\right) = 35\left(\frac{314.5}{245.5}\right) = 43.2\ \Omega \tag{2-6a}$$

Figure 2-2 shows this solution graphically.

2-7 ***American wire gauge.*** To assure the manufacture of conductors in sizes that will be suitable for all applications, the American Wire Gauge (AWG) has been developed to assign a number to a particular size of wire. These numbers start at #40, the smallest, which is assigned to a wire with a diameter of 3.145 mils. The gauge numbers then *descend* in order to #0000, the largest, with a diameter of 460 mils. The size of any wire larger than #0000 is expressed directly in its circular mil area.

Table 2.3. Properties of conductors

Size (AWG or MCM)	Area (cmil)	Solid or Stranded Number Wires	Solid or Stranded Diameter Each Wire (in.)	dc Resistance at 25°C, 77°F (Ω/1000 ft) Copper	dc Resistance at 25°C, 77°F (Ω/1000 ft) Aluminum
18	1,620	Solid	0.0403	6.51	10.7
16	2,580	Solid	0.0508	4.10	6.72
14	4,110	Solid	0.0641	2.57	4.22
12	6,530	Solid	0.0808	1.62	2.66
10	10,380	Solid	0.1019	1.018	1.67
8	16,510	Solid	0.1285	0.6404	1.05
6	26,240	7	0.0612	0.410	0.674
4	41,740	7	0.0772	0.259	0.424
3	52,620	7	0.0867	0.205	0.336
2	66,360	7	0.0974	0.162	0.266
1	83,690	19	0.0664	0.129	0.211
0	105,600	19	0.0745	0.102	0.168
00	133,100	19	0.0837	0.0811	0.133
000	167,800	19	0.0940	0.0642	0.105
0000	211,600	19	0.1055	0.0509	0.0836
250	250,000	37	0.0822	0.0431	0.0708
300	300,000	37	0.0900	0.0360	0.0590
350	350,000	37	0.0973	0.0308	0.0505
400	400,000	37	0.1040	0.0270	0.0442
500	500,000	37	0.1162	0.0216	0.0354
600	600,000	61	0.0992	0.0180	0.0295
700	700,000	61	0.1071	0.0154	0.0253
750	750,000	61	0.1109	0.0144	0.0236
800	800,000	61	0.1145	0.0135	0.0221
900	900,000	61	0.1215	0.0120	0.0197
1000	1,000,000	61	0.1280	0.0108	0.0177

Table 2.3 lists the standard sizes of wire from #18 to 1 million cmil. The resistance of both copper and aluminum conductors is given in **Ω**/1000 ft in Table 2.3. This table is of great assistance when determining conductor resistance of any size and length.

2-8 ***Resistances in series.*** When examining Eq. (2-5) for finding resistance, it is noted that resistance of a material is directly proportional to its length (the unit is given in feet). We could then say that every foot of the material is in series with the next and could merely add them together to obtain the total resistance. This leads to our definition of series resistances. If two or more resistances are connected in such a way that they carry the *same* current, they are in series. A simple way to determine if one resistor is in series with another is to ask, "Does the same current go through both resistors?" If the current has no other possible path, the conclusion is obvious.

For resistances in series, therefore, the total resistance is the sum of the individually connected series resistances or

$$R_T = R_1 + R_2 + R_3 + \cdots + R_N \tag{2-7}$$

where R_1 through R_N are individual resistances in ohms and R_T is the total resistance in ohms. Examples 2-6 and 2-7 illustrate the value of Eq. (2-7).

Example **2-5.** If three resistors of 12, 22, and 30 Ω are connected in series, what is the total resistance of the circuit? Refer to Fig. 2-3.

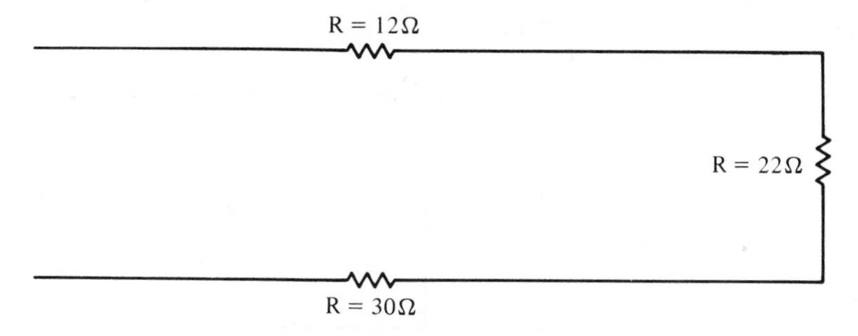

Fig. 2-3. Diagram for Ex. 2-5.

Solution

$$\begin{aligned} R_T &= R_1 + R_2 + R_3 \\ &= 12\ \Omega + 22\ \Omega + 30\ \Omega \\ &= 64\ \Omega \end{aligned} \tag{2-7}$$

Example **2-6.** If two conductors, each of which has 0.15 Ω resistance, are used to conduct current to a 20-Ω resistor, calculate the total resistance of the circuit. Refer to Fig. 2-4.

Solution

$$\begin{aligned} R_T &= R_1 + R_2 + R_3 \\ &= 0.15\ \Omega + 20\ \Omega + 0.15\ \Omega \\ &= 20.3\ \Omega \end{aligned} \tag{2-7}$$

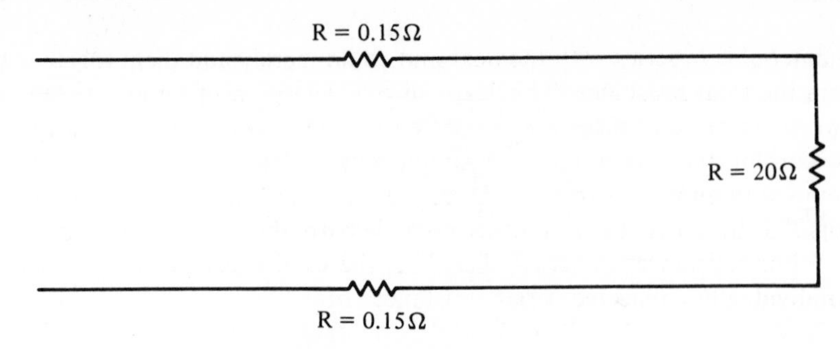

Fig. 2-4. Diagram for Ex. 2-6.

2-9 ***Resistances in parallel.*** What is meant by resistances connected in parallel? Parallel lines are defined as lines that never meet. Originally the idea of using the word "parallel" to describe a particular way of connecting resistances might stem from the idea that on a diagram they are drawn parallel to each other and then connected together. Figure 2-5 shows three resistors connected in parallel.

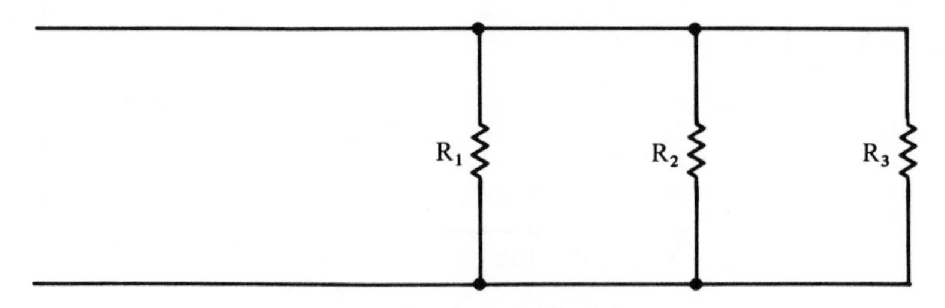

Fig. 2-5. A parallel circuit.

A proper definition might be "Two or more resistors are physically in parallel if they are connected together at both ends and have no other resistance between their terminals." Another test for parallel connections of resistors is to note if the *same* voltage exists *across* each resistor.

When calculating the total resistance of a number of resistances connected in parallel, we can again examine Eq. (2-5), which says that as the area of a resistor is increased its resistance decreases. It should also be obvious that as more resistances are connected in parallel, there are more paths for current flow, which means less opposition or less resistance to the flow of current.

Since resistances in parallel allow more current to flow, we add together the ability to conduct current rather than the ability to oppose current as when resistors are in series. This ability to conduct current is called, appropriately enough, *conductance* (which mathematically is the reciprocal of resistance), and the unit is called "mho" which, also appropriately, is ohm spelled backward.

The equation now must say, "The total conductance of a group of parallel resistors is the sum of the conductances."

$$\frac{1}{R_T} = \frac{1}{R_1} + \frac{1}{R_2} + \frac{1}{R_3} + \cdots + \frac{1}{R_N} \tag{2-8}$$

When there are only two resistors in parallel, Eq. (2-8) can be simplified as follows:

$$\frac{1}{R_T} = \frac{1}{R_1} + \frac{1}{R_2}$$

$$R_T = \frac{R_1 R_2}{R_1 + R_2} \tag{2-9}$$

When only two resistors are in parallel, Eq. (2-9) states that their combined (equivalent) resistance is equal to their product divided by their sum, as illustrated in Ex. 2-7.

Example **2-7.** Calculate the equivalent resistance of a 12- and 20-**Ω** resistor connected in parallel. Refer to Fig. 2-6.

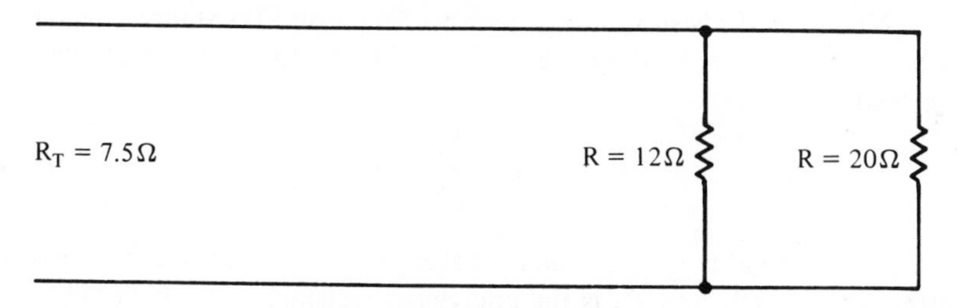

Fig. 2-6. Diagram for Ex. 2-7.

Solution

$$R_T = \frac{R_1 R_2}{R_1 + R_2} = \frac{(12)(20)}{32} = 7.5\ \Omega \tag{2-9}$$

Note from Ex. 2-7 and Fig. 2-6 that the equivalent resistance is less than either of the individual parallel resistances. As more devices are added in parallel, the circuit current must increase. This means the equivalent resistance must decrease. This is sometimes called an "*increase* in electrical *load.*"

Example 2-8. Calculate the total resistance of a 20-, 30-, and 35-Ω resistor all connected in parallel. Refer to Fig. 2-7.

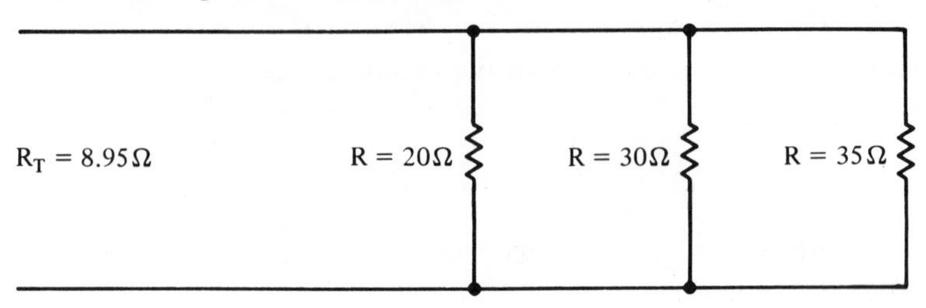

Fig. 2-7. Diagram for Ex. 2-8.

Solution

$$\frac{1}{R_T} = \frac{1}{R_1} + \frac{1}{R_2} + \frac{1}{R_3} = \frac{1}{20} + \frac{1}{30} + \frac{1}{35} \qquad (2\text{-}8)$$

$$= 0.05\ \Omega + 0.0333\ \Omega + 0.0286\ \Omega$$

$$= 0.1119\ \Omega \qquad 0.1119\ R_T = 1\ \Omega$$

$$R_T = \frac{1\ \Omega}{0.1119} = 8.95\ \Omega$$

It should be noted again from the above examples that the total resistance of a parallel circuit is always less than the smallest resistor.

When *equal* resistors are in parallel their total resistance calculation can be simplified by dividing the ohmic value by the number of parallel-connected resistors.

$$R_T = \frac{R_1}{N} \qquad (2\text{-}10)$$

where R_1 is the value of any equal resistor in Ω; N is the number of equal resistors in parallel; and R_T is the equivalent resistance of the circuit.

Example 2-9. What is the total (equivalent) resistance of five 25-Ω resistors connected in parallel?

Solution

$$\frac{1}{R_T} = \frac{1}{R_1} + \frac{1}{R_2} + \frac{1}{R_3} + \frac{1}{R_4} + \frac{1}{R_5} \qquad (2\text{-}8)$$

$$= \frac{1}{25} + \frac{1}{25} + \frac{1}{25} + \frac{1}{25} + \frac{1}{25} \qquad \text{(ohms)}$$

$$= \frac{5\ \Omega}{25}$$

$$5\,R_T = 25\ \Omega$$
$$R_T = 5\ \Omega$$

Alternate method using Eq. (2-10):

$$R_T = \frac{R_1}{N}\ \Omega \tag{2-10}$$
$$= \frac{25\ \Omega}{5} = 5\ \Omega$$

2-10 ***Series-parallel circuits.*** There are many times when resistors are connected in networks, or particular combinations of resistors, where some are in parallel and some are in series. To properly evaluate the electrical characteristics of a *series–parallel* circuit it is necessary to know the *equivalent resistance* R_T of the entire circuit. The equivalent resistance is that value of ohmic resistance that has the *same* resistance as the original circuit. For example, a dollar bill is equivalent in value to three quarters, two dimes, and a nickel, but the bill's appearance is not similar to the various coins.

To calculate equivalent resistance it is strictly necessary to observe the rules for combining resistors when they are in series or parallel with other. For example, in the combination of resistors shown in Fig. 2-8 we can make some conclusions regarding their connections:

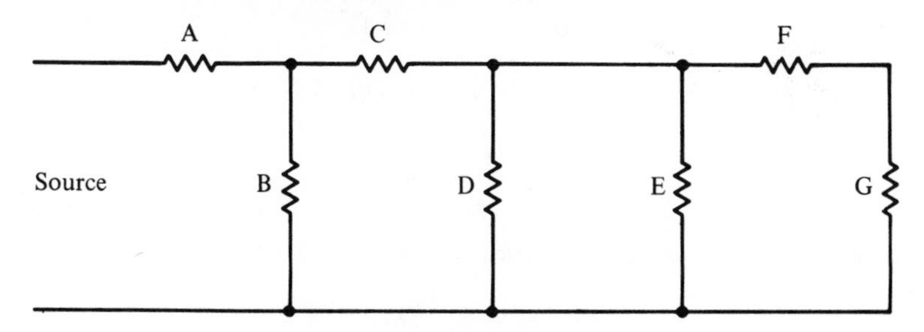

Fig. 2-8. A series-parallel circuit.

1. Resistors *F* and *G* are in series because the same current must flow through them.
2. Resistors *D* and *E* are in parallel with each other because there is no other resistance between them and the voltage across them is the same.
3. Resistors *A*, *B*, and *C* are not in series or parallel with any other single resistor.
4. Resistors *D* and *E* are in parallel with the sum of *F* plus *G*.
5. Resistor *C* is in series with the parallel combination of *D*, *E*, and the sum of *F* and *G*.

6. Resistor B is in parallel with the sum of C plus the parallel combination of D, E, and F plus G.
7. Resistor A is in series with the rest of the circuit.

It is very important when combining resistances to form equivalent that the *starting* point be at the *end* of the circuit (furthest from the source) and not the beginning (at the source). If we start at the beginning of the circuit we do not yet know what resistance value to add to A, since the rest of the circuit has not yet been evaluated.

Example ***2-10.*** Using ohmic values as given, calculate the equivalent resistance of circuit of Fig. 2-8.

$$A = 2\ \Omega \qquad E = 15\ \Omega$$
$$B = 12\ \Omega \qquad F = 7\ \Omega$$
$$C = 5\ \Omega \qquad G = 3\ \Omega$$
$$D = 20\ \Omega$$

Solution

$$R_{FG} = F + G = 7\ \Omega + 3\ \Omega = 10\ \Omega$$

$$R_{DE} = \frac{(D)(E)}{D + E} = \frac{(20\ \Omega)(15\ \Omega)}{35\ \Omega} = 8.57\ \Omega$$

but R_{DE} of 8.57 Ω is in parallel with R_{FG} of 10 Ω. Then

$$R_{DEFG} = \frac{(8.57\ \Omega)(10\ \Omega)}{18.57\ \Omega} = 4.62\ \Omega$$

and R_{DEFG} of 4.62 Ω is in series with R_C of 5 Ω, so

$$R_{CDEFG} = 4.62\ \Omega + 5\ \Omega = 9.62\ \Omega$$

but R_{CDEFG} of 9.62 Ω is in parallel with R_B of 12 Ω, so

$$R_{(B\text{-}G)} = \frac{(9.62\ \Omega)(12\ \Omega)}{21.62\ \Omega} = 5.32\ \Omega$$

but R_{BG} of 5.32 Ω is in series with R_A of 2 Ω, so

$$R_T = R_{A\text{-}G} = 5.32\ \Omega + 2\ \Omega = 7.32\ \Omega$$

This value of R_T or 7.32 Ω is called the "*equivalent resistance* of the entire circuit." For all practical purposes with respect to the source, all seven resistors could be replaced by a single resistor of 7.32 Ω.

PROBLEMS

2-1 Convert the following into mils:
(a) $\frac{1}{4}$ in.
(b) 0.1 in.
(c) $\frac{3}{8}$ in.
(d) 0.285 in.

2-2 Express the cross-sectional area of the following conductors in circular mils:
(a) $\frac{1}{8}$-in. diameter.
(b) 0.2576-in. diameter.
(c) 0.050 in. square (0.05″ × 0.05″).
(d) 0.06 in. diameter.
(e) 0.061 in. diameter.
(f) $\frac{1}{8}$ in. × $\frac{1}{4}$ in.

2-3 What is the resistance of 600 ft of copper wire having a diameter of 0.1 in.?

2-4 Using Eq. (2-5) calculate resistances of the following conductors:
(a) a heating element for an electric oven, made of 35-ft nichrome wire with a diameter of 30 mils.
(b) 250-ft #14 AWG copper wire.
(c) 850-ft #00 AWG copper wire.
(d) 300-ft #6 AWG aluminum wire.
(e) 2000-ft #14 AWG iron wire.

2-5 Using Table 2-3, calculate resistances of the following conductors:
(a) 400-ft #1 AWG copper.
(b) 400-ft #1 AWG aluminum.
(c) 250-ft #14 AWG copper.
(d) 850-ft #00 AWG copper.
(e) 1200-ft 300 MCM aluminum.

2-6 Calculate the total resistance of both #18 AWG copper conductors of a 100-ft cord.

2-7 What would be proper size copper wire for a two-wire 100-ft extension cord if its resistance must be less than 0.75 Ω?

2-8 Calculate the resistance of a circuit that has three resistors of 20-, 28-, and 45-Ω connected in series.

2-9 Calculate the total resistance of a series circuit comprising a 35-, a 20-, and two 10-Ω resistors and two #14 AWG conductors, each of which is 150 ft long.

2-10 Calculate the resistance of a circuit made up of a 25- and a 40-Ω resistor connected in parallel.

2-11 Calculate the resistance of a circuit consisting of a 15-, a 25-, and a 30-Ω resistor all connected in parallel.

2-12 Calculate the resistance of a circuit with five 20-Ω resistors connected in parallel.

2-13 Calculate the resistance of a 65- and 90-Ω resistor connected in parallel.

2-14 Calculate the resistance of the following circuit: a 5-Ω resistor connected in series with a 60- and 100-Ω resistor in parallel.

2-15 Find the equivalent resistance of the circuit shown in Fig. P2-15.

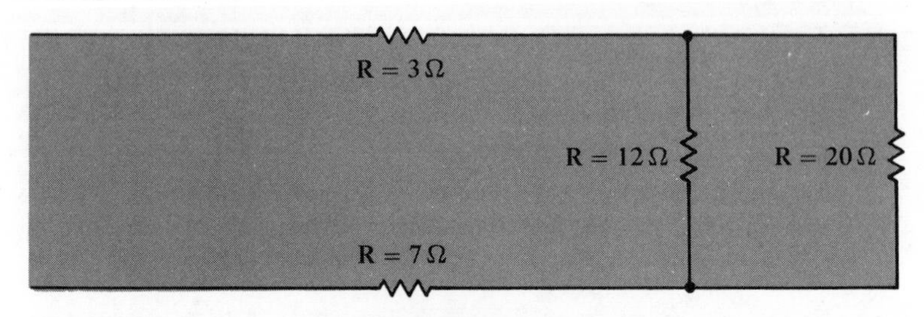

Figure P2-15

2-16 Find the equivalent resistance of the circuit shown in Fig. P2-16.

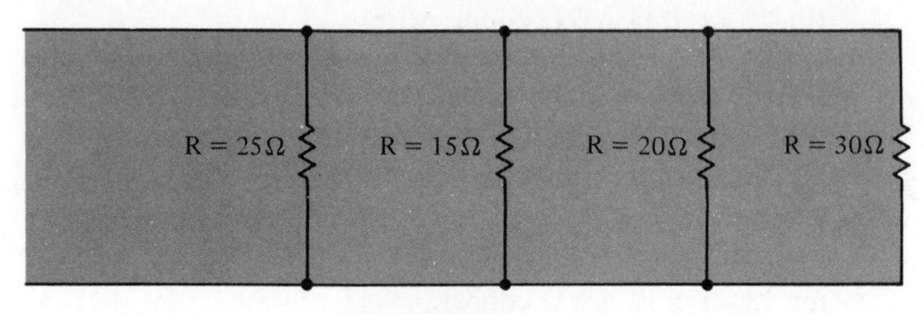

Figure P2-16

2-17 Find the equivalent resistance of the circuit shown in Fig. P2-17.

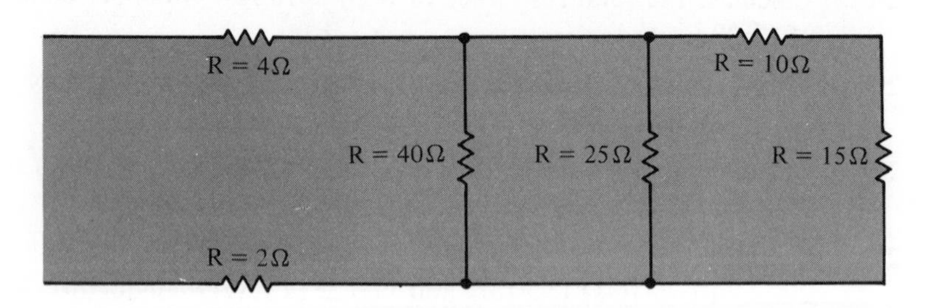

Figure P2-17

2-18 Find the equivalent resistance of the circuit shown in Fig. P2-18.

2-19 Calculate the equivalent resistance of a circuit that has a load of 7.5 Ω supplied with current by means of two #10 AWG copper conductors, each 150 ft long.

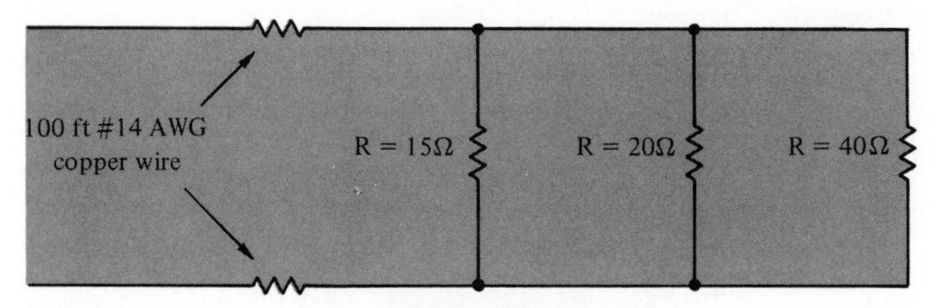

Figure P2-18

2-20 By use of Eq. (2-5) calculate the resistance of the following conductors:

(a) 400-ft #8 AWG copper. (b) 400-ft #8 AWG aluminum.
(c) 1200-ft #0 copper. (d) 1200-ft #0 aluminum.

2-21 If the resistance of both conductors supplying current to a load must be no more than 2% of the resistance of the load, by the use of Table 2-3 find suitable copper and aluminum conductors for the following loads:

(a) 5 Ω.
(b) ten 10-Ω resistors in parallel.
(c) three 20-Ω resistors in parallel.
(d) twenty 40-Ω resistors in parallel, connected in series with a 1.5-Ω resistor.

Chapter Three

The Electric Circuit

3-1 ***Ohm's law.*** The previous chapter emphasized that the electrical quantity called "resistance" is in reality completely physical. Two of its functions are to conduct current and to produce heat. In each case we must use the proper material in the correct dimensions for each particular situation.

The current flow in a circuit is measured in amperes (A). The electromotive force that causes the current to flow in an electric circuit is measured in volts (V). The unit of opposition to current flow is measured in ohms (Ω). These three units would have no relationship with each other unless they were defined. A very important basic relationship among these three units has been established and is called *Ohm's law*. This law states that the current flow is proportional to voltage but is inversely proportional to resistance. Ohm's law in symbol form is

$$I = \frac{E}{R} \tag{3-1a}$$

where I is the current in amperes, E is the voltage in volts, and R is the resistance in ohms.

Ohm's law enables us to calculate any of the three electrical quantities. For the purpose of calculations it also may be written

$$E = IR \tag{3-1b}$$

or

$$R = \frac{E}{I} \tag{3-1c}$$

The use of these equations enables us to calculate the third quantity if any two are already known.

Example 3-1. Calculate current that will flow in a 25-Ω resistor if a voltage of 120 V is applied. Refer to Fig. 3-1.

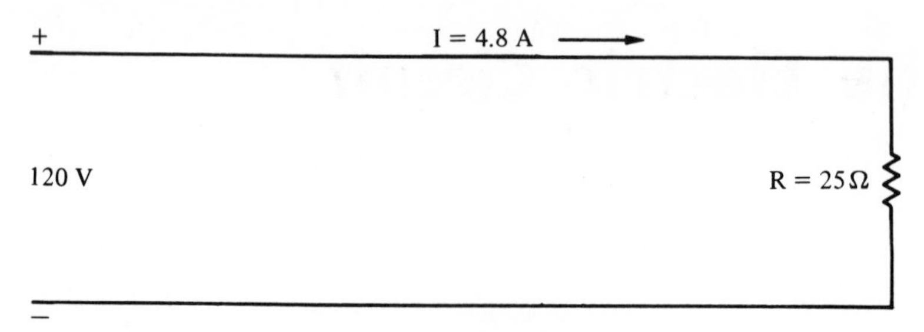

Fig. 3-1. Diagram for Ex. 3-1.

Solution

$$I = \frac{E}{R} = \frac{120\text{ V}}{25\ \Omega} = 4.8\text{ A} \tag{3-1a}$$

3-2 ***Kirchhoff's voltage law*** (KVL). When resistors are connected in series we can easily determine the total resistance using Eq. (2-7). Then according to Ohm's law we can determine the total current to the entire circuit for any given applied voltage. Since this original circuit has all its resistors in series, this current must go through every resistor and back to the voltage source.

An electric direct current (dc) power supply that produces a voltage will have one terminal of positive polarity and the other of negative. Conventional current flow assumes that current flows from the positive terminal of the supply and returns to the negative terminal. These polarities are easily found with a voltmeter since all dc measuring instruments have polarity indications on their terminals.

The question of the use of dc or alternating current (ac) will be investigated later. When resistances are the *only* devices used in a circuit, the electrical calculations and observations are the *same* whether a direct or an alternating current flows through them. The positive terminal of a dc supply will always be positive. Any one terminal of an ac supply will be alternately positive and then negative. This change of polarity will be uniform among all resistive elements in a purely resistive circuit. Therefore, we can assume a fixed polarity and a constant current direction for the purpose of circuit analysis. This will be valid as long as we are consistent.

Kirchhoff's voltage law (KVL) states that the net voltage in any continuous path of current must always be zero. When two or more resistors are connected in series there will be a voltage drop across each resistor. The sum of these voltage *drops* then must equal the voltage *rise* of the power supply. Hereafter in this text we shall use the letter V as a symbol for the voltage drop across a load but retain E as the symbol for the electromotive force (voltage rise) of the power supply. The polarity of the voltage drops in a series circuit also can be determined with a voltmeter. According to one of the Ohm's law equations a voltage drop across any resistor will be

$$V = IR \tag{3-1b}$$

One convenient method of writing a voltage equation for a circuit according to KVL is to follow three simple rules:

1. Indicate current direction with an arrow *out* of the positive terminal of power supply. If the positive terminal is unknown, assume one as positive.
2. As current goes through a resistor, place a plus sign where current enters a resistor and a minus sign where it leaves.
3. Start at any point in the circuit, in the direction of current flow, and add up all voltages. When going from a + to a −, this would be taken as a voltage *drop* and given a *minus* sign. When going from a − to a +, this would be a voltage *rise* by a power supply and given a *plus* sign.

KVL: The algebraic sum of all voltages in any closed path must be zero.

$$E - (V_1 + V_2 + V_3 + V_4) = 0 \tag{3-2a}$$

Kirchhoff's voltage law states that the current flow, as permitted by all resistances in the circuit, will cause the voltage drops to be equal to the voltage supply, or

$$E = V_1 + V_2 + V_3 + V_4 \tag{3-2b}$$

Example **3-2.** Connect a 10-, 20-, and 30-Ω resistor in series across a 120-V supply. Refer to Fig. 3-2. Calculate (a) the current and (b) the voltage drop across each resistor and (c) write the equation according to KVL for the circuit.

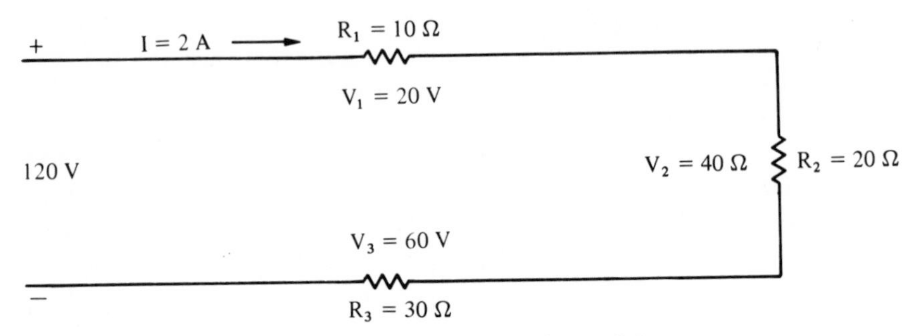

Fig. 3-2. Diagram and solution for Ex. 3-2.

Solution

(a)
$$R_T = R_1 + R_2 + R_3 \tag{2-7}$$
$$= 10 + 20 + 30 = 60\ \Omega$$
$$I = \frac{E}{R_T} = \frac{120\ \text{V}}{60\ \Omega} \tag{3-1a}$$
$$= 2\ \text{A}$$

(b)
$$V_1 = IR_1 = (2)(10) = 20\ \text{V} \tag{3-1b}$$
$$V_2 = IR_2 = (2)(20) = 40\ \text{V}$$
$$V_3 = IR_3 = (2)(30) = 60\ \text{V}$$

(c)
$$E - (V_1 + V_2 + V_3) = 0 \tag{3-2a}$$
$$120 - (20 + 40 + 60) = 0$$
$$120 - (120) = 0$$

3-3 ***Kirchhoff's current law*** **(KCL).** When resistors or any electrical devices are connected in *parallel* there will be *different* paths for current to flow in. If we examine any junction, regardless of how many current paths there might be, Kirchhoff's current law (KCL) states that the net current of any junction must equal zero. Stated in another way, the current *into* a junction must equal the current *out* of a junction. This is quite logical. The number of automobiles *into* a highway intersection must equal the number of cars *out* of the intersection or we have a traffic back-up. In an electric circuit there can be no current back-up. The current will flow instantly or not at all. Stated as an equation,

$$I_T - (I_1 + I_2 + I_3 + \cdots + I_N) = 0 \qquad (3\text{-}3a)$$

When using Kirchhoff's current law it is necessary to assign arrows to all currents that enter or leave a junction. When assigning these arrows we assume the current flows from − to + through the power supply and from + to − through an electrical load. The sum of these currents as designated by the arrows must be zero.

Example **3-3.** A parallel circuit consists of two resistors connected across a 120-V supply. The first resistance has 10 **Ω**, and the other resistor is unknown. The total current is 30 A. Refer to Fig. 3-3a. Calculate (a) the unknown current I and (b) the resistance in ohms of unknown resistor.

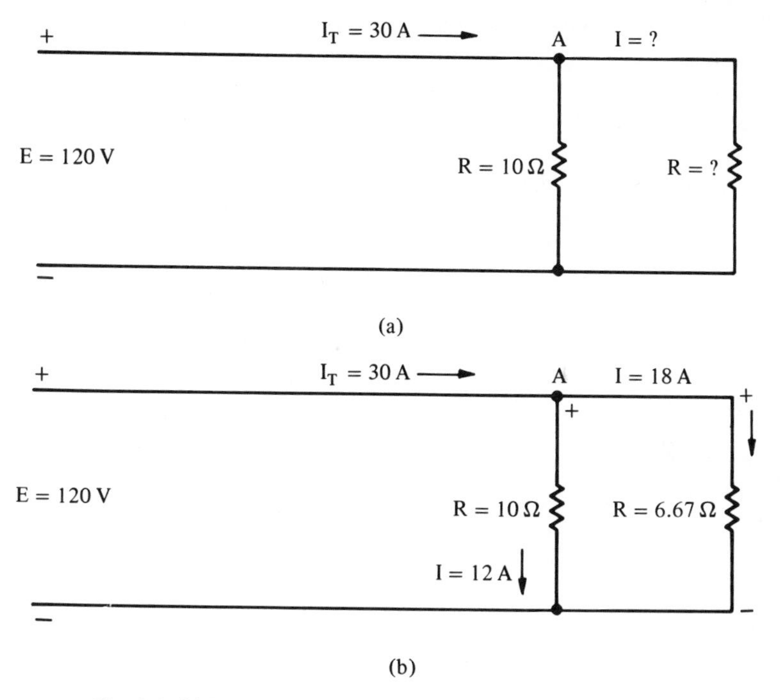

Fig. 3-3. (a) Data for Ex. 3-3. (b) Solution for Ex. 3-3.

Solution

(a) Current through a 10-**Ω** resistor

$$I = \frac{120\text{ V}}{10\ \Omega} = 12\text{ A}$$

For a complete labeled diagram of the circuit, refer to Fig. 3-3b.

$$\text{Current into junction } A = \text{current out of junction } A$$
$$30 = I + 12$$
$$I = 30 - 12$$
$$= 18\text{ A} \qquad (3\text{-}3b)$$

There can be no loss of current in and out of an electric junction.

(b)

$$R = \frac{E}{I} = \frac{120\text{ V}}{18\text{ A}} = 6.67\ \Omega$$

3-4 ***Power in an electric circuit.*** *Power* is defined as the *rate of doing work.* The size of an electric motor needed to raise an elevator is selected not only according to the weight we wish to raise but how fast we wish it to rise. Likewise, an electric heating element must be selected according to how fast we wish the heat to be produced.

Electric power is measured in watts, abbreviated W as a unit. This unit is named after James Watt, a Scotch inventor. It is equal to the product of the voltage multiplied by the current. The total power of a circuit is obtained by multiplying the total current by the voltage. The power of any one resistor in any circuit configuration can be calculated by multiplying the current through that resistor by the voltage across that resistor.

$$P = IE \qquad (3\text{-}4)$$

By combining this expression with Ohm's law we derive two other power equations:

$$P = IE \qquad (3\text{-}4)$$

But

$$I = \frac{E}{R}$$

Then

$$P = \frac{E}{R}(E) = \frac{E^2}{R} \qquad (3\text{-}5)$$

Substituting IR for E in the power equation,

$$P = I(IR)$$
$$= I^2R \tag{3-6}$$

Equations (3-5) and (3-6) state that the power of a resistor also can be calculated by dividing the square of the voltage by the resistance or by multiplying the resistance by the square of the current. This is logical because to increase the current in a resistor we must first increase the voltage. For example, doubling the voltage across a resistor would cause the current to double. This would increase the power to four times the original value.

Many instances will arise when the resistance of the conductors of a system between certain points will have to be taken into consideration. The voltage drop in these conductors will be of great concern, and this voltage drop will cause a loss of power in the order of V^2/R or I^2R. In these expressions V would represent the voltage drop in the conductors and I would represent the current the conductors must carry.

Example 3-4. A series circuit has a 30-, 40-, and 50-Ω resistor connected to a 240-V supply. Refer to Fig. 3-4. Calculate (a) the power of the circuit and (b) the power of each resistor by the three methods.

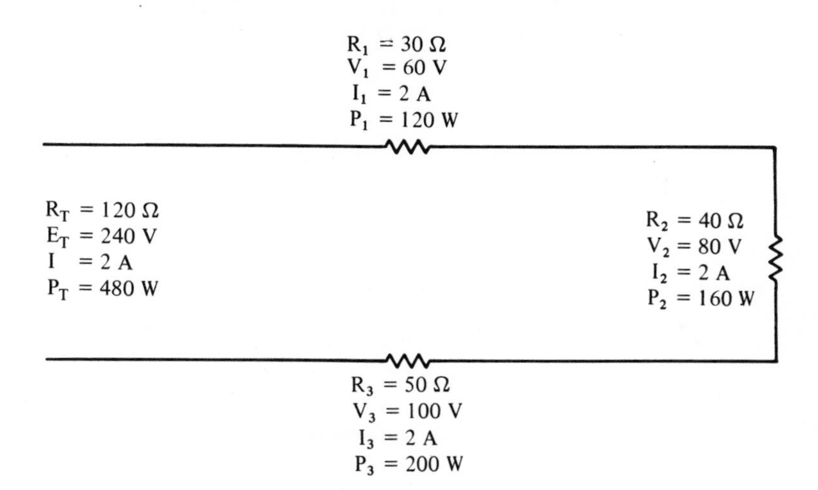

Fig. 3-4. Diagram for Ex. 3-4.

Solution

(a)

$$R_T = R_1 + R_2 + R_3$$
$$= 30 + 40 + 50 = 120\ \Omega \tag{2-7}$$

$$I = \frac{E}{R} = \frac{240 \text{ V}}{120 \, \Omega} = 2.0 \text{ A} \tag{3-1a}$$

Power of circuit

$$P_T = EI = (240)(2) = 480 \text{ W} \tag{3-4}$$

(b) Power of each resistor $= I^2R$
Power of the 30-Ω resistor $= (2)^2(30) = 120$ W
Power of the 40-Ω resistor $= (2)^2(40) = 160$ W
Power of the 50-Ω resistor $= (2)^2(50) = 200$ W
Total power $= 480$ W (3-6)

The power of each resistor can be found by calculating the voltage across each resistor and them multiplying by the current.

Voltage across each resistor $= IR$
Voltage across the 30-Ω resistor $= (2)(30) = 60$ V
Voltage across the 40-Ω resistor $= (2)(40) = 80$ V
Voltage across the 50-Ω resistor $= (2)(50) = 100$ V
Total voltage of supply $= 240$ V (3-1b)

Power of each resistor $= EI$
Power of the 30-Ω resistor $= (60)(2) = 120$ W
Power of the 40-Ω resistor $= (80)(2) = 160$ W
Power of the 50-Ω resistor $= (100)(2) = 200$ W
Total power $= 480$ W (3-4)

The power may be calculated by using Eq. (3-5): $P = E^2/R$.

Power of the 30-Ω resistor $= \dfrac{60^2}{30} = 120$ W

Power of the 40-Ω resistor $= \dfrac{80^2}{40} = 160$ W

Power of the 50-Ω resistor $= \dfrac{100^2}{50} = 200$ W

Total power $= 480$ W

It should be noted from Ex. 3-4 that the largest resistor has the greatest voltage drop and dissipates the most power.

Both power and voltage drop occur in proportion to resistance. This may be expressed as follows:

$$\frac{R_1}{R_T} = \frac{IR_1}{IR_T} = \frac{V_1}{V_T} = \frac{I^2R_1}{I^2R_T} = \frac{P_1}{P_T} \tag{3-6a}$$

Example 3-5. Three 10-Ω resistors in parallel are supplied with current by two #8 copper conductors. Each conductor is 150 ft long and the source voltage is 120 V. Refer to Fig. 3-5a. Calculate (a) the voltage at the load and (b) the power loss in the conductors.

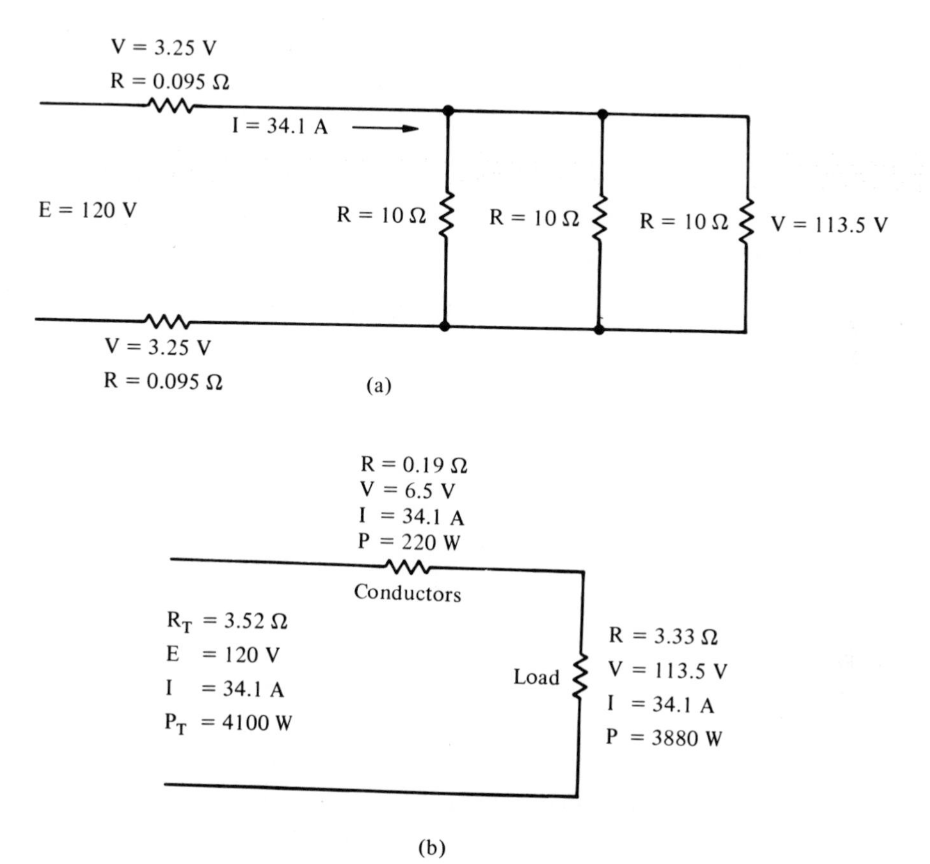

Fig. 3-5. Diagram and solution to Ex. 3-5.

Solution

Equivalent resistance of load

$$R_L = \frac{10}{3} = 3.33\ \Omega \qquad (2\text{-}10)$$

Equivalent resistance of both conductors

$$R_C = \frac{\rho l}{\text{cmil}} = \frac{(10.4)(300)}{128^2} = 0.19\ \Omega \qquad (2\text{-}5)$$

Note: The diameter of a #8 wire is 0.128 in. Equivalent resistance of circuit

$$R_T = R_L + R_C = 3.33 + 0.19 = 3.52\ \Omega \tag{2-7}$$

For the equivalent circuit refer to Fig. 3-5b.

$$\text{Circuit current } I = \frac{E}{R} = \frac{120\text{ V}}{3.52}\ \Omega$$

$$= 34.1\text{ A} \tag{3-1a}$$

(a) Voltage at load

$$IR_2 = (34.1)(3.33) = 113.7\text{ V} \tag{3-1b}$$

(b) Power loss in conductors

$$P = I^2R_C = (34.1)^2(0.19) = 220\text{ W} \tag{3-6}$$

where

$$V_{\text{loss in conductor}} = V_{\text{source}} - V_{\text{at load}}$$
$$= 120 - 113.5$$
$$= 6.5\text{ V}$$

or

$$\frac{V^2}{R} = \frac{6.5^2}{0.19} = 220\text{ W} \tag{3-5}$$

3-5 ***Electrical energy.*** The term *power* has been defined as the *rate* of doing work. A workman's salary is expressed as a *rate* of pay. It may be a certain rate per hour, week, month, or year, but wages earned cannot be determined until a period of time has been implied. Wages could equal the rate of pay in dollars per hour times the time worked in hours.

Similarly, a quantity of electrical energy cannot be evaluated until we include a period of time that a circuit has been operating at some rated value of power, expressed in watts. The unit of energy is the watt-hour (Wh), but since 1 Wh would be a very small amount of energy, the more common unit is the kilowatt-hour (kWh). One kilowatt-hour could be 1000-W for 1 hr, 500 W for 2 hr, or 10 W for 100 hr. It is the kilowatt-hour energy measurement whereby electric companies charge their customers. A kilowatt-hour of electrical energy can vary in price from 1 to 10 cents, depending on many factors such as total usage, demand for a particular period of time, location of customer facilities, type of equipment involved, and sometimes seasonal use. Summarized in equation form,

$$\text{Power } P = \frac{W}{t} \tag{3-7a}$$

where P is the power in watts; W is the energy in watt-hours; t is the time in hours; and energy

$$W = (P)(t) \tag{3-7b}$$

Example **3-6.** Calculate the cost of operating a heating element with a resistance of 8 Ω for an 8-hr day. The po'ver supply has a voltage of 230 V and the energy cost is 2 cents/kWh.

<u>*Solution*</u>

To find the current,

$$I = \frac{E}{R} = \frac{230 \text{ V}}{8\ \Omega} = 28.8 \text{ A}$$

To find the power,

$$P = EI = (230)(28.8) = 6620 \text{ W}$$

To find energy,

$$\text{kWh} = \frac{(P)(t)}{1000} = \frac{(6620)(8)}{1000} = 53 \text{ kWh}$$

To find the cost,

$$\left(\frac{\$}{\text{kWh}}\right)(\text{kWh}) = \left(\frac{\$0.02}{\text{kWh}}\right)(53 \text{ kWh}) = \$1.06$$

The above operation could be summarized in a single equation to find total cost for a given period:

$$\text{Total cost} = \frac{(P)(C)(t)}{1000} \tag{3-8}$$

where P is power consumed in watts; C is cost of energy in dollars per kilowatt-hour and t is time in hours.

Using Eq. (3-8) for above example,

$$\text{Total cost} = \frac{(6620)(0.02)(8)}{1000} = \$1.06$$

3-6 ***Power evaluation.*** When an electrical system consists of many devices of various types and for different purposes, it becomes virtually impossible to evaluate them on an individual basis according to their resistances. It also will become evident that most electrical devices are designed to operate at some particular voltage, which necessitates that all devices must be connected in parallel. Should the electrical load under consideration consist of 50 or 100 devices, each with a specific value of resistance, the calculation of the equivalent resistance of the total load would be a monumental task.

When evaluating the electrical load of a particular system, which could be a simple circuit or one floor of a building or the entire building, the current or power is a much easier parameter to use than the equivalent resistance.

Example 3-7. Calculate the current required to operate one hundred 50-W incandescent lamps. The voltage of the system is 120 V.

Solution

$$\text{Total power} = (100 \text{ lamps})(50\text{-W lamp}) = 5000 \text{ W}$$

$$\text{Current } I = \frac{P}{E} = \frac{5000 \text{ W}}{120 \text{ V}} = 41.5 \text{ A} \qquad (3\text{-}4)$$

PROBLEMS

3-1 Three resistors of 8, 12, and 20 Ω are connected in series to a 120-V supply. Calculate the voltage across each resistor.

3-2 Four equal resistors are connected in series across a 120-V circuit. The current is 10 A. Calculate (a) the resistance of each resistor and the (b) voltage across each resistor.

3-3 A series circuit consists of a 4-Ω and an unknown resistor. The current is 12 A through the circuit. The voltage of the supply is 120 V. Calculate the resistance of the unknown resistor.

3-4 Two equal resistors of 10 Ω each are connected in series to a 120-V supply. A switch is connected across one resistor so that when closed the resistance is short-circuited. Calculate circuit current when this switch is (a) open and (b) closed.

3-5 A 22-Ω resistor is to operate with a voltage of 90 V. If connected to a 115-V supply, what resistance must be connected in series?

3-6 A Christmas tree light set has eight 15-V lamps connected in series across a 120-V supply. If one more lamp is added so that there are nine lamps in series, calculate the voltage across each lamp.

3-7 A series circuit of three resistors of 7, 12, and 15 Ω, respectively, is connected to a 120-V supply. It is desired to limit the current to 2.5 A. What resistance must be added?

3-8 A series circuit that has a current of 2 A has two resistors. If it desired to have 70 V across one resistor and 50 V across the other, what is resistance of each resistor?

3-9 A 120-V supply furnishes current to a series circuit of three resistors. Resistor *A* has 20 V across its terminals. Resistor *B* has a current of 4 A and has the same resistance as *C*. Calculate the resistances of *B* and *C*.

3-10 The field winding of a generator has a resistance of 90 Ω. It is necessary to vary the current to a minimum of 1.5 A, with a variable series-connected resistor called a *rheostat*. The rated voltage is 230 V. Calculate (a) the maximum resistance value of this rheostat (a rheostat can be adjusted to zero resistance) and (b) maximum current through the field winding.

3-11 A series resistor of 80 Ω has a current of 2 A. If the power supply is 240 V, what resistance must be added to the circuit so that the current remains at 2 A?

3-12 A group of five series-connected 10-Ω resistors are connected in series with one 20-Ω resistor. What voltages are available from this circuit if it is connected to a 120-V supply?

3-13 A 12-V automobile ignition coil normally operates at 9 V by using a series resistor. Calculate its resistance if the coil has a resistance of 10 Ω.

3-14 A resistance of 50 Ω is in series with a 30-Ω resistor. The 30-Ω resistor has a switch connected across its terminals. If the supply voltage is 115 V, calculate the current when the switch is (a) open and (b) closed.

3-15 Three resistors of 12, 20, and 25 Ω, respectively, are connected in parallel to a 120-V source. Calculate the total current. (Solve by two methods.)

3-16 Two 20-Ω resistors are connected in parallel, and this combination is connected in series with a 30-Ω resistor. The voltage of the source is

240 V. Calculate (a) the total current in the circuit and (b) the voltage across each resistor.

3-17 Twenty 50-Ω resistors are connected in parallel to a 240-V supply. Calculate (a) the current to each resistor and (b) the total current of the circuit.

3-18 Two resistors are connected in parallel to a 120-V source. One resistor has four times the resistance of the other. The total current is 25 A. Calculate the current to each resistor.

3-19 A 25- and a 40-Ω resistor are connected in parallel to a 240-V supply. Calculate the total current.

3-20 A 40- and a 60-Ω resistor connected in parallel are required to have a voltage of 150 V. Calculate the series resistor necessary if the source voltage is 230 V.

3-21 A series circuit consists of three resistors, 20, 40, and 60 Ω, respectively. If the supply voltage is 240 V, calculate (a) the power of each resistor and (b) the total power of the circuit.

3-22 An electrical load consists of fifty 100-W incandescent lamps. The voltage is 115 V. Calculate the current required for this load.

3-23 A parallel circuit consists of a 10-, 20-, and 40-Ω resistor. The voltage is 115 V. Calculate (a) the power of each resistor and (b) the total power.

3-24 A group of lamps must operate at a voltage not less than 115 V. When operating at 115 V, the power consumed is 6500 W. Calculate (a) the minimum conductor (two wires) resistance permitted if the source voltage is 120 V, (b) the cost of operating lighting for an 8-hr day if the energy cost is 2 cents/kWh, and (c) the cost of power lost in conductors.

3-25 Calculate the cost of operating an electrical load for an 8-hr day. The energy cost is 2 cents/kWh. The load consists of (1) twenty-five 100-W lamps, (2) three 2-kW heaters, and (3) a motor with an input of 2400 W.

3-26 A series circuit consists of a 50-, 75-, and 100-Ω resistor. The source voltage is 120 V. Calculate (a) the voltage across each resistor and (b) the voltage across the 75-Ω resistor if another 75-Ω resistor is connected across its terminals.

3-27 Calculate the power of each resistor in the circuit shown in Fig. P3-27.

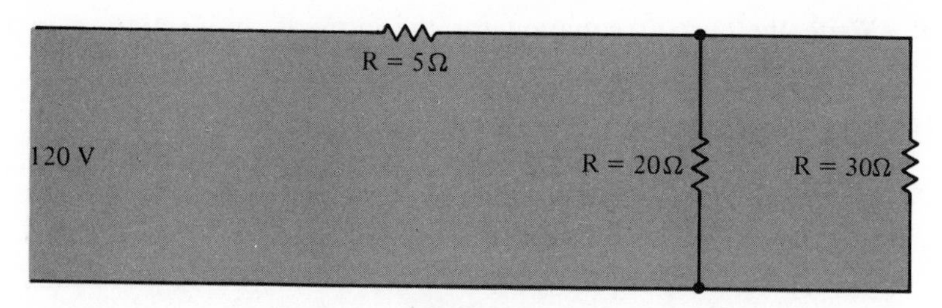

Figure P. 3-27

3-28 (a) Calculate the current in each of the parallel connected loads shown in Fig. P3-28.

(b) Calculate the resistance of load *A*.

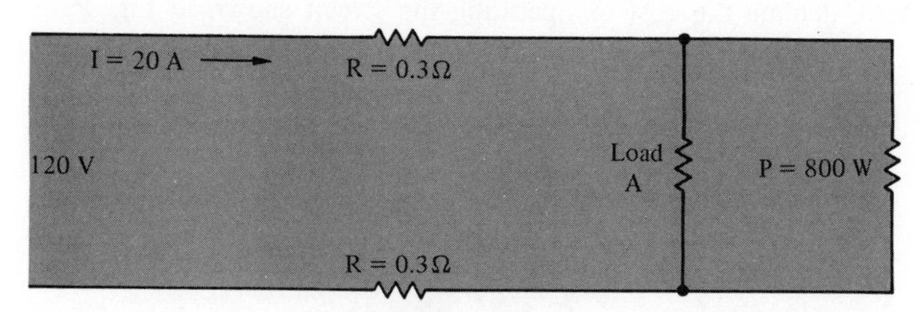

Figure P. 3-28

3-29 Write the current equation for the currents at junctions *A* and *B* in Fig. P3-29 according to Eqs. (3-3a) and (3-3b).

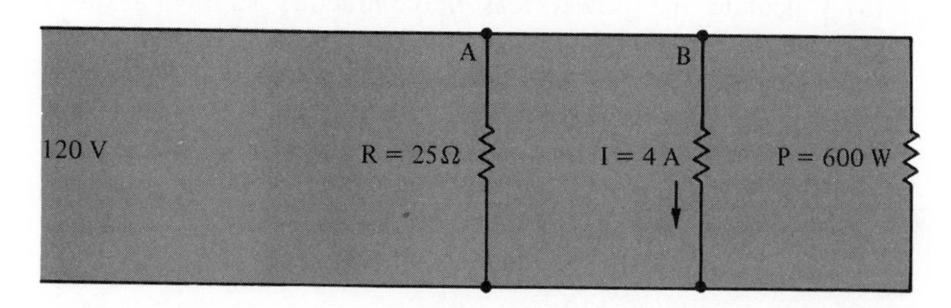

Figure P. 3-29

3-30 Write the current equation for the currents at junction A in Fig. P3-30 according to Eq. (3-3b).

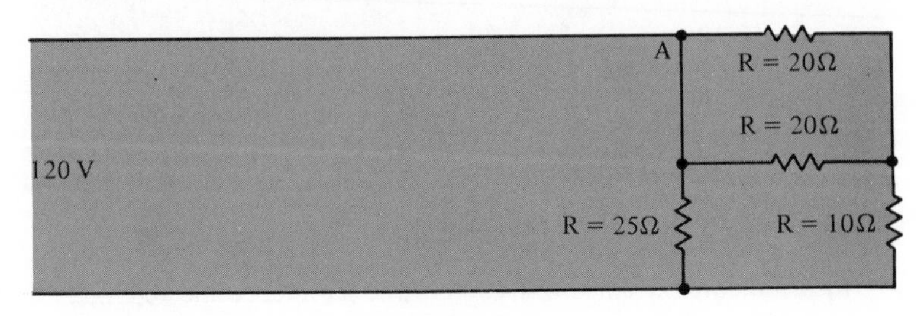

Figure P. 3-30

3-31 Calculate the cost of operating the circuit shown in Fig. P3-31 for twenty 8-hr days. The energy cost is 2 cents/kWh.

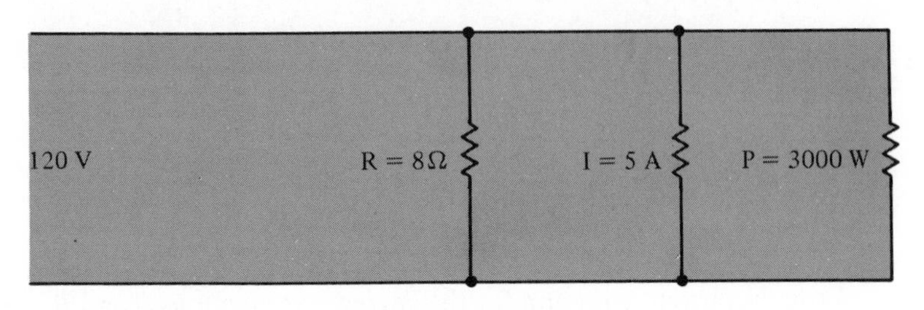

Figure P. 3-31

3-32 (a) Calculate the power loss in conductors supplying the 150-A load in Fig. P3-32.

(b) Calculate the cost of this power loss for 1 month, 20 days, 8 hr/day. The energy costs 2 cents/kWh.

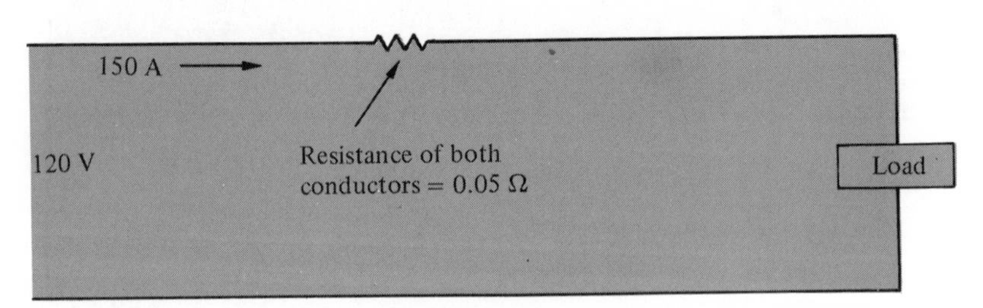

Figure P. 3-32

Chapter Four

Power in ac Circuits

4-1 ***Direct vs. alternating current.*** Thus far we have explored only current flow in series and parallel circuits. Little has been mentioned about whether this current is direct current (dc) or alternating current (ac). Before extending this subject any further a definition and a description of the practical use of each is in order.

The term dc is used to describe unidirectional current that flows in one direction. It may not always be of the same magnitude—it may change from 2 to 3 A and then back to 2 A in a steady cycle of change—but if it does not reverse direction, it would be called a dc.

To clarify some dc circuits, if the current is always at the same value it may be called "pure" dc. A battery is a good source of pure dc. Since the voltage of a battery is produced by chemical action, the voltage thus produced would not change its value for a given load; therefore, the current flow as a result of this voltage would not change its value.

Direct current *generators*, small or large, are the usual source of dc power. A more detailed study of electrical machinery is presented in Chapter 6. It will suffice for the present to state that the voltage produced by a dc generator is constant for a given load. Therefore, any current that flows would not vary, unless of course the circuit being supplied is changed.

Direct current systems are used in a variety of applications; these include welding, electroplating, and emergency lighting. Direct current motors have certain speed and torque characteristics that justify their use.

The term "alternating current" (ac) is used to describe a current that periodically changes its direction, and by so doing it must be *continuously* changing its magnitude. The only instant it is not changing is the instant when it actually reverses. This situation is similar to a ball thrown straight up in the air that is continuously moving, except for the instant when it stops going up and has not yet started to come down.

The operation of the entire electrical industry and hundreds of various electrical devices depend on the concept of current change in the device itself. The method by which this current is made to change continuously in the form of a *sine wave* is built into the design of the ac generator. This machine generates a voltage that *alternates*; therefore, the current that flows also will alternate.

Many electromechanical devices have been invented that cause a current to change from one value to another even though the voltage source does not change. However, this involves other equipment whereas the ac generator produces this variable voltage by rotation and thus produces a sine wave variation. The number of times a voltage goes through its cycle of change per unit time is called the *frequency* of the system. This is usually measured in cycles per second, such as 60 cps. A new unit was adopted recently called the *hertz* (Hz) that enables us to avoid any mention of a unit of time. A 60 cycles-per-second (cps) voltage can now be called 60 Hz.

Practically the entire United States now uses 60-Hz current. Previously, some areas used 50, 40, or 25 Hz and even dc. Obviously energy systems at different frequencies made system interconnection impossible. Electrical devices also had to be designed for each different frequency. This would be like different railroads having different gauges or spacing for their rails.

4-2 ***Alternating current in a resistance.*** Power in a resistor has been described as heat produced by the electron movement in the atoms of the material of which the resistor is made. It should be quite obvious that the heat produced does not vary with the instantaneous current that is continuously changing in value, provided we can average the changes that occur over a cycle. For example, a certain amount of heat would be produced by a direct current of 10 A in a resistor. If a current of 20 A flowed for 1 sec and then 5 A flowed for 2 sec, 15 A for 1 sec, and 5 A for 1 sec, and this 5-sec cycle repeated itself indefinitely, the *average* current over each 5-sec period would be 10 A. This average current would produce the same amount of heat as a steady or direct current of 10 A.

In Fig. 4-1 we can calculate this average value by finding the area under the graph of current values and dividing by the number of units, in this case time in seconds.

$$I = \frac{(20 \times 1) + (5 \times 2) + (15 \times 1) + (5 \times 1)}{5} = 10 \text{ A}$$

The speed with which a current changes is of no importance; this particular cycle could take place in 0.001 sec and still give the same results. The current we have graphed is a direct current, since it does not reverse.

However, if in the next 5-sec period the current actually reversed and repeated the same changes, the same heat would be produced by the reversed current even though on a graph we would call it a negative current. In this case we must recall that one expression for power or heat in a resistor is I^2R. If current reversed, either periodically or permanently, then $-I$ would represent a current flow in the opposite direction. We could now write this power relationship as $(-I)(-I)R = I^2R$. When alternating current flows in a resistor it produces the same heat effect as a direct current, if our unit of measurement of the ampere is the same. To assure this equality we identify the ac ampere as I_{rms}, or the *effective* ampere.

The term "rms" is an abbreviation for *root-mean-square*. This is an average obtained by taking very small increments of current, squaring each value, adding them all together, dividing by the number of increments, and

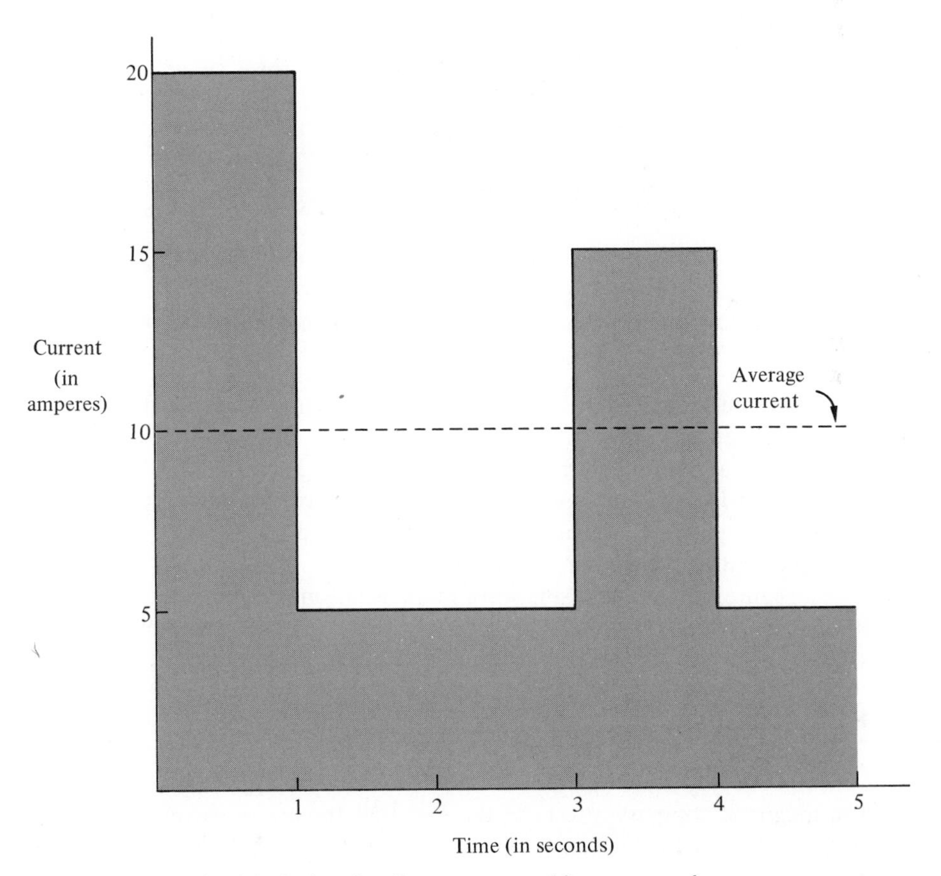

Fig. 4-1. A changing direct current and its average value.

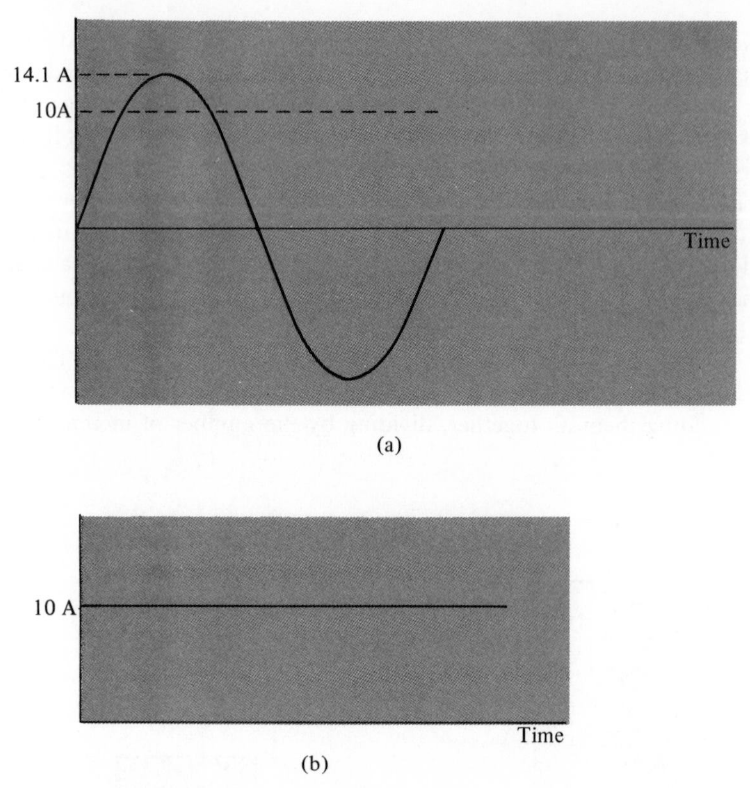

Fig. 4-2. (a) Maximum and effective values of an ac. (b) A pure dc of 10 A. Each of these currents will cause the same heating effect in a resistance.

extracting the square root. This task is all done for us when we read the ac amperes on an ac ammeter.

Since alternating currents are usually varying sinusoidally, I_{rms} will be equal to 0.707 of the maximum value of its sine wave. Throughout this book the symbol I means I_{rms}, which is effectively equal to I_{dc}, as far as producing the same heat in the same resistor is concerned.

Figure 4-2a shows a sine wave of an ac. Its maximum value is 14.1 A. Its effective value is (14.1)(0.707) or 10 A. Also, a dc of 10 A is shown in Fig. 4-2b.

4-3 ***Magnetism.*** In Chapter 1 the statement was made that "when a conductor is moved through a magnetic field an electromotive force is produced," but we were not told how to produce a magnetic field. The magnetic compass attests to the fact that the entire earth possesses a magnetic field with this strange force leaving its North Pole and entering its South Pole.

Here we introduce another fundamental fact of electricity, that an electric current in a conductor produces a magnetic field around that conductor.

Let us keep account of these basic facts of electricity thus far encountered:

1. **Current flow in a conductor is proportional to the voltage applied and inversely proportional to the opposition (Ohm's law).**
2. **An electric current in a conductor produces a magnetic field around that conductor.**
3. **Moving a conductor through the magnetic field will generate a voltage in that conductor.**

A source of a magnetic field, then, is an electric current. The strength of the field is represented by magnetic lines of force. The *number* of lines around the conductor depend on the *current* and the medium through which the lines are set up. Iron or any alloy containing iron is an excellent producer of magnetism and is used extensively to generate magnetic circuits. A *magnetic circuit*, like the electric circuit, must have a *complete* path for the magnetic field or lines of magnetism. This magnetic field is called *flux*; its symbol is ϕ. The force that causes the magnetic flux is called *magnetomotive force* or *mmf*. The mmf can be evaluated in ampere-turns because the field flux depends both on the current and on the turns of the coil carrying the current that set up a magnetic field within the coil.

As a coil of wire is formed, a small current going through many turns can produce the same flux as a large current going through a few turns. This product of current and turns is called ampere-turns and is abbreviated NI.

Figure 4-3 shows a typical magnetic core and circuit with its exciting winding. Magnetic cores of course may take on any shape dependent on the use. Some magnetic circuits have a closed core while others have an air gap. The air gap is necessary to produce a north and a south pole. Obviously, to move a conductor through a magnetic field there must be an air gap.

Since magnetic flux has a direction, we define the north pole (if there is one) as the pole where the flux leaves and the south pole where flux enters. Between the two poles must be a nonmagnetic material such as air, wood, or aluminum. If the current in the winding is reversed, the poles will reverse. Fleming's right-hand rule states, "Let the fingers of the right hand point in the direction of the current flow, then the thumb will point in the north direction of the magnetic flux."

The ability of a magnetic circuit to establish flux lines is called its *permeability*. This is merely a ratio or number without units that is the

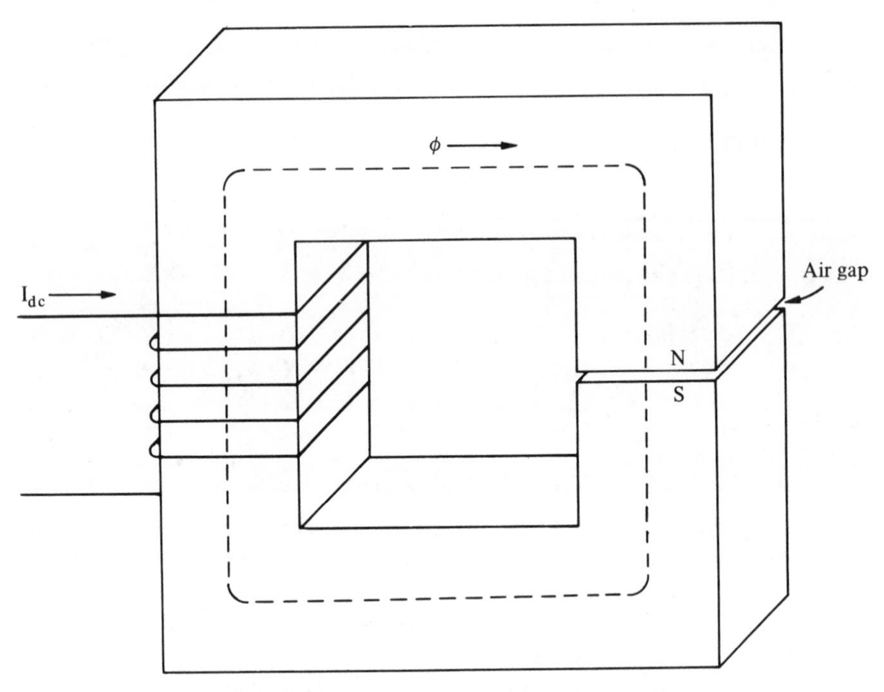

Fig. 4-3. Typical magnetic circuit. I_{dc} is the North and South poles constant. I_{ac} is the North and South poles alternate.

permeability of a material compared to the permeability of air (unity). (Permeability in this respect is similiar to the specific gravity of liquids, which is the density of a material compared to the density of water, which is also unity.) All nonmagnetic materials, including air, have a permeability of unity.

4-4 ***Inductance.*** To properly understand ac circuits, the student must become familiar with an electrical quantity not previously mentioned, called *inductance*. We shall describe it simply as the property of a coil of wire to oppose any change of current in it. This coil may or may not be wound on an iron core. The equation for inductance is

$$L = \frac{N^2 \mu A}{l} \tag{4-1}$$

where L is inductance as measured in henrys, N is number of turns of wire, and μ is permeability or a measure of the magnetic circuit's ability to become magnetized. A and l are physical dimensions representing the area and the length of the magnetic circuit. This equation is not given to evaluate the inductance as much as to note those factors affecting its magnitude. Obviously, the number of turns N and their presence on an iron core (high μ) contribute greatly to inductance.

A single length of wire has one turn. If a coil of wire wound on an iron core is removed and simply spread out to its entire length, its property of inductance would be very small, since N and N^2 would be unity. The property of resistance of the wire, of course, would be the same in either case, regardless of turns.

Magnetism is needed for the operation of many electrical devices from both dc and ac sources. In electrical circuitry we must be able to evaluate the effect of inductance when a magnetic circuit is required. In this text we shall not design magnetic circuits, we shall only discover how to change the magnetism and observe the results from such change.

4-5 ***Inductive reactance.*** When dc passes through a coil possessing the property of inductance, magnetism is produced in its iron core. This magnetism is steady and changes only when the current is turned on or off. These changes are called *transient* periods and are of such short duration that they can be neglected. However, if ac flows in the coil, the magnetic flux produced by this current is *continuously* changing as well as *reversing*. Now think about rule No. 3 (Sec. 4-3), that moving a conductor in a magnetic field will generate a voltage in that conductor. You may say that the conductors of the coil on the iron core are *not* moving. True, but the magnetic flux is! It is continuously changing in an orderly manner *dictated* by the alternating current. If the flux is changing through a coil of wire, it produces the same effect as if the conductors of the coil moved physically through a constant magnetic field excited with a direct current.

This action means that the coil, influenced by the changing magnetic flux, will have a voltage induced in itself. This self-induced voltage is in opposition to the applied voltage because of Lenz's law, which states that "any action in a circuit will produce a reaction in such a direction as to oppose the cause that started the action." Figure 4-4 shows how this might be represented in diagram form.

Figure 4-4a shows two batteries in opposition. Figure 4-4b shows an induced voltage opposing an ac source. The voltage in the enclosure in Fig. 4-4b represents an induced voltage in the windings of the inductive load.

The evaluation of this opposing induced emf becomes very difficult. However, if we represent this reaction as equivalent to *opposition* to current flow in ohms, it becomes very simple, provided the current that flows is sinusoidal. The current supplied by all commercial power systems, classified as ac, varies in a sine wave pattern and meets this condition.

In Fig. 4-4a if the 2-V battery were replaced with a resistor, some particular value of current could cause a 2-V drop and also leave us 4 V to apply to a load. This means that we could replace the induced emf in Fig.

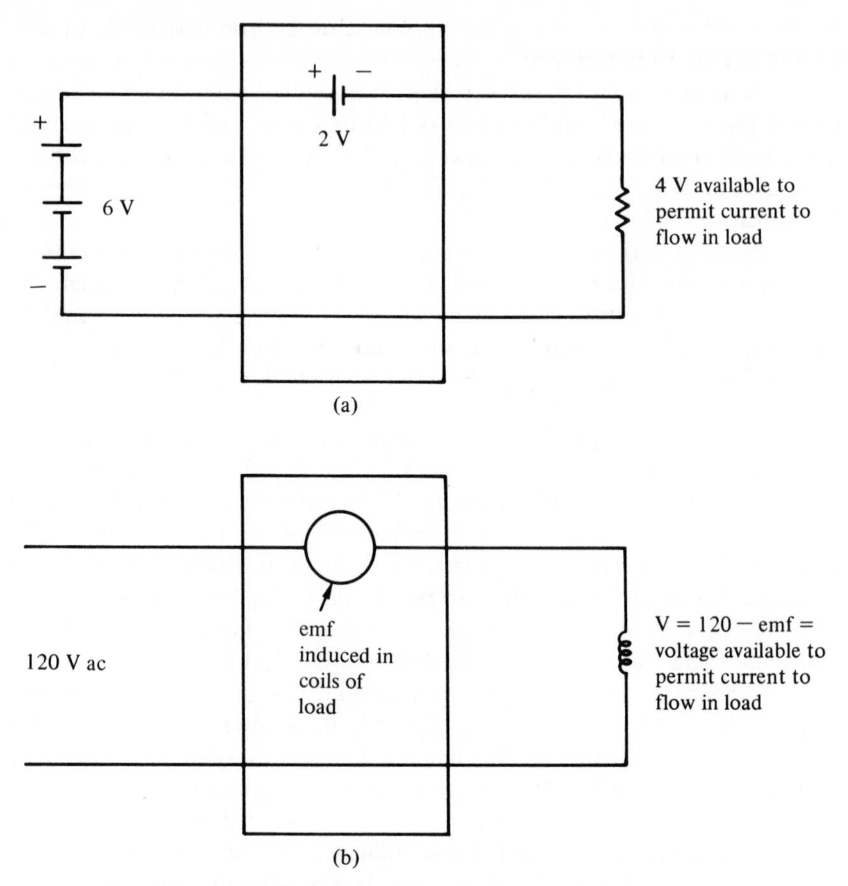

Fig. 4-4. (a) A 6-V dc supply showing an opposition voltage reducing available voltage. (b) A 120-V ac supply connected to an inductive load, showing effect of reactance in the coil.

4-4b with an ohmic value to cause a voltage drop. The ohmic value we use is called an *inductive reactance* and mathematically is equal to

$$X_L = 2\pi f L \tag{4-2}$$

where X_L is inductive reactance in a coil of wire measured in ohms; 2π is a mathematical constant used when the current is sinusoidal; f is the frequency of the supply voltage in cps or Hz; and L is the inductance of the coil as measured in henrys. Note that when dc is flowing through a coil, the reactance is zero because the frequency is zero. In a more practical approach it could be said that a dc sees no opposition due to inductance when flowing through an inductance. This relates quite properly to our definition of inductance (Sec. 4-4). An exception to this would be when the dc is changing, and this opposition known as inductive reactance appears only while the dc is changing.

Example ***4-1.*** Calculate current flow in a coil with 0.2-H inductance if connected to a 60-Hz, 120-V power supply.

Solution

$$X_L = 2\pi f L \qquad (4\text{-}2)$$

$$= (6.28)(60)(0.2)$$

$$= 75.4\ \Omega$$

$$I = \frac{E}{X_L}$$

$$= \frac{120\ \text{V}}{75.4\ \Omega}$$

$$= 1.59\ \text{A}$$

4-6 ***Power in an inductance.*** In Ex. 4-1 we may ask what effect this 1.59 A has in the coil. The answer must be that it is producing varying magnetism, since there is theoretically no resistance to produce any heat. In most electrical devices a current can provide only these functions: heat or magnetism, and usually both. We may now ask whether it consumes power to produce this magnetism.

Figure 4-5a shows the sine waves of the voltage and current in an inductor. Multiplying these two sine waves together ($P = EI$) gives the double frequency power curve. It shows that during one-half of the period power is positive, flowing to the inductor from the power supply. For the other half period, power is negative, flowing back to the power supply. This is logical since we have described the inductor as a voltage generator and in such a role it would not be able to furnish power. This means the average net power consumed by the inductor for one cycle is zero.

In order for current and voltage sine waves to produce no power when multiplied together, their displacement must be 90° apart. This places them one-fourth of a cycle apart. The vector counterparts of the sine waves are shown in Fig. 4-5b.

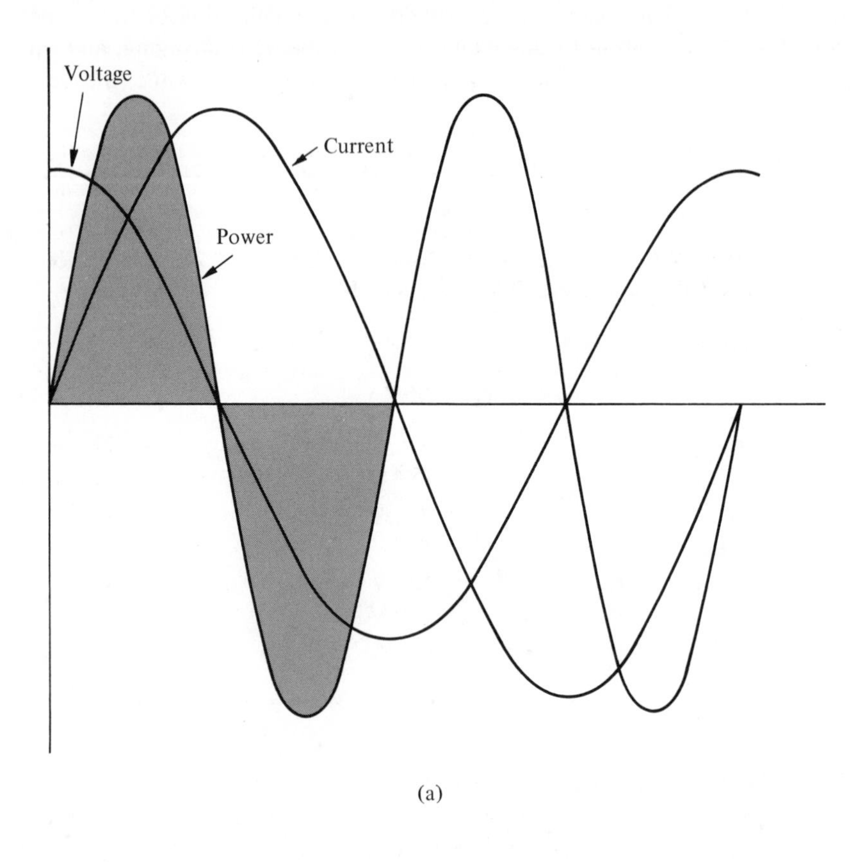

(a)

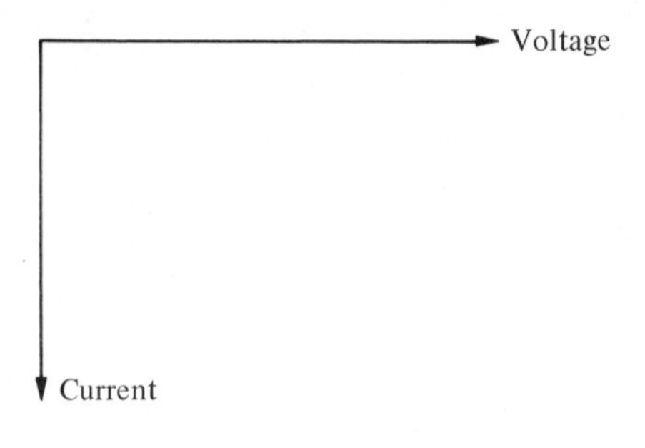

(b)

Fig. 4-5. (a) Sine waves of voltage, current, and power in an inductance. (b) Vector relationship of voltage and current in an inductance.

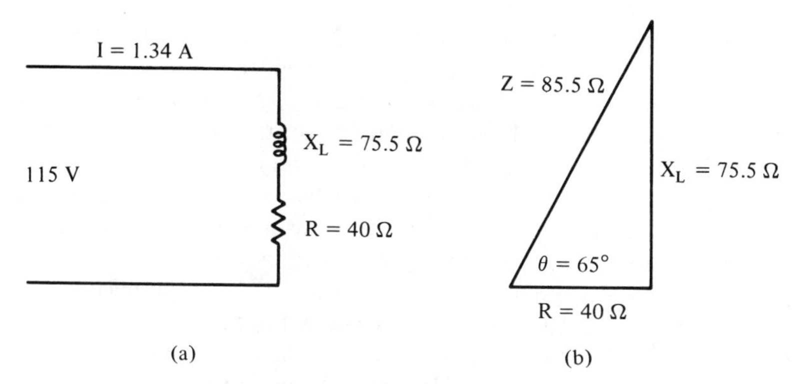

Fig. 4-6. (a) Circuit diagram of resistance and inductive reactance in series. (b) Vector addition of resistance and inductive reactance in series.

Example **4-2.** Calculate the impedance of the coil of Ex. 4-1 if it has 40 Ω resistance. Refer to Fig. 4-6.

Solution

$$Z = \sqrt{R^2 + X_L^2} \tag{4-3}$$

$$= \sqrt{40^2 + 75.4^2} = \sqrt{7300}$$

$$= 85.5\ \Omega$$

The practice of using vectors for representing electrical quantities that are at some angle with respect to each other greatly simplifies such calculations.

The previous discussion assumed an *ideal* inductor having no resistance. This is impossible because all wires have resistance. Whether the resistance of a coil might be small enough to be neglected would depend on the relative magnitude of resistance with respect to reactance. The resistance of a coil must be added to its reactance. This addition is not performed arithmetically but vectorially in agreement with the 90° angle of current displacement with respect to voltage. Therefore, resistive and inductive ohms are added at right angles. Their sum, called *impedance*, is abbreviated Z. Impedance represents the total opposition to current flow *created by* a coil.

$$Z = \sqrt{R^2 + X_L^2} \tag{4-3}$$

where R is the ac resistance in ohms; X_L is the inductive reactance in ohms; and Z is the impedance or total opposition in ohms.

4-7 *Power factor*. Now that we have included the resistance of the inductor, it follows that power must be expended in the resistance of the inductive circuit. In inductive devices current cannot be in phase with the voltage because of the reactance present. The amount of angular displacement (actually, time) between voltage and current depends on the ratio or relative amount of resistance to reactance. Because the current and voltage do not reach maximum at the same instant (like two men tugging on a load from different directions), the real power of the circuit (the in-phase product of voltage and current) must be $E \times I$ multiplied by some factor less than one. This factor compensates for the phase displacement between voltage and current. This factor is called the *power factor* of the load and is equal to the *cosine* of the angle between the voltage across and the current drawn by the load. The power factor is also equal to the ratio of load resistance to load impedance. The angle identified as θ in Ex. 4-1 is called the power factor (pf) angle.

$$\text{pf} = \cos\theta = \frac{R}{Z} \tag{4-4}$$

4-8 *Power in an impedance*. Before considering circuits made up of many devices, let us first investigate the technique of power calculation in a single device containing both reactance and resistance.

Example 4-2 illustrates the vector addition of resistive and inductive ohms when in series or in one device. Angle θ is the actual angle that will exist between current and voltage when current flows through the coil. The inductance causes the current to lag behind the voltage by 90°, and the resistance tends to keep current in phase with voltage. The result is a compromise, and because of the relative parameters of this particular coil the current will lag behind the voltage by 65°. We measure these phase angles in degrees. One cycle is 360°. For a 60-Hz voltage, one cycle has a duration of $\frac{1}{60}$, or 0.0167 sec. Therefore, 65° would represent $\frac{65}{360}$ of 0.0167 sec, or 0.003 sec. Translated, this means that the current would reach a maximum value 0.003 sec after the voltage. This short period of time might appear to be inconsequential, but the current and voltage will continue to maintain this constant relationship. When multiplied together to obtain the power, this phase (or time) displacement between current and voltage must be considered. Our power equation then must be modified to include power factor and

$$P = EI\cos\theta = EI\,\text{pf} \tag{4-5}$$

where pf is the cosine of the angle between the voltage E and current I.

Example 4-3. Calculate: (a) The power factor of the coil of Ex. 4-2, (b) the power factor angle, and (c) the power of the coil if it is connected to a 115-V, 60-Hz supply. Refer to Fig. 4-6.

Solution

(a)
$$\text{pf} = \frac{R}{Z} = \frac{40\ \Omega}{85.5\ \Omega} = 0.465 \qquad (4\text{-}4)$$

(b)
$$\text{pf} = \text{power factor angle with cosine of } 0.465 = 65^\circ$$

(c)
$$I = \frac{E}{Z} = \frac{115\ \text{V}}{85.5\ \Omega} = 1.34\ \text{A}$$

$$P = EI\ \text{pf} = (115)(1.34)(0.465) = 71.5\ \text{W} \qquad (4\text{-}5)$$

or

$$P = I^2R = (1.34)^2(40) = 71.5\ \text{W}$$

4-9 ***Capacitors.*** There is another physical element that can permit an electric current to pass through it: a *capacitor*. A capacitor produces neither heat nor magnetism. It does, however, have several extremely useful characteristics. If any current passes through it, the capacitive current will *lead* the voltage across it by 90°. Since the current through a pure inductor lags its voltage by 90°, these two currents will be 180° apart or directly opposite. Capacitance is defined as the property to *oppose any change in voltage*.

Capacitors also have the ability to store electrical energy and are used in all sizes and shapes in electronic circuits. This role is important since they will not conduct a dc. Capacitors limit or oppose ac because the voltage magnitude is constantly changing. (See above definition.)

Physically a capacitor is merely two conducting materials separated by an insulator. This insulator may be simply wax paper or a heavy sheet of polyethylene film, depending on its voltage rating.

When connected to a dc power supply, current cannot pass through, but because of the large area of conducting material, the two surfaces charge up to the supply voltage in a short time interval. A capacitor can hold this charge for an indefinite period of time after being disconnected from the supply voltage. It will discharge through an external circuit if one is provided. The capacitor discharge takes place very quickly. A photographic flash unit is a good example of this case.

When connected to an ac power supply the plates will charge alternately positive and negative and give the illusion of current flow through the insulating medium. Because of the charge-discharge cycle, the current must lead its voltage by 90°; therefore, its power factor is zero and the capacitor consumes no power.

The magnitude of the alternating current limited by a capacitor is dependent on its physical size, the insulating material or dielectric, and how much area of the conducting materials it contains. We can evaluate its opposition in ohms to ac by the equation

$$X_C = \frac{1}{2\pi fC} \qquad (4\text{-}6)$$

where X is its capacitive reactance measured in ohms; C is its capacitance in farads; and f is the frequency in hertz of the ac applied to the capacitor.

A farad is a very large unit; therefore, capacitance is usually measured in microfarads. Then the equation may be written

$$X_C = \frac{10^6}{2\pi fC} = \frac{159 \times 10^3}{fC} \qquad (4\text{-}7)$$

where C is capacitance measured in microfarads (μF), or millionths of a farad.

Examples 4-4 and 4-5 illustrate how capacitors can limit current flow and reduce the circuit current to an inductive load. Example 4-5 also illustrates a simple technique of adding currents that are out of phase with each other.

Remember two important rules:

1. **When impedances or reactances are in series they must be added, mathematically or by vectors.**
2. **When impedances or reactances are in parallel we must add their *currents*, mathematically or by vectors.**

The current through a capacitor can be calculated in the same manner as the resistor or the inductor by dividing the voltage across the capacitor by its capactive reactance in ohms, or

$$I = \frac{E}{X_c} \qquad (4\text{-}8)$$

Example **4-4.** Calculate the current that will flow in a 50-μF capacitor if it is connected to a 120-V, 60-Hz power supply.

Solution

$$X_C = \frac{(159)(10^3)}{fC} \tag{4-7}$$

$$= \frac{(159)(10^3)}{(60)(50)} = \frac{(159)(10^3)}{(3)(10^3)}$$

$$= 53\ \Omega$$

$$I = \frac{E}{X_C} = \frac{120\text{ V}}{53\ \Omega} = 2.26\text{ A} \tag{4-8}$$

Example **4-5.** A magnetic device has 20 Ω total impedance. Its resistance is 12 Ω. It is connected to a 120-V, 60-Hz power supply. A 75-μF capacitor is then added in parallel to the circuit. Calculate (a) the current in the magnetic device (see Figs. 4-7 and 4-8), (b) the circuit current after the capacitor is added (see Figs. 4-9 and 4-10), and (c) the power of the circuit in (a) and (b).

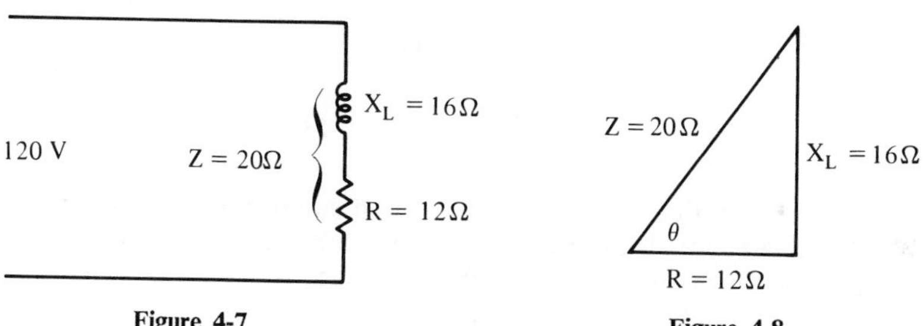

Figure 4-7

Figure 4-8

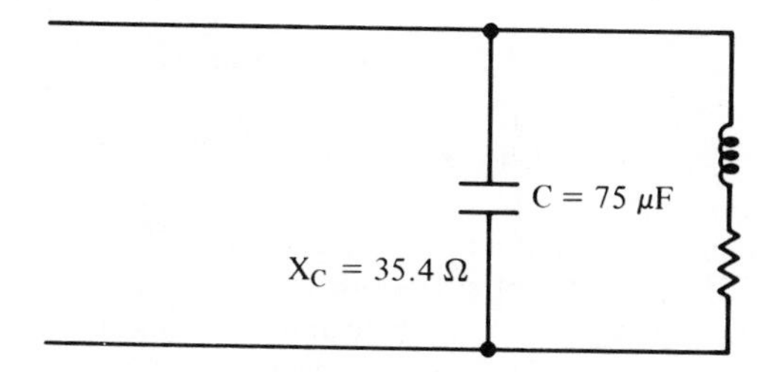

Figure 4-9

Solution

(a)

$$\text{pf} = \frac{R}{Z} = \frac{12\ \Omega}{20\ \Omega} = 0.6 \tag{4-4}$$

Angle θ is the angle whose cosine is 0.6, or 53°.

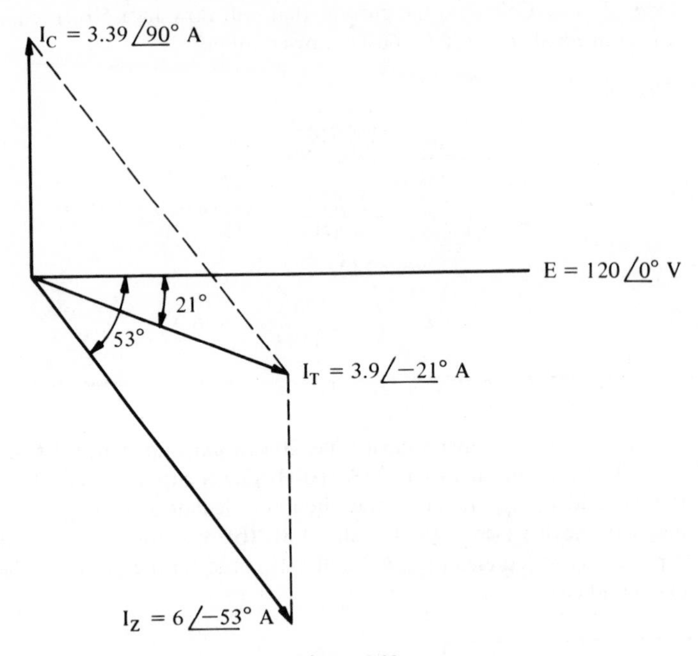

Figure 4-10

$$I = \frac{120\text{ V}}{20\ \Omega} = 6\angle -53^\circ\text{ A}$$

Note that the method of notation shown for I indicates both the magnitude and phase of the current that lags behind the voltage by 53°.
(b) The total current with the capacitor added is

$$X_C = \frac{(159)(10^3)}{fC} = \frac{(159)(10^3)}{(60)(75)} \quad (4\text{-}6)$$

$$= 35.4\ \Omega$$

$$\text{Current through capacitor} = I_C = \frac{E}{X_C} = \frac{120\text{ V}}{35.4\ \Omega} = 3.39\angle 90^\circ\text{ A} \quad (4\text{-}8)$$

Again, the above notation shows that this 3.39-A capacitor current must lead the voltage by 90°. Using a protractor and a scale, these two currents, $6\angle -53^\circ$ and $3.39\angle +90^\circ$ A, can be added graphically as shown in Fig. 4-10.

The resultant total current lags behind the voltage by 21° and has a magnitude of 3.9 A and is expressed as shown. The cosine of 21° is 0.93. This is the power factor of the circuit.
(c) The power of the circuit in (a) is

$$P = EI\text{ pf} = (120)(6)(0.6) = 430\text{ W} \quad \text{or}$$
$$P = I^2R = (6)^2(12) = 430\text{ W}$$

The power of the circuit, with a capacitor, in (b) is

$$P = EI\cos\theta = (120)(3.9)(\cos 21°)$$
$$= (120)(3.9)(0.93)$$
$$= 430 \text{ W}$$

Note in Ex. 4-5 that adding the capacitor (which dissipates no power) to the circuit does not change the power. The addition of a capacitor, however, reduces the circuit current from 6 to 3.9 A and raises the pf of the circuit. The importance of this property will become apparent in later discussions.

If done carefully the graphical method of Ex. 4-5 can be quite accurate when adding currents of different angles. In this case, when worked out mathematically, the total current is 3.86 A. The graphical addition shows 3.9 A.

4-10 ***The power triangle.*** Many of the electrical problems we will meet involve an electrical load made up of many different devices, all connected in parallel. Some may be resistive, such as incandescent lighting or heating elements. Some may be part inductive and part resistive, such as fluorescent lighting or motors. If a load is capacitive it probably would consist of pure capacitors added to an existing system to compensate for an inductive load. This actually will reduce the magnitude of the total current to the system, as in Ex. 4-5b.

To evaluate such a load by determining the ohmic value of each device is virtually impossible. This is because of the possibility of resistive and inductive ohms in the same device. Since the whole is the sum of its parts, we shall add up our electrical loads by considering the current and power of a system. We shall recognize that part of this current is in phase with the voltage (resistive) and part is lagging behind the voltage by 90° (inductive) and part may be leading the voltage by 90° (capacitive). Section 2-8 illustrates how this is done when the load has only resistive components connected in parallel.

An *impedance triangle* is a means of determining all of the parameters of a device if any two of the four are known. The four parameters are resistance, reactance, impedance, and power factor. These form a right triangle. (Refer to Fig. 4-11a.)

By multiplying all sides of this triangle by the current (R and X_L are in series and therefore carry the same current) we get a *voltage triangle*, as in Fig. 4-11b. IX_L is voltage across inductance; IR is voltage across resistance; and IZ is the sum of the two, or voltage of supply.

Now refer to Fig. 4-11c. If the voltage triangle is multiplied again by

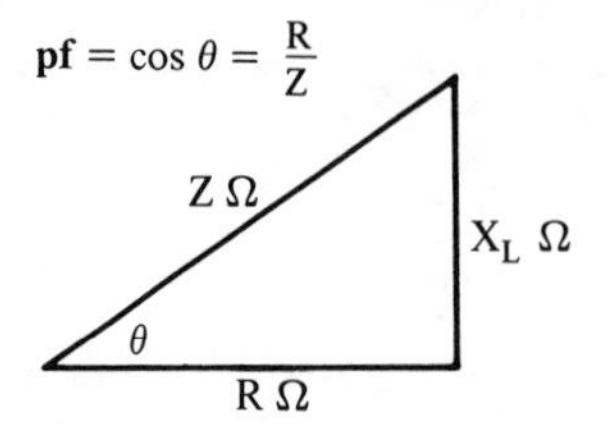

(a) Impedance Triangle

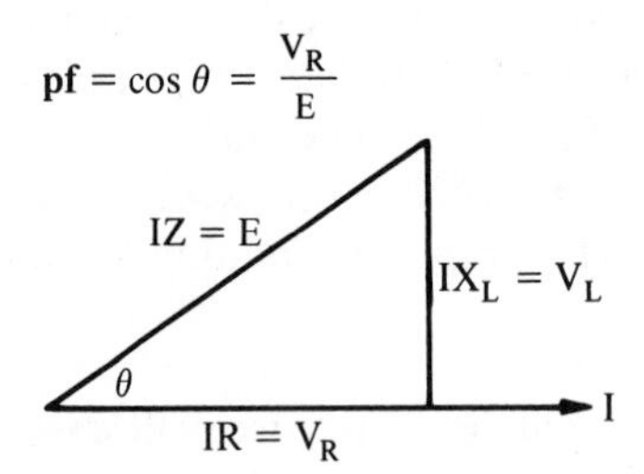

(b) Voltage Triangle

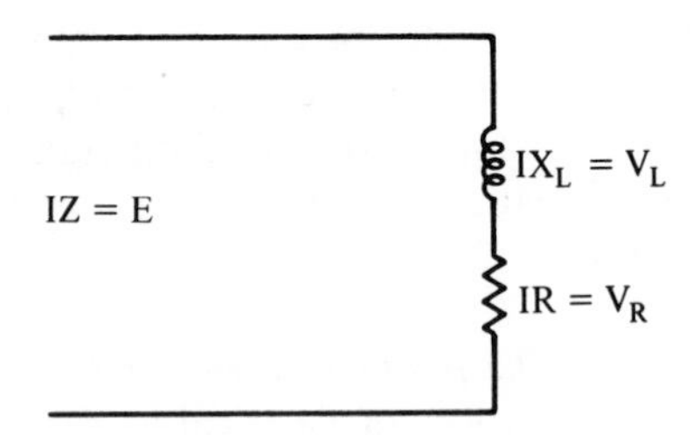

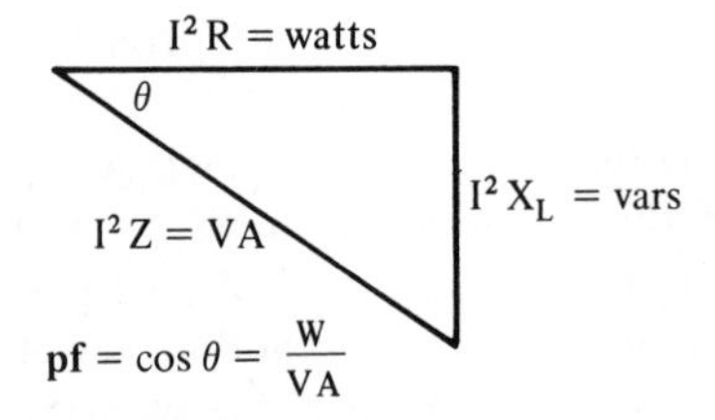

(c) Power Triangle

Figure 4-11

the current, the result will be a *power triangle*. The horizontal side is power expressed in watts; the vertical side is reactive power, which actually has no named unit but is referred to as *var*. This is the abbreviation of *volt-amperes reactive*. The hypotenuse is the product of the voltage and the current. This unit is sometimes called *apparent power* or simply *volt-amperes*, abbreviated VA.

This power triangle now has the same shape, and therefore the same angles, as the original impedance triangle. If any two sides, or any combination of one side and one acute angle, are known, the rest of the triangle can be solved. To be consistent with the vector picture that always shows the current *lagging behind* the voltage in an inductive load, the power triangle is drawn with the vars going vertically down. If they were capacitive vars they would be drawn upward.

From the power triangle the relationships between power and volt-amperes can be shown by

$$\text{VA} = \frac{W}{\cos\theta} = \frac{W}{\text{pf}} \tag{4-9}$$

Also, the relationship between volt-amperes and vars can be shown by

$$\text{vars} = (\text{VA})(\sin\theta) \tag{4-10}$$

The sine of θ, or the sine of the power factor angle, is sometimes called the reactive factor.

Example **4-6.** Calculate the power factor of a 40-kW load that has a current of 250 A and operates at 220 V. Draw the power triangle. Refer to Fig. 4-12.

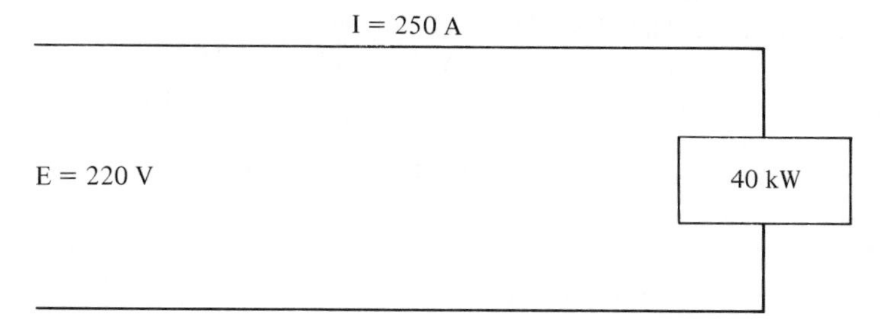

(a) Diagram for Example 4-6

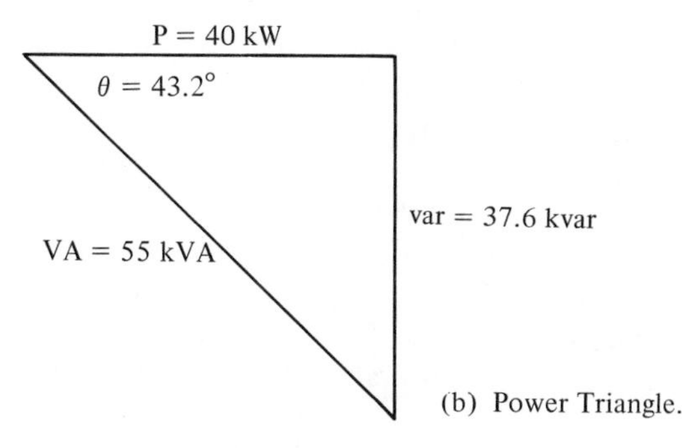

(b) Power Triangle.

Figure 4-12

Solution

$$P = EI\,\text{pf} \tag{4-5}$$

$$\cos\theta = \text{pf} = \frac{P}{EI} = \frac{40{,}000}{(220)(250)} = 0.73$$

The angle between the power and the volt-amperes is the angle whose cosine is 0.73 or 43.2°.

$$\text{kVA} = \frac{\text{kW}}{\cos\theta} = \frac{40}{0.73} = 55\ \text{kVA} \tag{4-9}$$

$$\text{vars} = \text{VA}\,(\sin\theta) \qquad \sin 43.2° = 0.684 \tag{4-10}$$

$$= (55)(0.684) = 37.6\ \text{kvars}$$

4-11 ***Addition of loads with different power factors.*** In large buildings, electrical loads must be parceled out in sections to assure proper distribution. The slogan "Don't put all your eggs in one basket" is good advice when carrying eggs. It can also guide us to good electrical design by not allowing all of the electrical load to depend on one set of conductors or one set of fuses.

When two or more electrical loads are resistive, the power factors are all unity and the true power is equal to the volt-amperes. Thus 35 kW plus 50 kW plus 20 kW is simply 105 kW. The current for each resistive load or the total resistive load is obtained simply by dividing the power by the system voltage.

Let us consider three 25-kW loads, each with a different power factor: unity, 0.8, and 0.65, respectively. Obviously the total power is 75 kW, but the current cannot be obtained by dividing the 75 kW by the system voltage because more current is required by the 25-kW loads whose power factors are less than one. We also must recognize that the kVA of each of these loads is different and has a different phase angle from its individual 25 kW of power. Because of these angles the kVA of the three loads cannot be added together arithmetically but must be added by the use of vectors. This can be done easily by adding horizontal and vertical components of triangles. We add the horizontal sides (kW) and then add the vertical sides (kvar). These sums form two sides of a new resultant right triangle whose hypotenuse will be the total kilovolt-amperes. The power factor of the entire system can then be calculated from the resultant triangle. This procedure is illustrated in Ex. 4-7.

Example 4-7. Three 25-kW loads have different power factors, 1.0, 0.8, and 0.65, respectively. Calculate (a) the overall power factor (see Figs. 4-13 to 4-15), (b) the total kVA of the system, and (c) the total current if the voltage is 230 V.

Solution

(a) 25 kW

Load 1 pf $= 1.0$ $\quad \theta = 0°$ $\quad \cos\theta = 1.0$

kvars $= 0$ $\quad$ kVA $=$ kW $= 25$

Load 2 pf $= 0.8$ $\quad \theta = 36.9°$ $\quad \cos\theta = 0.8$

$$\text{kVA} = \frac{\text{kW}}{\cos 36.9°} = \frac{25}{0.8} = 31.3 \text{ kVA}$$

$$\text{kvar s} = \text{kVA}(\sin 36.9°) = (31.3)(0.6) = -18.75 \text{ kvars}$$

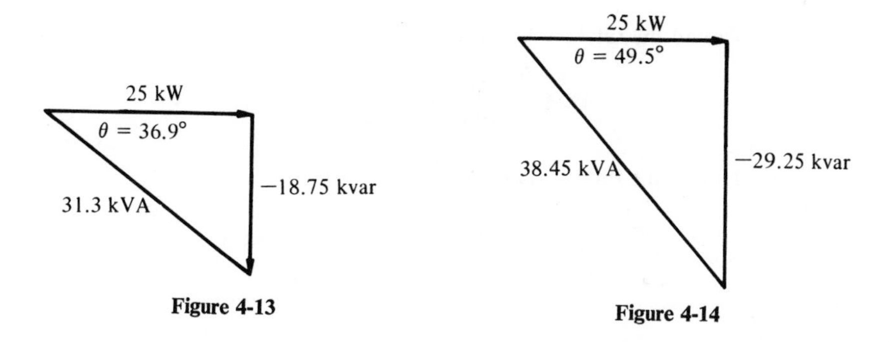

Figure 4-13 **Figure 4-14**

Minus sign indicates lagging kvars because of lagging current.

Load 3 pf $= 0.65$ $\quad \theta = 49.5°$ $\quad \cos\theta = 0.65$

$$\text{kVA} = \frac{\text{kW}}{\cos 49.5°} = \frac{25}{0.65} = 38.45 \text{ kVA}$$

$$\text{kvars} = \text{kVA} \sin(49.5°) = (38.45)(0.76) = -29.25 \text{ kvars}$$

Adding the three loads by adding horizontal and vertical components of the three triangles, produces the results shown in Fig. 4.15. The i-f power factor angle of entire system. It may be measured graphically or

$$\tan\theta = \frac{48 \text{ kvars}}{75 \text{ kW}} = 0.64$$

and angle θ is then 32.6°. The power factor of the entire system is cos 32.6° or 0.842.

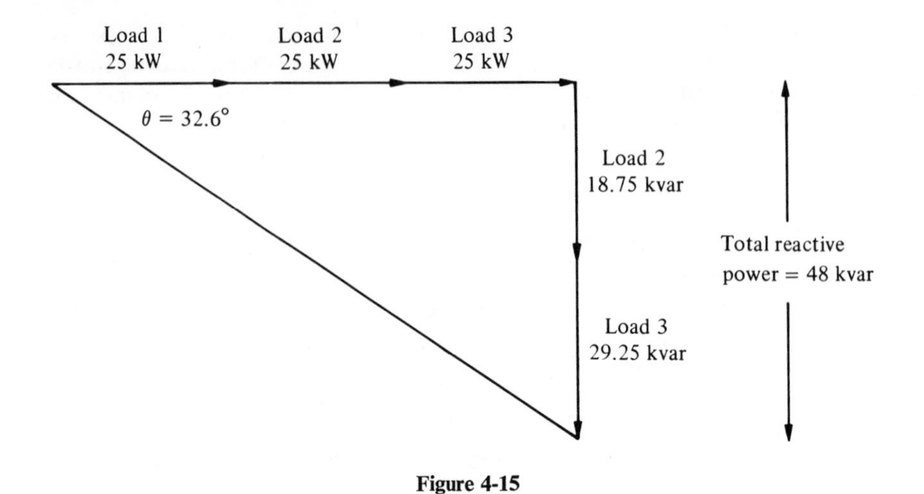

Figure 4-15

(b) To find the total kVA of the entire system,

$$\text{kVA} = \frac{\text{kW}}{\cos\theta} = \frac{75}{0.842} = 89.2 \text{ kVA}$$

(c) To find the total current to all three loads,

$$I = \frac{(\text{kVA})(1000)}{E} = \frac{89{,}200}{230} = 388 \text{ A}$$

4-12 ***Power factor correction.*** In Sec. 4-9 the capacitor was described as a device whose current leads its voltage. Since a capacitor has little or no resistance, the current drawn by it will lead its voltage by almost 90°. The net power drawn by a capacitor is zero. The reactive current multiplied by the applied voltage would then be properly entitled vars, as applied to the power triangle. The capacitor consumes no power but receives power for half of each cycle and returns it back to the supply during the other half of the cycle. This makes its power factor zero and its phase angle 90°. The current through a capacitor is 180° or opposite to that of an ideal inductor in ac circuits.

This feature alone makes the capacitor useful in power systems by introducing *leading* vars into the system. These directly oppose the *lagging* vars caused by highly inductive devices. This reduction in vars brings the apparent power (kVA) of the system closer in magnitude to the true power (kW), raising the power factor closer to unity and requiring less current. The physical size of almost all equipment and installations in an electrical system depends on the current that is required. Good design practice, therefore, calls for continual vigilance in this matter of approaching unity pf insofar as possible.

Power companies often assess surcharges to the cost of power when the power factor falls below some particular point, usually 0.8 or 0.85. When correction is made, therefore, it is not economically feasible to correct to unity but rather from pf's well below 0.8 to approximately 0.8 pf.

The measurement of capacitance and capacitive reactance is explained in Sec. 5-9. When capacitors are connected in parallel their capacitances are added:

$$C_1 + C_2 + C_3 = C_T \tag{4-11}$$

Equation (4-11) should be obvious. To increase current flow in capacitors to accomplish the desired result, more devices must be connected in parallel.

Example 4-8 illustrates the use of capacitors in the correction of power factor.

---

*Example **4-8.*** A 50-kW load operates with a power factor of 0.6 lagging from a 230-V, 60-Hz power supply.

(a) Calculate the current.
(b) It is required to improve the power factor to 0.85. What is the kvar rating and capacitance of capacitors necessary (see Figs. 4-18 and 4-19)?
(c) What is the current after power factor correction (see Figs. 4-16 and 4-17)?

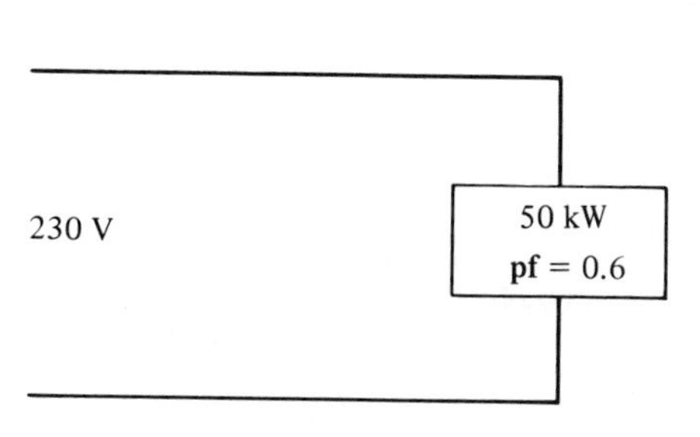

Figure 4-16

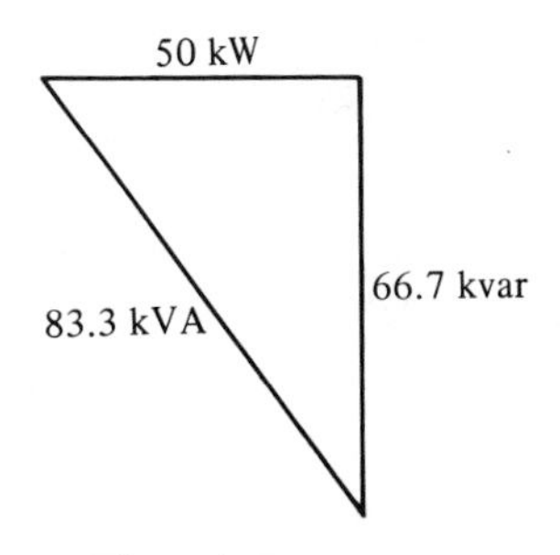

Figure 4-17

Solution

(a)

$$\text{kVA} = \frac{\text{kW}}{\text{pf}} = \frac{50}{0.6} = 83.3 \text{ kVA}$$

$$\text{kvars} = (\text{kVA})(\sin\theta) = (83.3)(0.8) = 66.7 \text{ kvars}$$

$$I = \frac{(\text{kVA})(1000)}{E} = \frac{(83.3)(1000)}{230} = 362 \text{ A}$$

(b) To correct pf to 0.85, the original 50-kW load must have an angle of 31.75° with its kVA (cos 31.75° = 0.85).

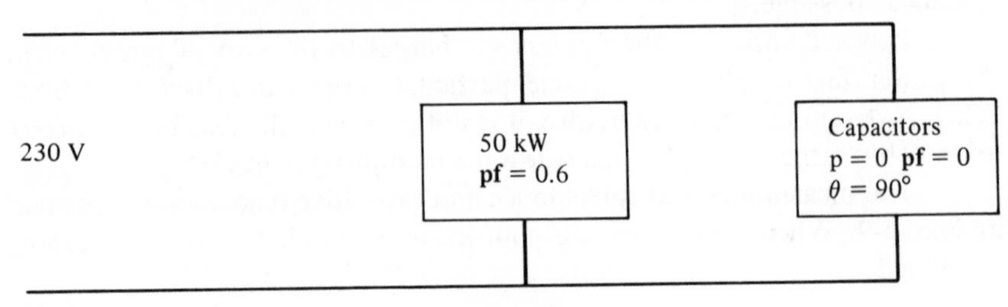

Figure 4-18

Adding the two loads so that the pf will be 0.85 must result in a new triangle with 31.75° between the kW and the kVA.

$$\text{kVA} = \frac{\text{kW}}{\text{pf}} = \frac{50}{0.85} = 58.8 \text{ kVA}$$

$$\text{kvars} = (\text{kVA})(\sin\theta) = (58.8)(0.526) = 30.9 \text{ kvars}$$

The power triangle with the capacitors has only 30.9 lagging kvars as compared to 66.7 lagging kvars in the original load. This means the capacitors must contribute 66.7 − 30.9 = 35.8 leading kvars. Capacitors used for this purpose are rated in kvars.

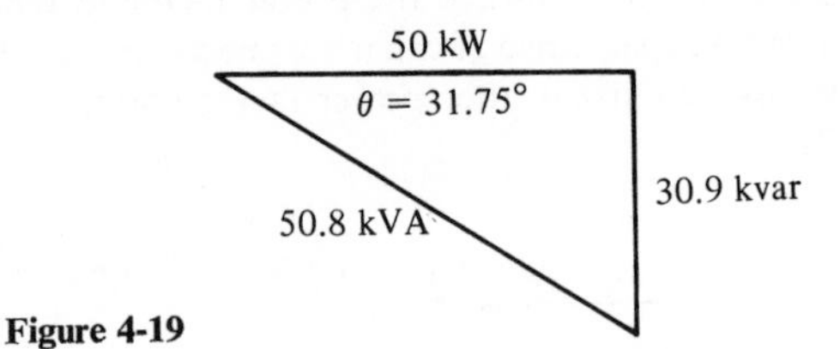

Figure 4-19

To calculate the capacitance of capacitors as measured in microfarads,

$$I_C = \frac{(\text{kvars})(1000)}{E} = \frac{(30.9)(1000) \text{ vars}}{230 \text{ V}} = 134.5 \text{ A}$$

$$X_C = \frac{I_C}{E} = \frac{134.5}{230} = 0.585\ \Omega$$

$$= \frac{(159)(10^3)}{fC} \quad \text{or} \quad C = \frac{(159)(10^3)}{fX_C}$$

$$C = \frac{(159)(10^3)}{(60)(0.585)} = 4530\ \mu\text{F}$$

or approximately 5000 μF.

This would be a very large capacitor, but since capacitors are additive when connected in parallel, we could use fifty 100-μF capacitors in parallel to satisfy this situation or a synchronous capacitor rated at 35.8 kVA.

(c) Calculate the current to load after adding capacitors

$$I_T = \frac{(\text{kVA})(1000)}{E} = \frac{(58.8)(1000)}{230} = 256 \text{ A}$$

The original current = 362 A.

This is 106 A less than the original load or a reduction of about 30%. This means all current-carrying equipment would be 30% smaller if the pf were corrected to 0.85.

4-13 ***Three-wire systems.*** We have seen the importance of current consideration in good electrical design. The lower the current for a particular situation, the more efficient its operation. Also, the cost of the initial installation is reduced. One way to reduce current for a given load is to increase the voltage.

The selection of a higher voltage, resulting in a lower current for a given amount of power, has its limitations. Many years ago, mostly because of consumer safety, 110–120 V was selected as a compromise for the operation of household equipment. This puts serious limitations on the distances circuits may be extended.

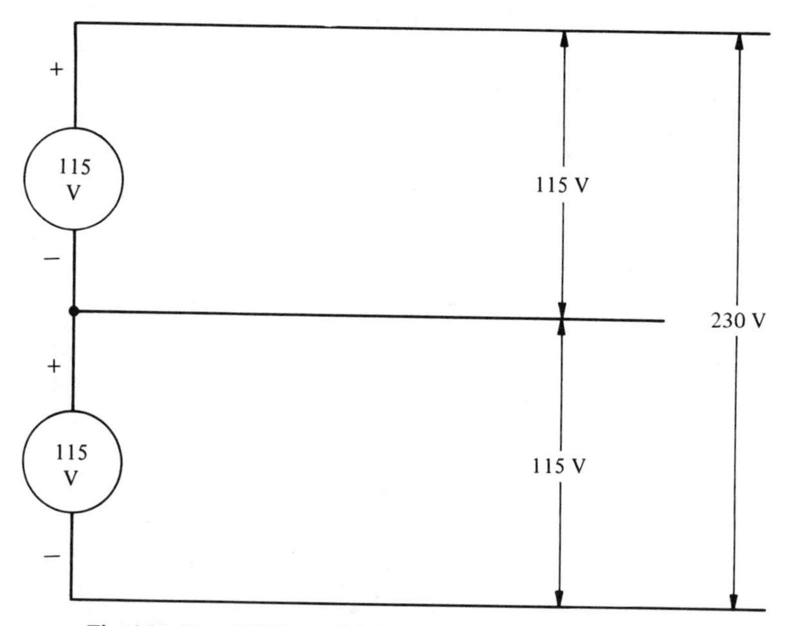

Fig. 4-20. Two 115-V supplies in series to obtain an Edison three-wire system.

This problem is somewhat alleviated by the use of the Edison Three-Wire System. This basically involves the use of connecting two 115-V power supplies in series in such a way that their polarities cause the voltages to be additive. The common point is extended to form what is called the *neutral conductor*. This can be done with both dc and ac power supplies. Let us investigate how this system functions. Figure 4-20 shows how the Edison Three-Wire System is obtained.

The system now is actually a 230-V system, but the neutral conductor, sometimes called *the third wire*, enables us to use 115 V, in addition to 230-V equipment. Extreme caution must be observed to be certain that this neutral conductor is never fused nor permitted to become disconnected in any way. Should this occur, the voltages across the individual loads will not be stable at 115 V. The smaller load (greatest number of ohms and lowest current) will have a higher voltage. The larger load will have less than 115 V. Of course, if loads were balanced (not always possible), a neutral would not be necessary. However, it always must be included in the system and remain unfused and unswitched.

Example 4-9. A 10-kW, 115-V load is supplied from a 115-V, two-wire supply. Assume the load to be resistive, with pf = 1.0.
(a) Calculate the current (see Fig. 4.21).

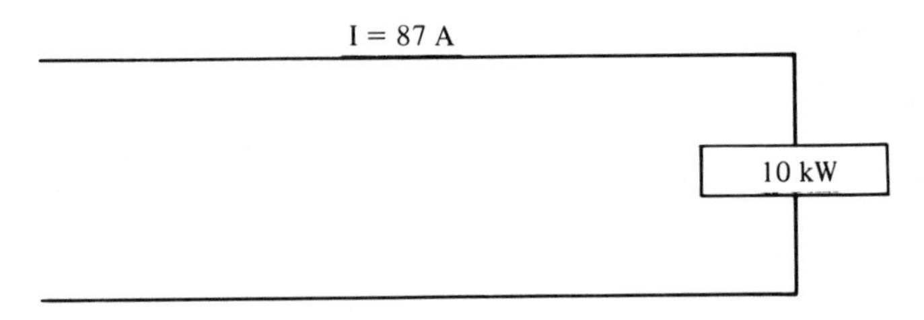

Figure 4-21

(b) Connect the same load to a 115–230-V, three-wire supply so that the load is balanced. Calculate the current in each conductor (see Fig. 4-22).

Solution

(a)

$$I = \frac{\text{kW}(1000)}{E}$$

$$= \frac{(10)(1000)\text{ W}}{115\text{ V}}$$

$$= 87\text{ A}$$

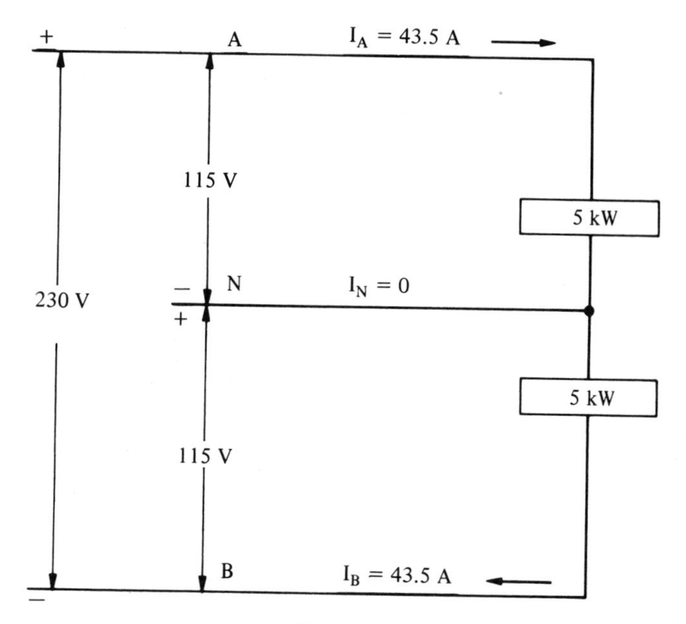

Figure 4-22

(b)

$$I_A = \frac{(\text{kW})(1000)}{E} = \frac{(5)(1000)\text{ W}}{115\text{ V}}$$

$$= 43.5\text{ A}$$

$$I_B = \frac{(\text{kW})(1000)}{E} = \frac{(5)(1000)\text{ W}}{115\text{ V}}$$

$$= 43.5\text{ A}$$

$$I_N = 43.5 - 43.5$$

$$= 0$$

Since the current in the neutral conductor is zero in Ex. 4-9, the system can be considered to be 230 V, with the two loads in series. Since they are equal in resistance the voltage would divide, 115 V for each load.

Note that the use of 230 V instead of 115 V has reduced the feeder current in half. This means a 50% reduction in the size of service equipment.

Example ***4-10.*** A 3-kW load is connected to one side and a 7-kW load to the other side of the same three-wire supply as in Ex. 4-9. Calculate the current in each conductor (see Fig. 4-23).

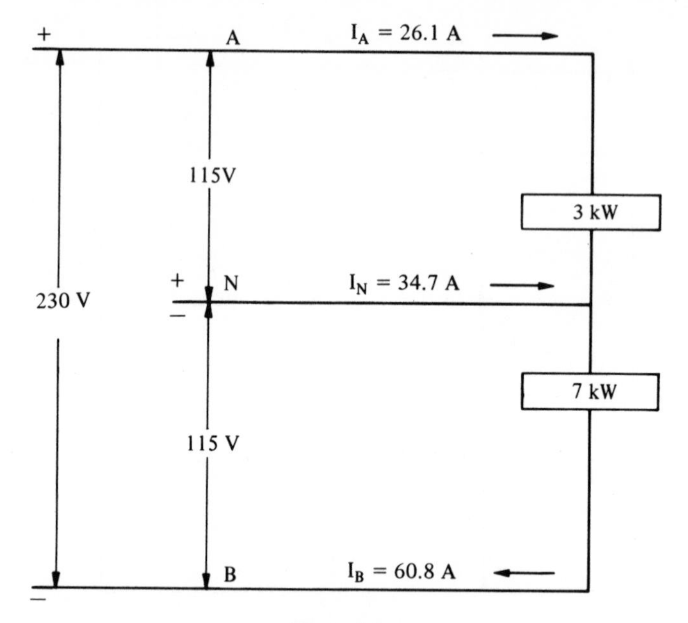

Figure 4-23

Solution

$$I_A = \frac{(\text{kW})(1000)}{115} = \frac{(3)(1000)\text{ W}}{115\text{ V}}$$

$$= 26.1\text{ A}$$

$$I_B = \frac{(\text{kW})(1000)}{115} = \frac{(7)(1000)\text{ W}}{115\text{ V}}$$

$$= 60.8\text{ A}$$

$$I_N = 60.8\text{ A} - 26.1\text{ A} = 34.7\text{ A}$$

flowing from supply to load.

When loads are unbalanced the neutral always will carry the unbalanced current. The current to the load from the supply always must equal the current from the supply to the load. In this case:

$$I_B = I_A + I_N = 26.1\text{ A} + 34.7\text{ A} = 60.8\text{ A}$$

Example ***4-11.*** In Ex. 4-10, open the neutral conductor and calculate the voltage across each load (see Fig. 4-24).

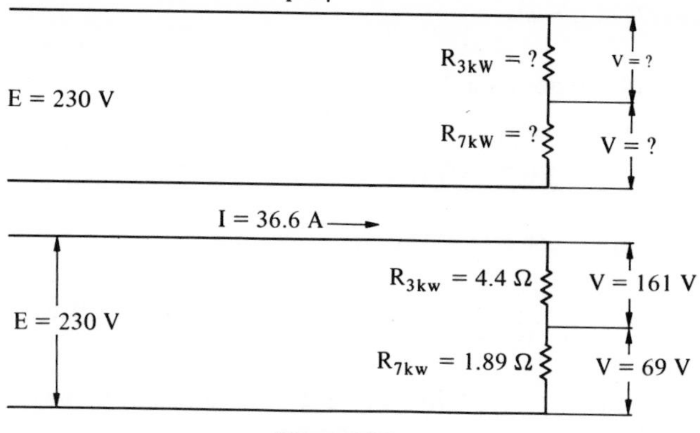

Figure 4-24

Solution

The 3-kW load is rated at 115 V. Its resistance cannot change. Therefore,

$$R_{3\text{kW}} = \frac{V^2}{W} = \frac{(115\text{ V})^2}{3000\text{ W}} = 4.4\ \Omega$$

Similarly, a 7-kW load is rated at 115 V. Its resistance cannot change and

$$R_{7\text{kW}} = \frac{(115\text{ V})^2}{7000\text{ W}} = 1.89\ \Omega$$

When connected to a 230-V system with the neutral open,

$$R_T = 4.4\ \Omega + 1.89\ \Omega = 6.29\ \Omega$$

With no neutral to carry unbalanced current there can be only one current in this series circuit.

$$I = \frac{E}{R_T} = \frac{230\text{ V}}{6.29\ \Omega} = 36.6\text{ A}$$

The voltage across the 4.4-Ω load or original 3-kW load is $V = IR$ or (36.6)(4.4) = 161 V. This is $\frac{161}{115} \times (100) = 140\%$ of the load's rated voltage. In the case of a heating appliance it may burn out in a short time.

The voltage across the 6.29-Ω or original 7-kW load is $V = IR$ or (36.6)(1.89) = 69 V.

This would not cause any damage to a heating appliance or lighting equipment, but a motor would draw excessive current and possibly burn out.

Example 4-11 also may be solved by the voltage division principle. The voltages in a series circuit divide in direct ratio to the resistance. The larger the power the smaller the resistance; therefore, the voltages divide in inverse ratio to the power of the individual loads.

Voltage across a 3-kW load

$$= \left(\frac{7\ \text{kW}}{7\ \text{kW} + 3\ \text{kW}}\right)(230\ \text{V}) = (0.7)(230) = 161\ \text{V}$$

Voltage across a 7-kW load

$$= \left(\frac{3\ \text{kW}}{7\ \text{kW} + 3\ \text{kW}}\right)(230\ \text{V}) = (0.3)(230) = 69\ \text{V}$$

To avoid this possibility of severe voltage unbalance, the neutral conductor of any system is never fused nor switched.

PROBLEMS

4-1 A circuit consisting only of resistance units requires 15 kW. Calculate (a) the current if power supply is 230-V dc, and (b) the current if power is provided from a 230-V, 60-Hz supply.

4-2 A coil has a 0.2 H inductance and 30 Ω resistance. Applied voltage is 115 V, 60 Hz. Calculate (a) the power consumed by the coil, (b) the pf of the coil, and (c) the current flow in the coil.

4-3 A 50-μF capacitor and a 100-Ω resistor are connected in series to a 115-V, 60-Hz source. Calculate (a) the impedance of the circuit and (b) the current flow.

4-4 A coil has 20 Ω resistance and 0.15 H inductance. Calculate (a) the impedance of the coil at 60 Hz, (b) the current that would flow if connected to 115 V dc, (c) the current that would flow if connected to a 115-V, 60-Hz power supply, and (d) the current that would flow if connected to a 115-V, 40-Hz power supply.

4-5 A coil with negligible resistance, a capacitor, and a resistance are all connected in series to a 115-V, 60-Hz source. The current through the circuit is 5 A and is in phase with the voltage of the supply. The power used by the circuit is 575 W. Calculate (a) the ohmic value of the resistor and (b) the impedance of the entire circuit.

4-6 If a voltage of 230 V is applied to a load, the current is 55 A. The angle between the voltage and the current is 35°. Calculate the power of the load.

4-7 The power used in a circuit is 6500 W, the voltage is 230 V, and the current is 43.5 A. Calculate the angle between the current and the voltage.

4-8 An impedance coil has 6 Ω resistance. When connected to a 115-V, 60-Hz supply, the current is 10 A. Calculate (a) its reactance and (b) the power used by the coil.

4-9 The power input to a single-phase, 230-V motor is 1200 W. At what pf is it operating if it draws 7 A?

4-10 Calculate the apparent power (VA) of a circuit if the power is 2500 W and the reactive power is 1000 vars.

4-11 The current to a 230-V, single-phase motor is 15 A. If its pf is 0.75, what is its power input?

4-12 A 50-Ω resistor is in parallel with a highly inductive coil. The coil has negligible resistance and 30 Ω reactance. The voltage is 115 V, 60 Hz. Calculate (a) the total current, (b) the pf of the entire load, and (c) the total power.

4-13 An electrical device has its electrical quantities measured and shows the following: voltage, 115 V; current, 9.5 A; and power, 900 W. The frequency is 60 Hz. Calculate (a) the pf of this device, (b) its resistance, (c) its reactance, (d) the size capacitor that must be added in parallel so that the current will be minimum, and (e) the minimum current in (d).

4-14 A four-lamp fluorescent luminaire is designed to operate at 115 V and requires 2.5 A. Each lamp uses 40 W, and the ballast supplying all four lamps requires an additional 25 W. (a) At what pf does the unit operate? (b) Calculate the size capacitor to be added in parallel to this lighting unit to correct its pf to 0.8.

4-15 Two impedances are in parallel, connected to a 115-V, 60-Hz supply. One consumes 250 W at 0.65 pf; the second, 500 W at 0.9 pf. Calculate (a) the total power, (b) the total current, and (c) the pf of the entire circuit.

4-16 An inductive load has a pf of 0.85, lagging. Capacitors rated at 15 kvars are added in parallel and this results in unity pf. The supply voltage is 230 V. Calculate (a) the power of the load, (b) the current before adding capacitors, and (c) the current after adding capacitors.

4-17 The coil of an ac-operated relay requires 0.4 A at 115 V to operate properly. Under these conditions the coil takes 15 W. If this relay is to operate on 208 V, what value of resistance must be added in series to assure current will not exceed 0.4 A?

4-18 Five 100-W incandescent lamps are each designed to use 100 W at 115 V. Calculate (a) the hot resistance of each lamp and (b) the total resistance of the five lamps when connected in parallel.

4-19 (a) A 20-kW incandescent (resistive) lighting load is connected to a 115–230-V, three-wire system. Load is balanced. Calculate the current in each conductor.

(b) Calculate the currents if lighting is fluorescent with a pf of 0.8.

4-20 The 20-kW load of Prob. 4-19 is connected so that 14 kW is on one side of the three-wire system and 6 kW is on the other side.

(a) Calculate the current in each conductor.

(b) Calculate the currents if lighting is fluorescent with a pf of 0.8.

4-21 A 30-kW load with a pf of 0.65 is connected to a 115–230 V, three-wire system and is balanced.

(a) Calculate the current.

(b) If the pf is to be corrected to 0.85, what value of capacitors in kvars is necessary?

4-22 A 20-kW resistive load is supplied from one load center and a 30-kW load with a pf of 0.65 is supplied from another. The voltage of the system is 115–230 V. The loads are balanced. Calculate (a) the current required for both loads, (b) the value of capacitors required (in kvars) if the pf is to be corrected to 0.85, and (c) the current after capacitors are added.

4-23 A three-wire, 115–230-V lighting load at 15 kW is operating with 6 kW on one 115-V side and 9 kW on the other. Calculate (a) the neutral current and (b) the voltages on each load if the neutral wire should be disconnected.

Chapter Five

Three-Phase Systems

5-1 ***Three phase vs. single phase.*** Our studies of alternating current have been referred to from time to time as "single phase." This leads to the question, "Are there systems other than single phase?" Perhaps we should examine the shortcomings of a single-phase system before investigating any others.

When a sinusoidal current and voltage are multiplied together the resultant power is pulsating. The power curve also goes through zero twice each cycle. When the power factor of the load is less than unity, part of the power curve is below zero, or negative. This represents the reactive load returning power back to the source during part of the cycle.

This presents no serious problem when the resistive circuits are those producing light or heat. However, for electric motors this pulsating power waveform makes motor design difficult. A single-cylinder gasoline engine is able to provide uniform torque only with the aid of a large flywheel to keep the engine running smoothly when no power stroke is available.

A properly connected three-phase supply provides a less pulsating power pattern. Like the multicylinder gas engine, a three-phase motor provides a more constant torque.

It appears to be somewhat facetious to compare a gas engine to an electric motor, but why not? They perform the some function. They do, however, have a different form of input power.

Three-phase system are in use practically throughout the United States. We shall prove that the use of three wires in a three-phase system can provide 173% more power than the two wires of a single-phase system. This means less copper required for conductors, smaller power distribution equipment, and less associated losses.

A three-phase system is one that has three sources of power with a

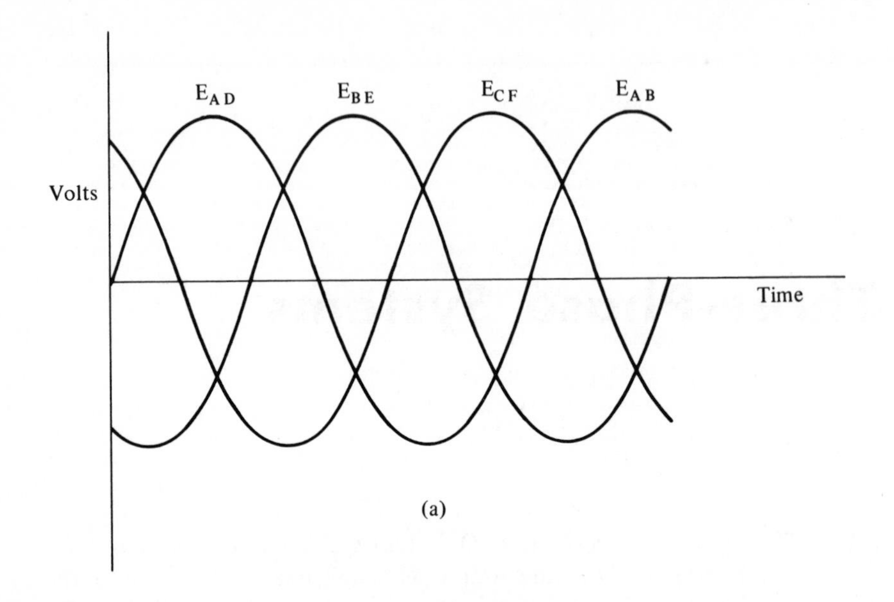

(a)

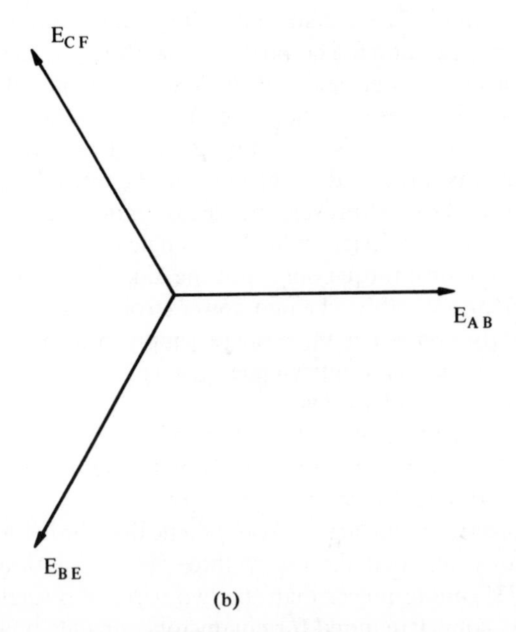

(b)

Fig. 5-1. (a) Sine waves of three equal-phase voltages of a three-phase generator, 120° apart. (b) Vector representation of the same three voltages.

time interval between each. Figure 5-1a shows the three sine waves of the three voltages equidistant from each other and their vector counterparts.

The generation of these three voltages is accomplished quite easily by a three-phase generator. It is only necessary to locate three windings 120° apart on the generator. As these windings rotate through the magnetic field, their induced voltages will be 120° apart. Note that 120° is one-third of a complete cycle, which represents $\frac{1}{60}$ sec for a 60-Hz generator. Therefore, 120° represents $\frac{1}{3}$ of $\frac{1}{60}$ or $\frac{1}{180}$ sec. The maximum value of each voltage is $\frac{1}{180}$ sec apart.

5-2 ***Connecting a three-phase supply in delta.*** A three-phase system may be defined as three single-phase voltages connected together. It is possible to operate a three-phase system by connecting a load across each phase or voltage source. This is illustrated in Fig. 5-2, but six conducting wires are necessary, and one of the principal advantages, that of saving copper, has vanished.

By connecting the three voltages properly the number of conductors between the generator and the load can be reduced to three. Before we elaborate on the procedure for these connections of the individual phases, let us investigate a similar situation of combining voltage sources, but with dc or batteries.

Batteries have polarities, + and −. If two 12-V batteries are connected in series, there are two possibilities, as shown in Fig. 5-3. To show the addition of voltages, vectors are drawn either in phase with each other or 180° out of phase. Also shown are their vector additions. These are the only two

Fig. 5-2. Each phase of a three-phase generator connected to a separate load.

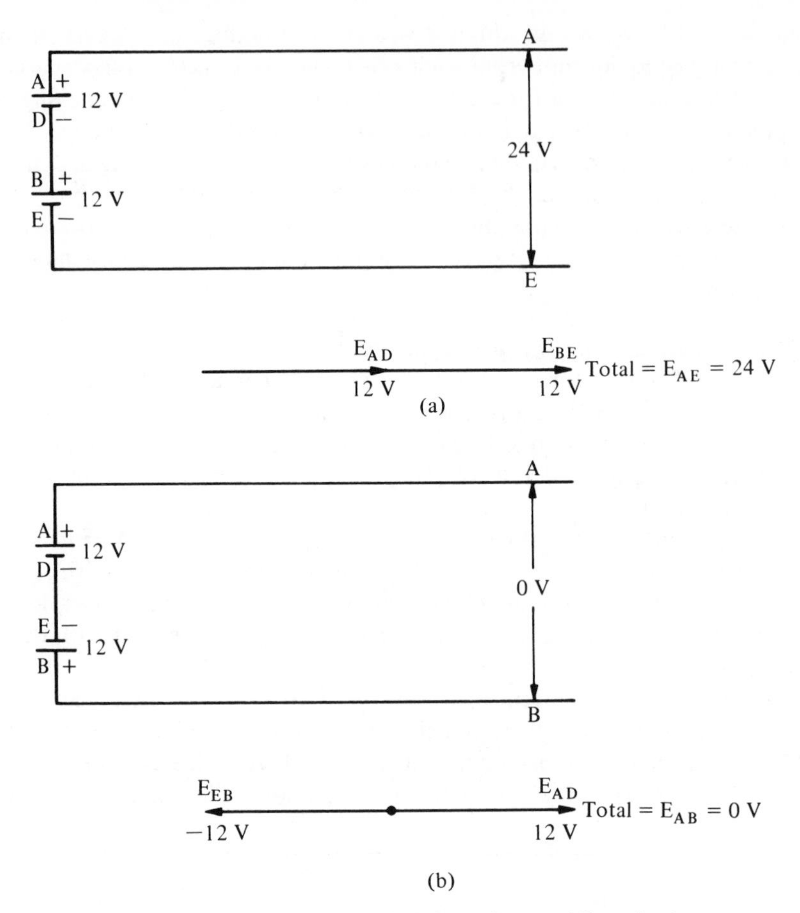

Fig. 5-3. (a) Two 12-V batteries in series, voltages adding—in phase. (b) Two 12-V batteries in series, voltages subtracting—180° out of phase.

possibilities. The sum of the two 12-V battery voltages can be either 0 or 24 V, depending on how they are connected.

To become familiar with the three-phase terminology of using double subscripts for voltages, we may call one battery voltage E_{AD} and the other E_{BE}.

In Fig. 5-3a we shall say that the voltage from A to E is E_{AD} plus E_{BE}. In Fig. 5-3b the voltage from A to B is $E_{AD} + E_{BE}$. Note that $E_{BE} = E_{EB}$.

We have stated that any two voltages of the three-phase generator are 120° apart. However, these two voltages also have two possible ways to be connected in series. One way they are 120° apart and must be so added vectorially to find their sum. If connected in series by reversing one voltage with respect to the other, they will be 60° apart and their addition must consider this angle. These two possibilities are the principal factors to consider when connecting three-phase supplies.

Figure 5-4 shows these two possibilities and their vector addition. It is noted that 230 V in series with 230 V must be 230 V, if they are 120° apart as in Fig. 5-4a. If phase *BE* is reversed, then the two voltages are 60° apart and their sum is 2(230)(cos 30°) or $230\sqrt{3} = 398$ V, as shown in Fig. 5-4b.

If in Fig. 5-5, *D* is connected to *B*, the voltage between *A* and *E* will be 230 V, as the vector addition illustrates. Next, if *E* is connected to *C*, the voltage from *A* to *F* should be zero. The vectors show that E_{AE} and E_{CF} are equal and opposite; therefore, their sum is zero. Now we can connect *A* and *F* since no voltage exists between them. Then we extend three wires from *A*, *B*, and *C* to obtain a three-phase supply from a delta-connected source. Note from the vectors that the voltage between any two of the *A*, *B*, and *C* terminals is equal to 230 V and each is 120° from any other.

Should any one of the phases be reversed, a delta connection would be impossible. The vectors of Fig. 5-6 show that twice phase voltage or 460 V would result between *A* and *F*.

Figure 5-7 illustrates the same connections as Fig. 5-5, which shows why the word "delta" is used to describe this particular configuration.

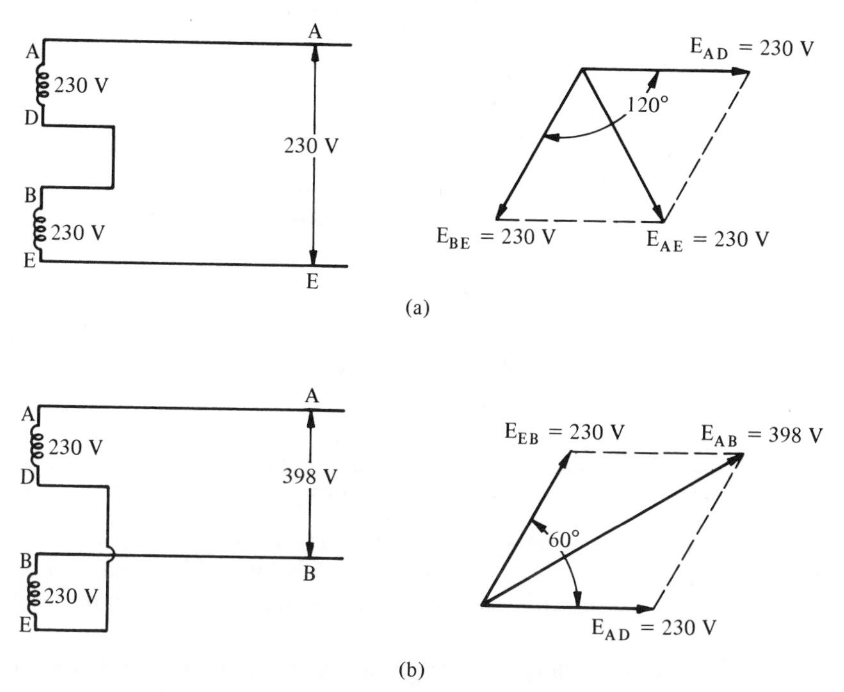

Fig. 5-4. The only two possibilities of resultant voltages when two phases of a three-phase generator are connected in series.

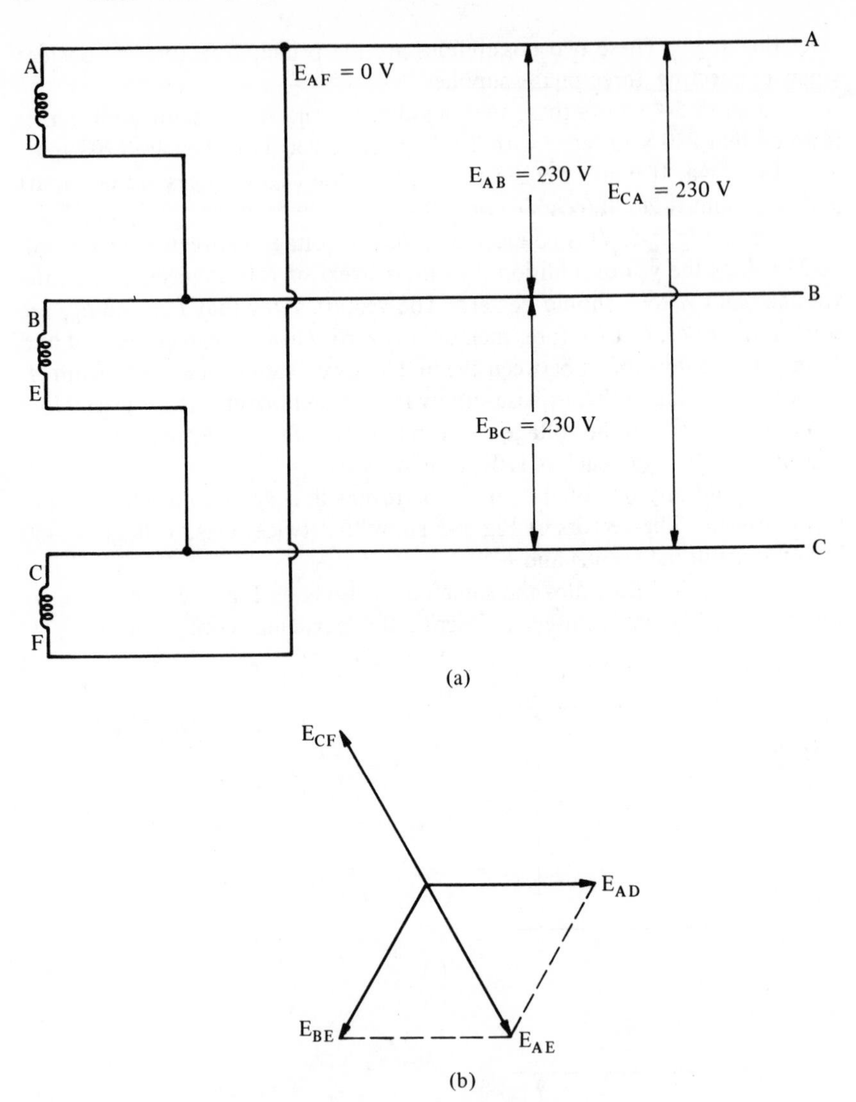

Fig. 5-5. (a) Correct connections of the three phases of a three-phase generator to obtain a delta. (b) Vectors of correct delta connection.

Figure 5-8 shows the incorrect delta connection according to our *A*, *B*, and *C* system of identification. The important point of this operation is not to close the delta unless the voltage measured at E_{AF} in Figs. 5-6a and 5-8 is zero.

Hereafter we shall refer to the terminals *A*, *B*, and *C* as a three-wire, three-phase power supply without any further consideration of the source connections.

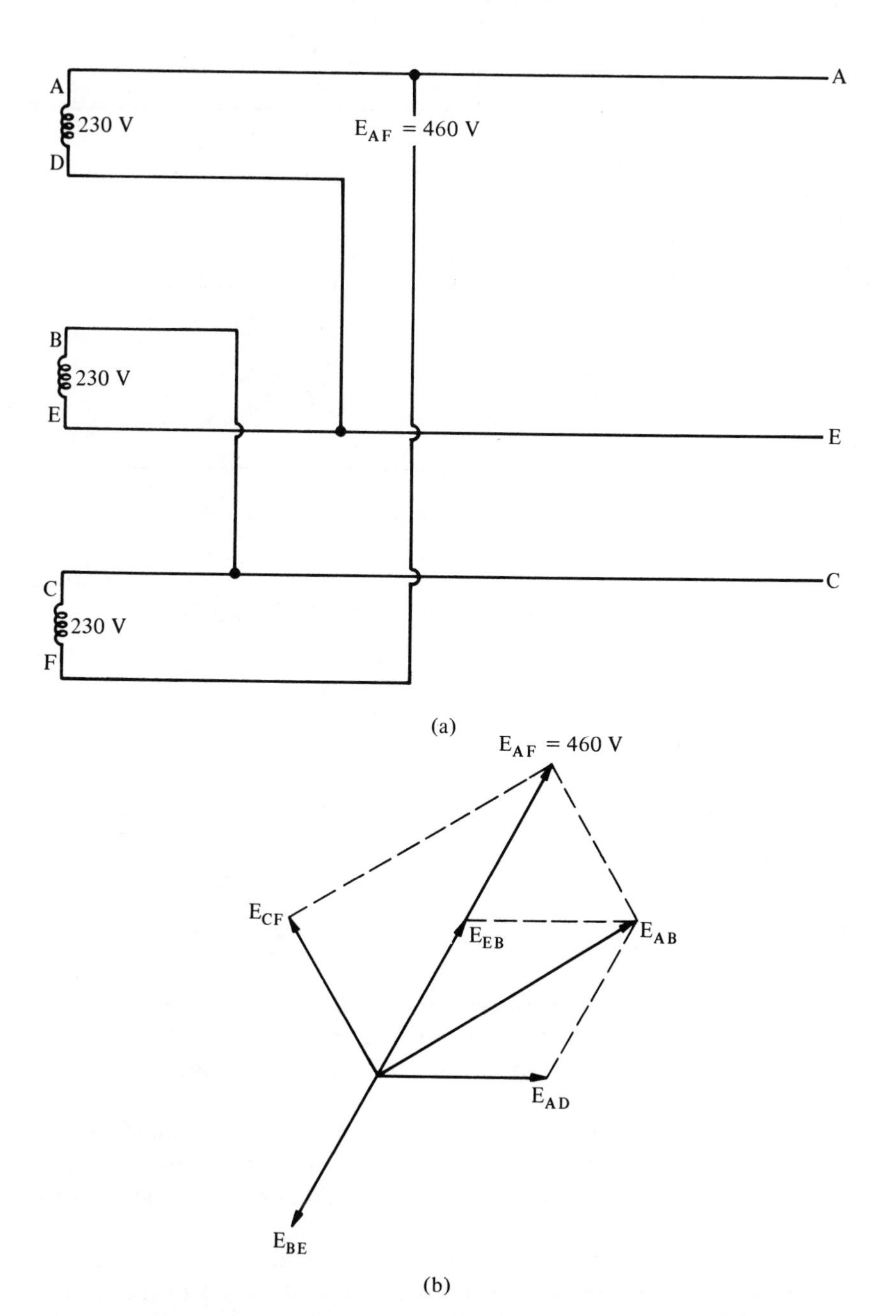

Fig. 5-6. (a) Incorrect connections of a three-phase generator. Phase *BE* is reversed. (b) Vectors of voltage additions of incorrect delta.

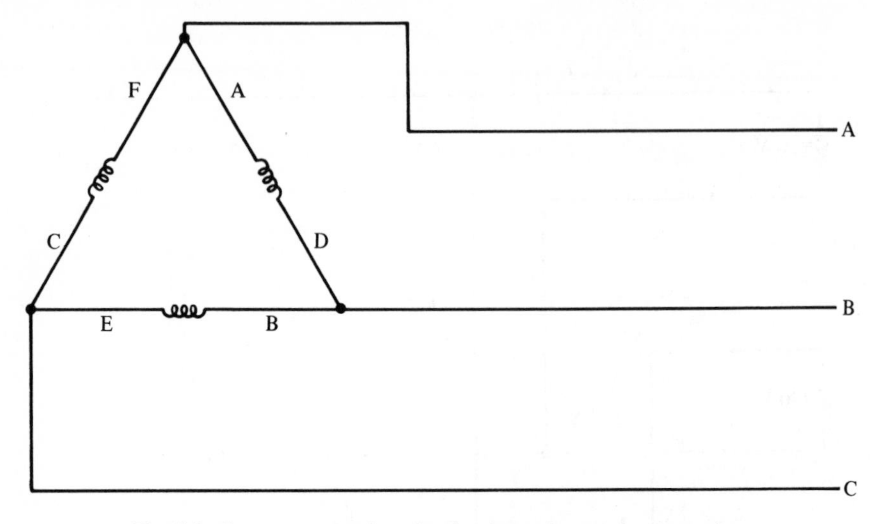

Fig. 5-7. Correct connections for the three phases of a three-phase generator to obtain a three-wire delta-connected supply.

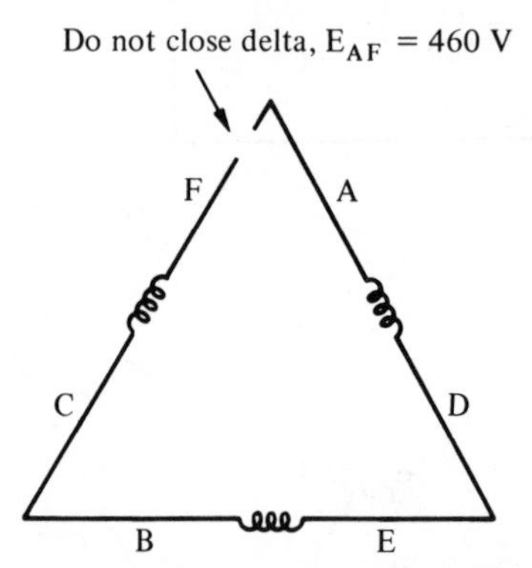

Fig. 5-8. Incorrect connections for a delta-connected generator.

5-3 ***Power in a balanced delta-connected load.*** From Sec. 4-8, power in a single-phase system is shown to be $EI \cos \theta$ or EI pf. This equation still holds for a three-phase system, with some modification. Let us see how.

The term *balanced delta-connected load* means three equal resistances or impedances connected in a delta to a three-wire, three-phase supply. (A four-wire supply could be used, but neutral would not be necessary. This is discussed later.)

The term *phase current* means the current in one phase or branch of a three-phase load. Line current refers to the current in one of the lines or connecting wires from supply to load.

The voltages that cause phase and line currents to flow are E_{AB}, E_{BC}, and E_{CA}, shown vectorially in Fig. 5-9 in the same location as derived in Fig. 5-5a and b.

Refer to Fig. 5-9a. The phase currents that flow in each 10-Ω resistor are

$$I_{AB} = \frac{230}{10} = 23 \text{ A}$$

$$I_{BC} = \frac{230}{10} = 23 \text{ A}$$

$$I_{CA} = \frac{230}{10} = 23 \text{ A}$$

Since these currents flow in pure resistors they must be in phase with their respective voltages. Strictly observing double subscript notations of the three currents we also draw the vectors of their negative values of current.

If we should connect an ammeter in line A it would read the vector sum of the currents I_{AC} and I_{AB}. Inspection of the vector diagram shows that these currents are 60° apart; therefore, their sum must be one of them multiplied by $\sqrt{3}$ or 1.732. In the same way the current in line B is the sum of I_{BA} and I_{BC}. Also, the current in line C is the sum of I_{CA} and I_{CB}.

This addition of phase currents to obtain line currents could be expressed according to Kirchhoff's current law, which states that the current into a junction is equal to the current away from that junction.

The arrows in Fig. 5-9 indicate that I_A is the current into the junction at A. The current away from the junction is the current from A to C, identified as I_{AC}, plus the current from A to B. This current is identified with the arrow I_{BA}, which points into the junction; therefore, we simply reverse the subscript and call it I_{AB} and assume it goes away from the junction.

Note that only one set of three voltages occur in a delta-connected load. They are of the same magnitude as the line voltages, or the voltage between any two line wires.

This leads to the following relationships in a balanced delta-connected load:

$$E_L = E_p \tag{5-1}$$

$$I_L = I_p\sqrt{3} \quad \text{or} \quad I_p = \frac{I_L}{\sqrt{3}} \tag{5-2}$$
$$= 23\sqrt{3} = 40 \text{ A}$$

The power per phase $P_p = E_p I_p \cos\theta$. The power in the entire system is

$$P_T = (3E_p I_p)(\cos\theta) \tag{5-3}$$
$$= (3)(230)(23)(1) = 15{,}900 \text{ W}$$

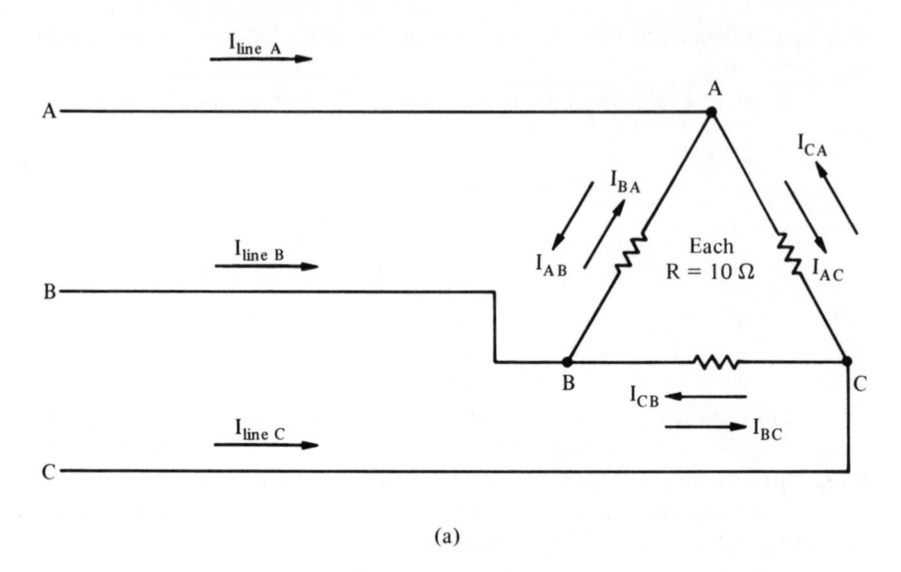

(a)

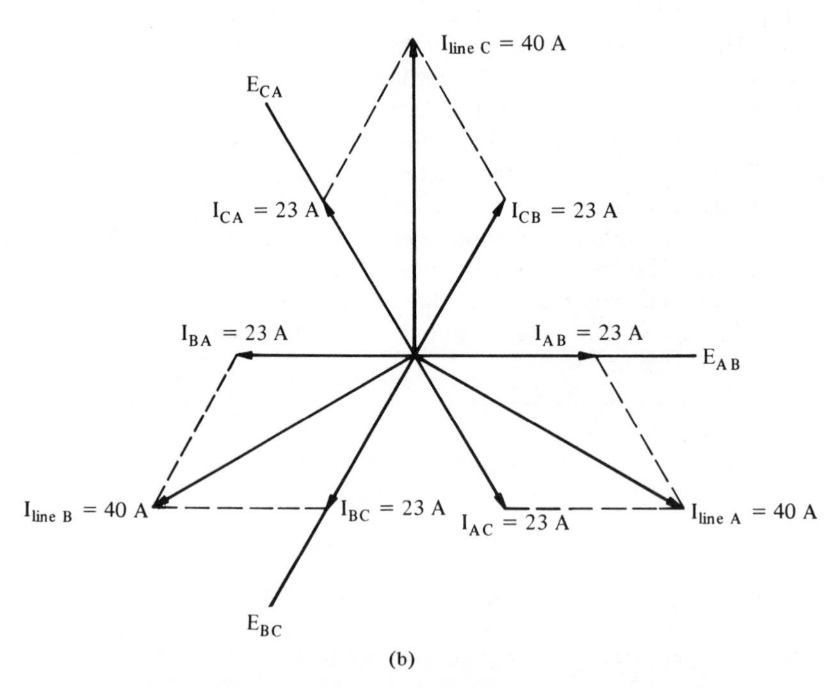

(b)

Fig. 5-9. (a) Balanced delta-connected resistive load. (b) Vector showing relationships between phase and line currents.

Using line values,

$$P_T = (3E_L) \times \left(\frac{I_L}{\sqrt{3}}\right) \cos\theta$$

$$= \sqrt{3}(E_L I_L) \text{ pf} \qquad (5\text{-}4)$$

$$= \sqrt{3}(230)(40)(1) = 15{,}900 \text{ W}$$

Equations (5-1) through (5-4) are most important equations in three-phase, delta-connected calculations.

In these equations E_L refers to the voltage between any two lines; E_p is the voltage across one of the phases, or loads; I_L is the current in any line wire; and I_p is the current in one of the phases or loads.

It should be noted that in the diagram of Fig. 5-9 the arrows denoting line current direction are all shown in the same direction. However, the vectors representing these three arrows are shown on the vector diagram 120° apart, and their sum then must be zero. This is in agreement with the single phase supply and load: The net current from a power supply to a load must be zero.

Example 5-1. Three 10-kW resistive loads are connected in delta to a 230-V, three-phase supply. Calculate (a) the current in each line and (b) the total power.

Solution

(a)

$$I_p = \frac{P_P}{E_p} = \frac{10{,}000 \text{ W}}{230 \text{ V}} = 43.5 \text{ A}$$

$$I_L = (\sqrt{3})(I_p) \qquad (\sqrt{3})(43.5) = 75.5 \text{ A}$$

(b)

$$P_T = 3P_p = 3(10{,}000) = 30{,}000 \text{ W}$$

or

$$P_T = (\sqrt{3})(E_L I_L) \text{ pf}$$

$$= (\sqrt{3})(230)(75.5)(1) = 30{,}000 \text{ W}$$

Example 5-2. A balanced 15-kW, three-phase load with a power factor of 0.75 is connected to a 230-V, three-phase supply. Calculate the current in each line.

Solution

$$P_T = (\sqrt{3})(E_L I_L)(\text{pf}) \qquad (5\text{-}4)$$

Then

$$I_L = \frac{P_T}{(E_L)(\sqrt{3})(\text{pf})} = \frac{15{,}000}{(230)(\sqrt{3})(0.75)} = 50.2 \text{ A}$$

5-4 ***Power in an unbalanced delta-connected load.*** The previous discussion and equation for power in a three-phase system assumes a balanced condition, that is, an identical load in each phase. A three-phase motor is that type of load. However, it is often practical to use a three-phase supply to provide for single-phase lights or heating elements. These single-phase loads are treated individually and, provided we connect the same number of identical electrical load units to each phase of the supply, the system will still be balanced and the equation

$$P = (\sqrt{3})(E_L I_L)(\text{pf}) \qquad (5\text{-}4)$$

is still valid.

Even if we do this the loads may become unbalanced, perhaps because some of the connected load is not in use. Then the balanced relationships no longer hold true, and we have an unbalanced three-phase load.

Figure 5-10 shows three different purely resistive loads (all unity power factor) connected in delta so that line voltage is applied to each load.

The current in line A is the sum of I_{AC} and I_{AB}, 30 and 20 A respectively. These currents are 60° apart, and advanced trigonometry is necessary for an exact solution. With a sharp pencil and a protractor it can be solved quite easily graphically. From Fig. 5-10 the current I_A is 43.5 A.

The current in line B is the sum of I_{BA} and I_{BC}, 20 and 40 A 60° apart. This adds up vectorially to 53 A in line B. Likewise, the current in line C is the sum of I_{CA} and I_{CB}, 30 and 40 A 60° apart; graphically, this is derived as 61 A in line C.

The power of the system cannot be calculated using the three-phase equation because the line currents are all different and are *not* 120° apart. The power must be calculated on a per phase basis ($E_p I_p \cos\theta$) and added together to obtain total power.

$$\begin{aligned}
\text{Power in phase } AC &= (230)(30)(1) = 6{,}900 \text{ W}\\
\text{Power in phase } CB &= (230)(40)(1) = 9{,}200 \text{ W}\\
\text{Power in phase } BA &= (230)(20)(1) = \underline{4{,}600 \text{ W}}\\
& \qquad\qquad\qquad\quad 20{,}700 \text{ W}
\end{aligned}$$

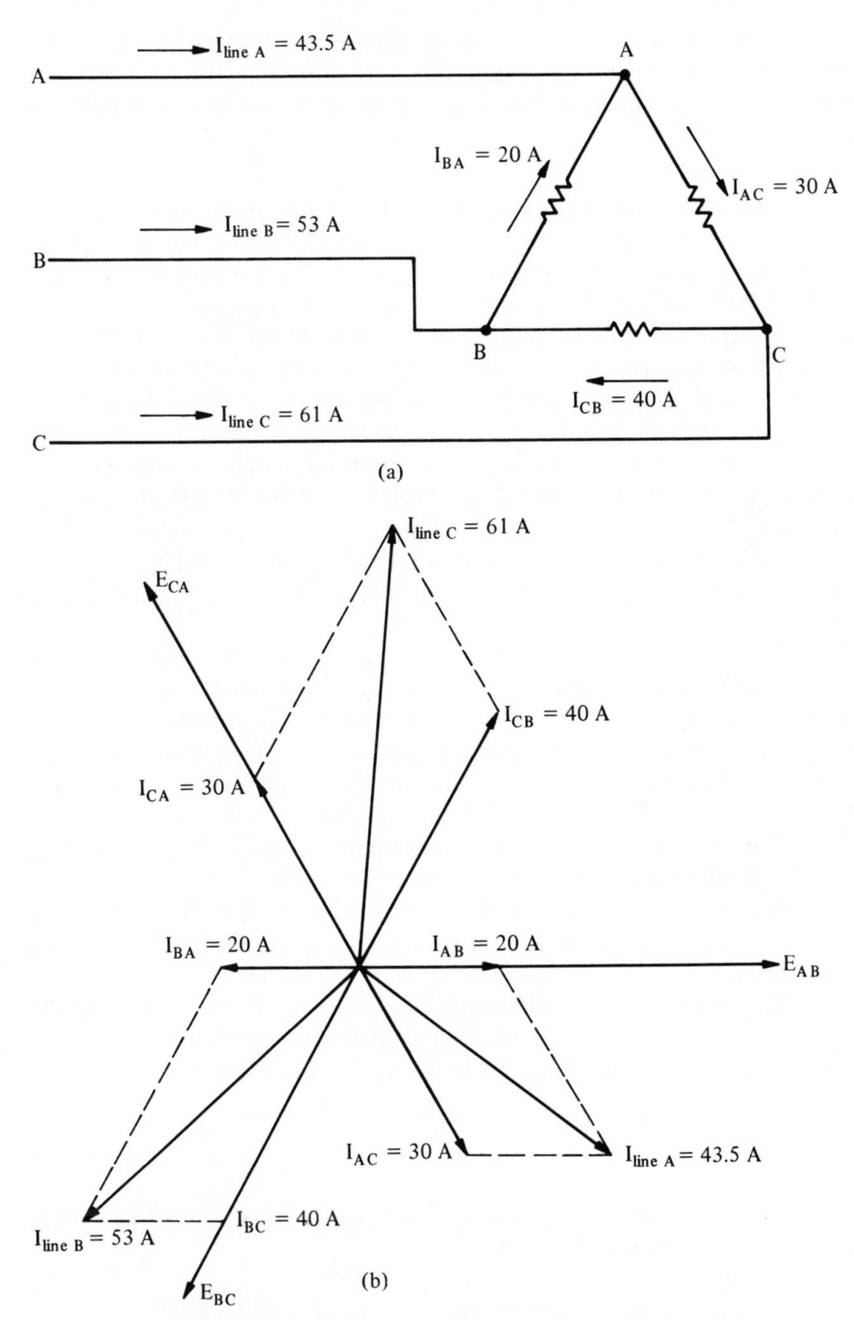

Fig. 5-10. (a) Three different resistance loads connected in delta, unbalanced load. (b) Vector addition of phase currents to indicate line-current magnitudes.

Upon examination of the three line current vectors it can be seen that when considering their magnitudes and directions their sum is zero, again proving Kirchhoff's current law that the net current in a circuit is zero.

5-5 ***Connecting a three-phase system in wye.*** When both single-phase and three-phase systems are to be made available from the same power supply, a four-wire system may be used. This is called the *wye* connection. Refer to Fig. 5-11a, which shows three voltages from a three-phase generator or three transformers. Each phase has a voltage of 120 V. Connect D to E. The voltage between A and B is (120)(2) cos 30° $= \sqrt{3} \times$ (120 or 208 V). Now connect E to F, making D, E, and F all common. The voltage between B and C also is $\sqrt{3} \times$ (120 or 208 V). From the vectors it can be determined that the voltage from C to A is the sum of E_{CF} and E_{DA}, which is also 208 V. Referring to the terminals A, B, and C, we again have a three-phase system because the voltages between any two line terminals is 208 V and line voltages all are displaced 120° from each other. These *line* voltages are E_{AB}, E_{BC}, and E_{CA}.

Since many devices require single-phase, 120-V ac, this neutral connection makes that voltage available from any line terminal A, B, or C to the neutral N. Voltages E_{AN}, E_{BN}, and E_{CN} are called *phase* voltages.

Figure 5-12 gives the vector representation if phase B were reversed. Note the unbalanced line voltages. Voltage E_{CA} is not equal to the other two, E_{AE} or E_{EC}; and they are not 120° from each other.

Figure 5-13 shows the same connections as Fig. 5-11a and illustrates how this configuration obtained its name.

Hereafter we shall refer to the four terminals A, B, C, and N as a four-wire, three-phase system without any further considerations of the source connections.

The phase and line relationships for a wye-connected source are different from those of the delta. The voltage between any two lines is the sum of two of the phases, displaced by 60° (Fig. 5-11b). Therefore,

$$E_L = E_p\sqrt{3} \tag{5-5}$$

The current that flows in any phase must be equal to the line current since they are in series. Therefore,

$$I_L = I_p \tag{5-6}$$

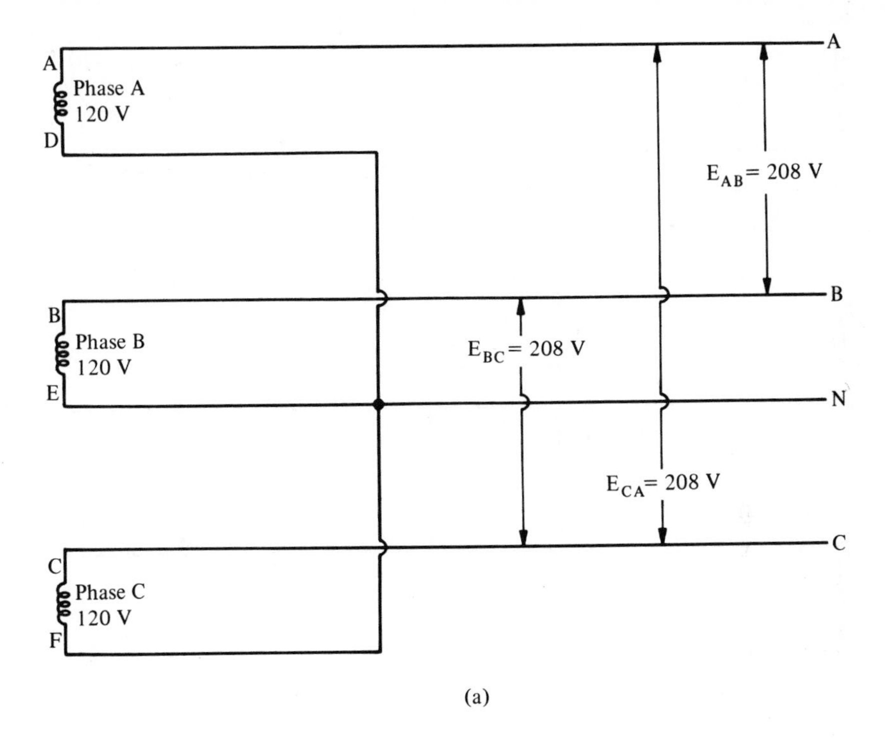

(a)

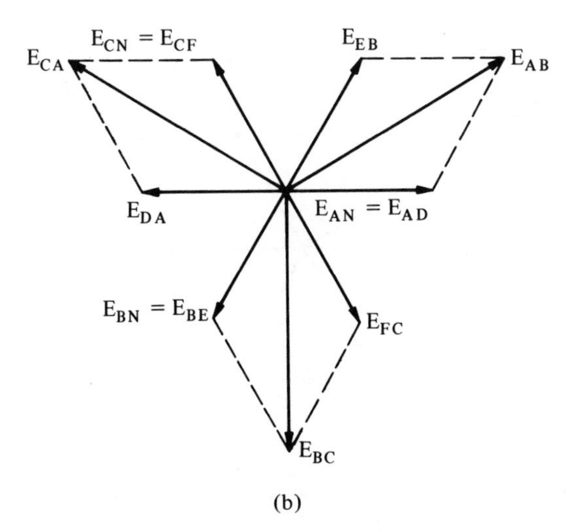

(b)

Fig. 5-11. (a) Correct connections of three-phase supply to obtain a wye. (b) Vectors of correct wye.

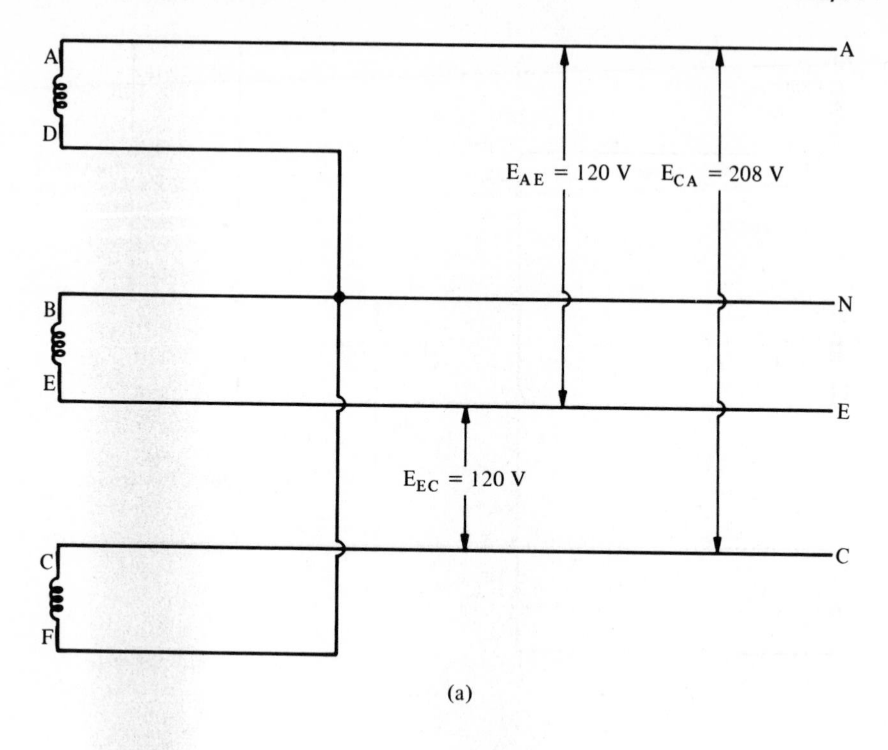

(a)

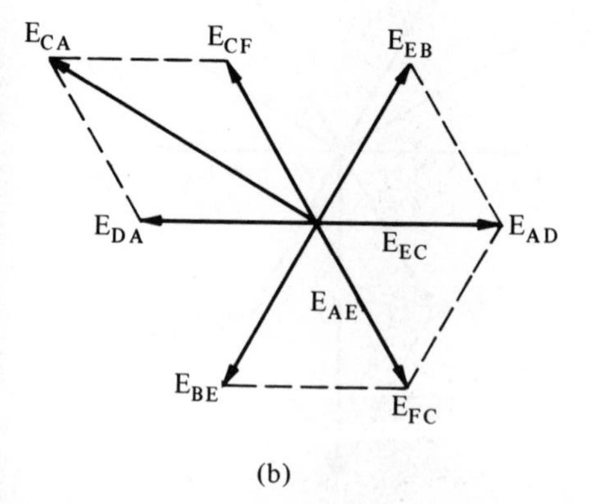

(b)

Fig. 5-12. (a) Incorrect connections for a wye-connected supply. (b) Vectors of incorrect wye.

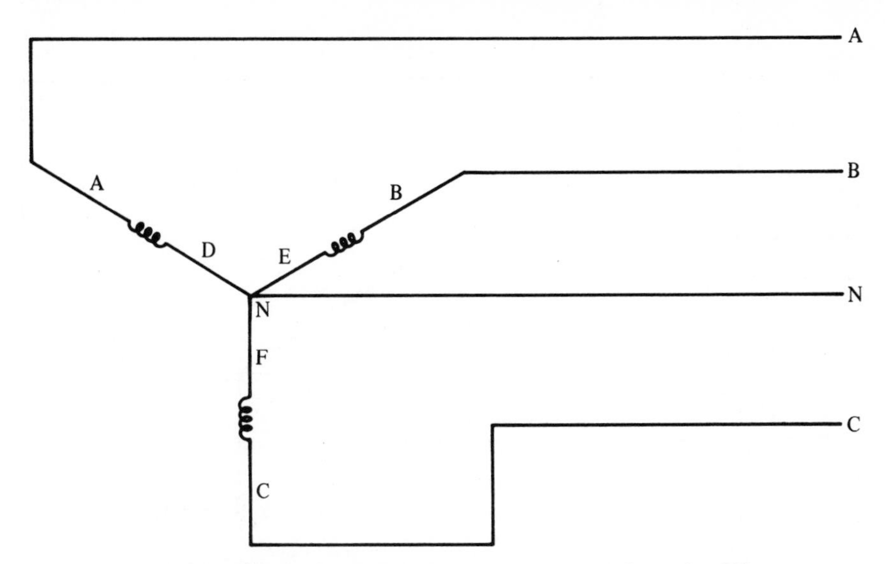

Fig. 5-13. Correct connections for a wye-connected, four-wire, 120–208-V, three-phase supply.

5-6 ***Power in a balanced wye-connected load.*** A wye-connected, three-phase load usually requires a wye-connected supply with four wires. Later we shall discover the important function of the fourth wire or neutral as described in Fig. 5-11. The criteria for selecting a wye or delta connection are somewhat involved and will be explored in a later chapter.

Figure 5-14 shows the system with balanced loads of 2500 W in each phase.

Since the line voltage is

$$E_L = E_p\sqrt{3} \tag{5-5}$$

then the phase voltage or voltage across each load is

$$E_p = \frac{E_L}{\sqrt{3}} = \frac{208 \text{ V}}{1.732} = 120 \text{ V}$$

The phase current or current through each load is

$$I_p = \frac{W}{E_p \cos\theta} = \frac{2500}{(120)(1)} = 20.8 \text{ A} \tag{4-5}$$

Figure 5-14b shows the vectors of the line voltages broken down into phase voltages with the neutral N common to all three. Since the phase currents are equal to line currents in a wye, they are shown in phase with their respective phase voltages (resistive load, pf $= 1.0$). Since these three

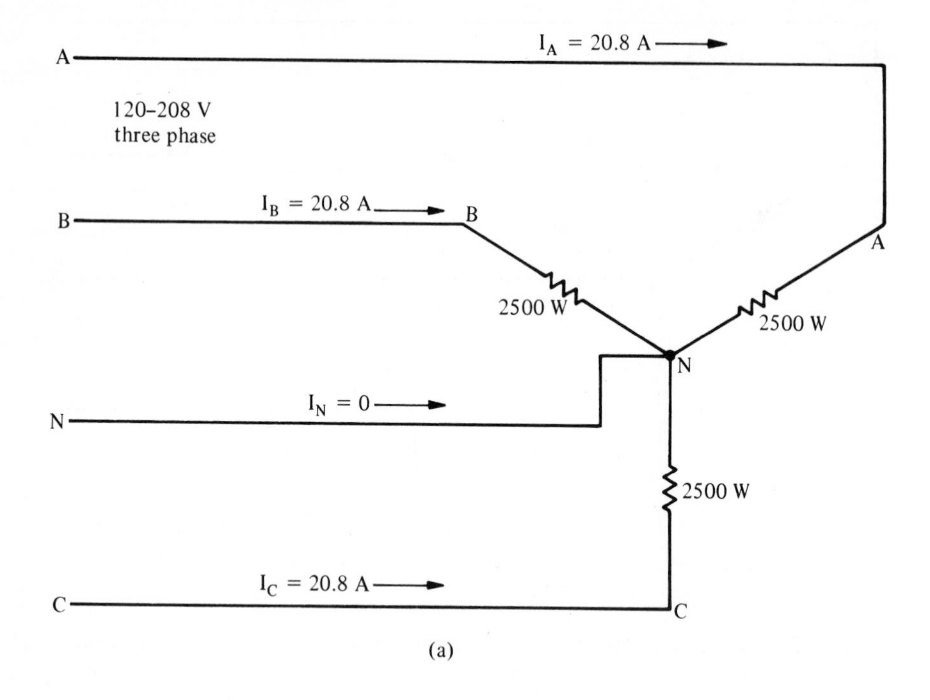

(a)

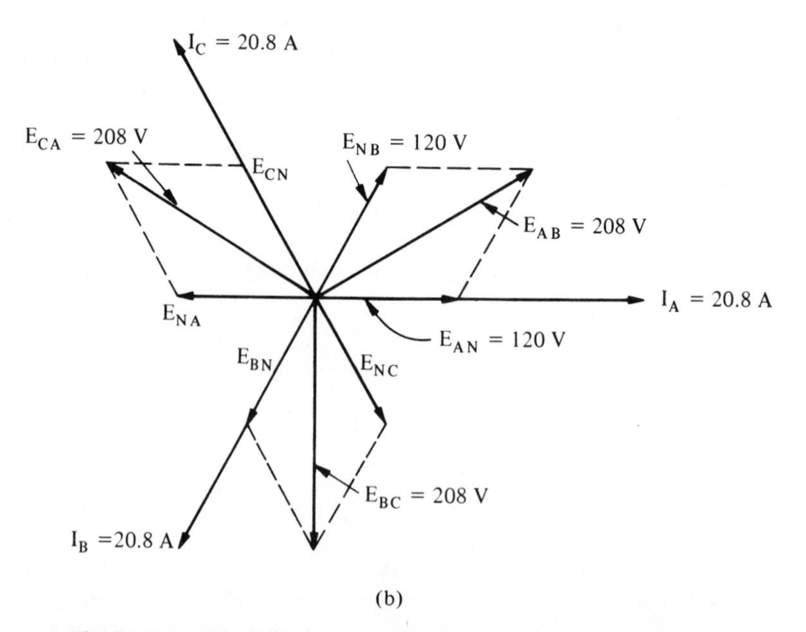

(b)

Fig. 5-14. (a) Balanced wye-connected load on a 208-V, four-wire supply. (b) Vectors of phase and line voltages and currents.

currents are equal and are 120° apart, their sum is zero; therefore, no current can flow in the neutral wire.

The definition of phase power factor is, in fact, the cosine of the angle between phase voltage and phase current, whether a wye or a delta configuration.

The power of the three-phase load of Fig. 5-14 can be considered to be the sum of individual powers of the three phases. That is,

$$P_T = (3E_P I_P)(\cos\theta) \quad \text{(same as delta)} \tag{5-3}$$

However, Eqs. (5-5) and (5-6) state

$$E_P = \frac{E_L}{\sqrt{3}}$$

$$I_P = I_L \quad \text{for a wye connection.}$$

Substituting these value in Eq. 5-3, we get

$$\begin{aligned} P_T &= (3)\left(\frac{E_L}{\sqrt{3}}\right)(I_L \cos\theta) \\ &= \sqrt{3}(E_L I_L)(\cos\theta) \end{aligned}$$

This is the same equation derived for the delta, as shown in Eq. (5-4). This means that when using line voltages and line currents the power can be calculated from line values regardless of whether the load is a delta or a wye. This is of great assistance when calculating three-phase power. The power of the load of Fig. 5-14 therefore is

$$\begin{aligned} P_T &= (3E_P I_P)(\cos\theta) \\ &= (3)(120)(20.8)(1) = 7500 \text{ W} \end{aligned} \tag{5-3}$$

or

$$\begin{aligned} P_T &= \sqrt{3}(E_L I_L)(\cos\theta) \\ &= (1.732)(208)(20.8)(1) = 7500 \text{ W} \end{aligned} \tag{5-4}$$

Example 5-3. A balanced 40-kW fluorescent lighting load is connected to a 277/480-V, four-wire, three-phase supply. If the power factor is 0.85, calculate the current in each line.

Solution

$$I_L = \frac{P_T}{(E_L)\sqrt{3}\,(\text{pf})} = \frac{40{,}000}{(480)(1.732)(0.85)} = 56.5 \text{ A} \tag{5-4}$$

To illustrate the contribution of each phase,

$$P_P = \frac{P_T}{3} = \frac{40 \text{ kW}}{3} = 13.33 \text{ kW}$$

$$I_P = \frac{P_P}{E_P \cos\theta} = \frac{13.330}{(277)(0.85)} = I_L = 56.5 \text{ A}$$

5-7 ***Power in an unbalanced wye-connected load.*** To cause the load of Fig. 5-14 to become unbalanced we shall change two of the loads to 2000 and 1500 W, respectively. The total power is now 6000 W, as shown in Fig. 5-15a.

$$I_A = I_{AN} = \frac{2000 \text{ W}}{(120)(1)} = 16.7 \text{ A}$$

$$I_B = I_{BN} = \frac{2500 \text{ W}}{(120)(1)} = 20.8 \text{ A}$$

$$I_C = I_{CN} = \frac{1500 \text{ W}}{(120)(1)} = 12.5 \text{ A}$$

To simplify the vector diagram only phase voltages are shown, but in the same position as Fig. 5-14b. The phase currents are drawn in phase with their correct voltages according to their proper magnitude.

It is obvious that the vector addition of these three currents is not zero; therefore, the system is said to be unbalanced. Were it not for the neutral conductor to carry the unbalanced current back to the supply the respective phase voltages would not be stable at 120 V each. This situation is like that described in Sec. 4-13 for a single-phase, three-wire system.

To find the magnitude of the neutral current it is necessary to add the three currents vectorially, and their sum must be the neutral current. The construction in Fig. 5-15b does this by first adding I_A and I_C and then adding that resultant to I_B. Graphically, the neutral current is about 7.3 A. Worked out mathematically the neutral current is 7.32 A.

Example 5-4. Three 120-V loads are connected to a 208/120-V, three-phase, four-wire supply.

Load A is 5 kW of incandescent lighting.
Load B is 3 kW of fluorescent lighting with an 80% power factor.
Load C is three 500-W heaters.

(a) Calculate the current in each line.
(b) Determine graphically the neutral conductor current.

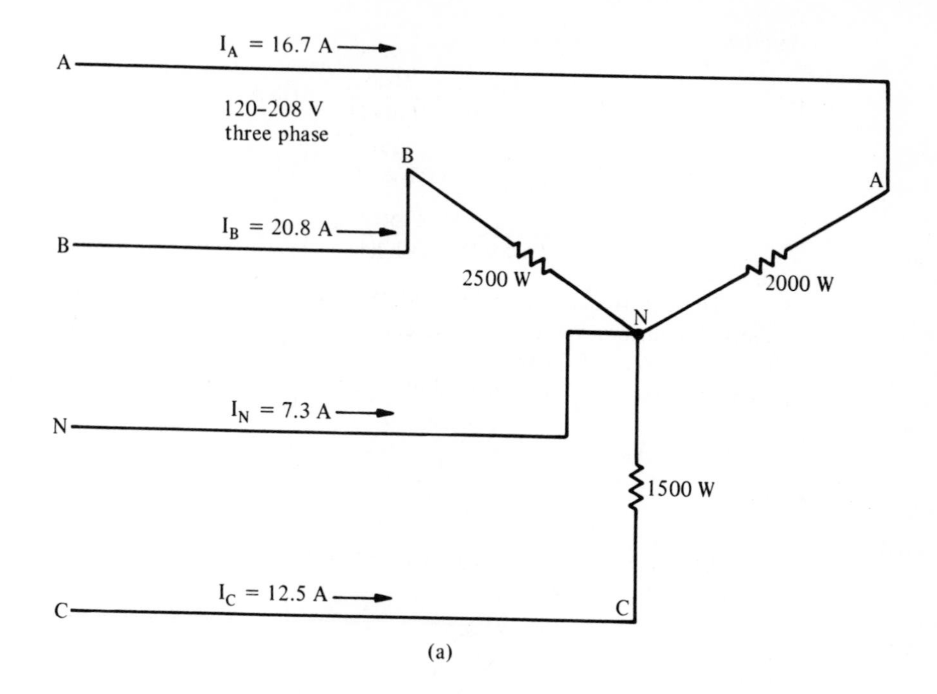

(a)

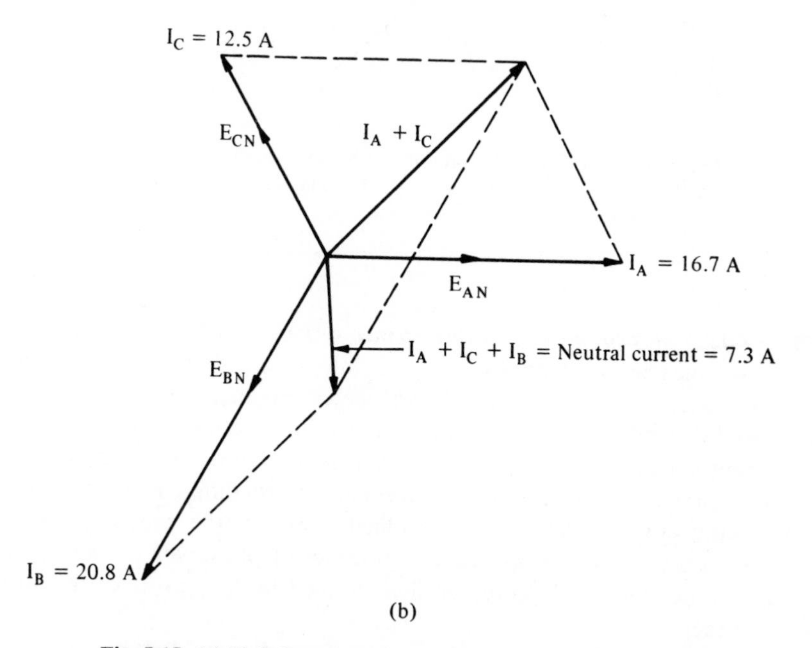

(b)

Fig. 5-15. (a) Unbalanced wye-connected load. (b) Vector addition of line currents to show neutral current.

Solution

(a)

$$I_A = \frac{P_P}{E_P \cos \theta} = \frac{5000}{(120)(1)} = 41.6 \text{ A}$$

$$I_B = \frac{P_P}{E_P (\text{pf})} = \frac{3000}{(120)(0.8)} = 31.2 \text{ A}$$

$$I_C = \frac{P_P}{E_P \cos \theta} = \frac{1500}{(120)(1)} = 12.5 \text{ A}$$

These currents were calculated as phase currents, but in a wye-connected load they are also the *line* currents. The power of the system is 5 + 3 + 1.5 kW, or 9.5 kW.

(b) See Fig. 5-16 for a graphical solution.

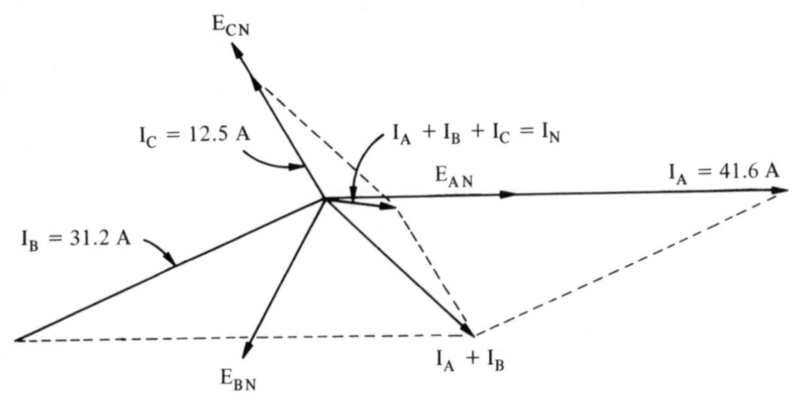

Fig. 5-16. Graphical solution for a obtaining neutral current in Ex. 5-4b. I_N = approximately 6.5 A. The mathematical solution is $I_N = 6.78$ A.

5-8 ***Power measurement of a three-phase resistive load.*** When both phase voltages and currents are in phase, the phase power is obviously their direct product. Therefore, if a voltmeter and an ammeter are properly connected to read phase values and their readings are multiplied, the actual phase power and total power in watts can be obtained. However, we have seen how inductive and capacitive loads cause phase differences between the voltage and current and when multiplied together (the product) become apparent power or volt-amperes. The product of phase voltage and phase current cannot be called power until multiplied by the cosine of the angle between them.

Wattmeters, however, are designed to recognize and to take this power factor angle, if one exists, into account. A wattmeter can multiply the current passing through its current coil by the voltage impressed on its voltage coil and then multiply this product by the cosine of the angle between them. According to our definition of power, this is the number of watts used by the circuit being measured.

A three-phase system actually consists of three separate single-phase loads, but when many different types of electrical loads are combined, the identity of each phase is lost. In our previous discussion of balanced and unbalanced loads we were always able to determine *line* currents and voltages. To properly measure three-phase power, we must be able to do so only by having access to the three line wires of the three-phase system.

An ingenious method of measuring three-phase power has been devised by using two single-phase wattmeters and adding their readings. This is done by connecting the wattmeters as shown in Fig. 5-17. Note that each voltage coil must be connected to the line that has no current coil.

A three-phase wattmeter would have its four coils connected in the same configuration but would indicate on only one meter scale. A three-phase watt-hour meter would have the same connections.

Figure 5-18 shows two wattmeters measuring the resistive load of Fig. 5-9. Note that all common connections are connected to the line side. Note also that the current of W_1 lags its voltage by 30° while the current of W_2 leads its voltage by 30°. Should one wattmeter read back scale it is only necessary to reverse the current connections.

If we follow the diagram and the vectors of Fig. 5-18 we can determine what the wattmeters will indicate.

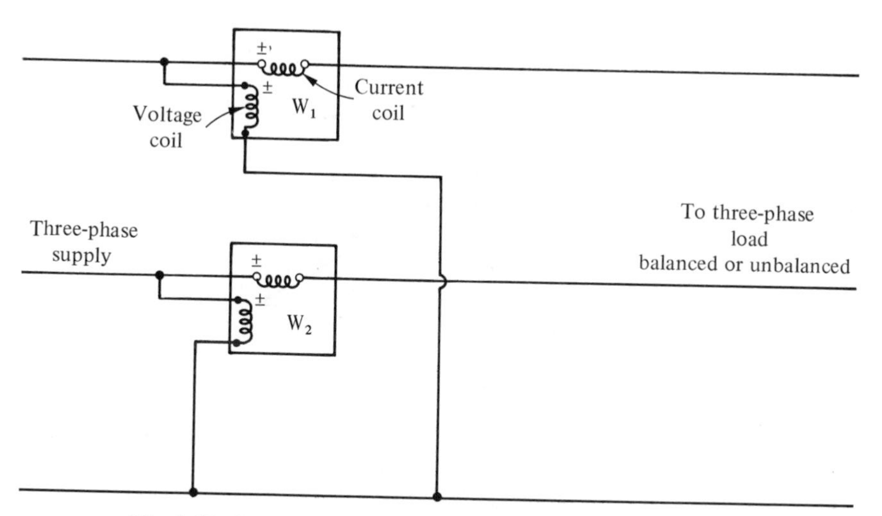

Fig. 5-17. Connections of current and voltage coils of two single-phase wattmeters to measure three-phase power.

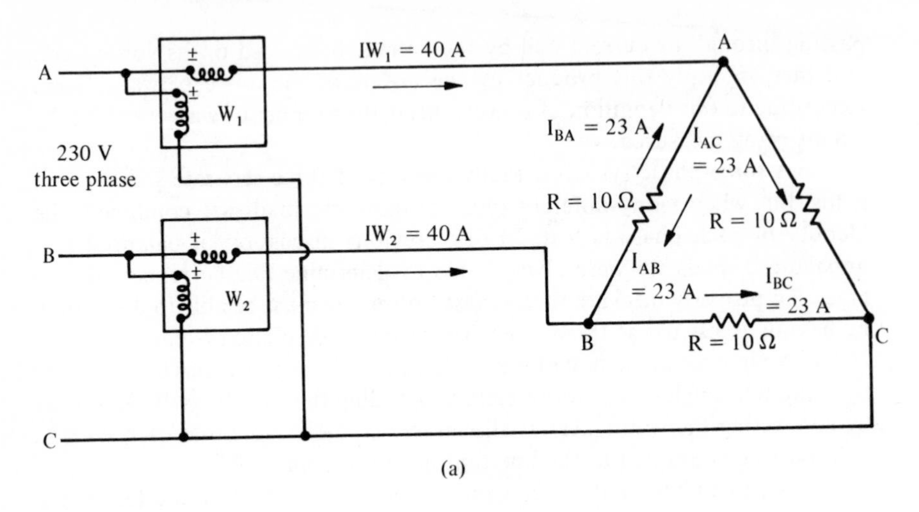

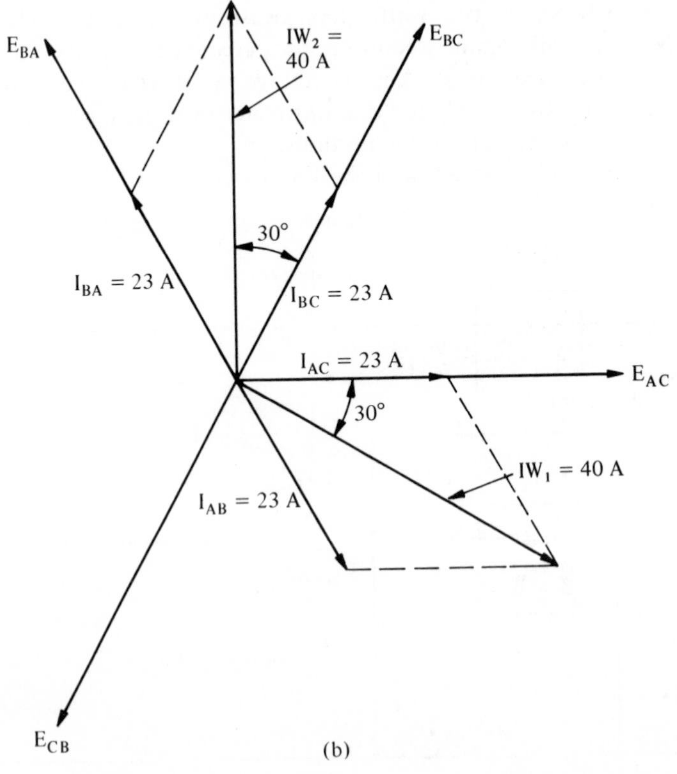

Fig. 5-18. (a) Two wattmeters measuring a three-phase resistive load. (b) Vector showing wattmeter voltages and currents.

Wattmeter 1 line current is 30° behind its line voltage E_{AC}:

$$\begin{aligned} W_1 &= (230)(40)(\cos 30°) \\ &= (230)(40)(0.866) = 7950 \text{ W} \end{aligned}$$

Wattmeter 2 line current is 30° ahead of its line voltage E_{BC}:

$$\begin{aligned} W_2 &= (230)(40)\ (\cos 30°) \\ &= (230)(40)(0.866) = 7950 \text{ W} \\ W_1 + W_2 &= 15{,}900 \text{ W} \end{aligned}$$

This is in agreement with Sec. 5-3.

5-9 ***Power measurement of a three-phase inductive load.*** Since each wattmeter in the previous discussion indicates the same number of watts, it would seem to be logical to only use one wattmeter and to double its reading. However, the vectors of Fig. 5-18 show a lead or lag situation that must be explored further in conjunction with an inductive load, which also creates a phase shift.

To do this we shall use a balanced load of three 10-Ω impedance loads in delta but assume they are partially inductive so that their power factor is 0.8. This means that each phase current lags its respective phase voltage by 37°.

Figure 5-19 shows the same magnitudes of voltages and currents except that the 37° angle is included. Reading off the vector diagram we state what each wattmeter "sees." Wattmeter 1 line current is 67° behind its line voltage E_{AC}:

$$\begin{aligned} W_1 &= (230)(40)(\cos 67°) \\ &= (230)(40)(0.391) = 3600 \text{ W} \end{aligned}$$

Wattmeter 2 line current is 7° behind its line voltage E_{BC}:

$$\begin{aligned} W_2 &= (230)(40)(\cos 7°) \\ &= (230)(40)(0.99) = 9100 \text{ W} \\ W_1 + W_2 &= 12{,}700 \text{ W} \end{aligned}$$

Let us check these results:

$$\begin{aligned} P_T &= (\sqrt{3})(E_L I_L)(\text{pf}) \qquad (5\text{-}4) \\ &= (\sqrt{3})(230)(40)(0.8) \\ &= 12{,}700 \text{ W} \end{aligned}$$

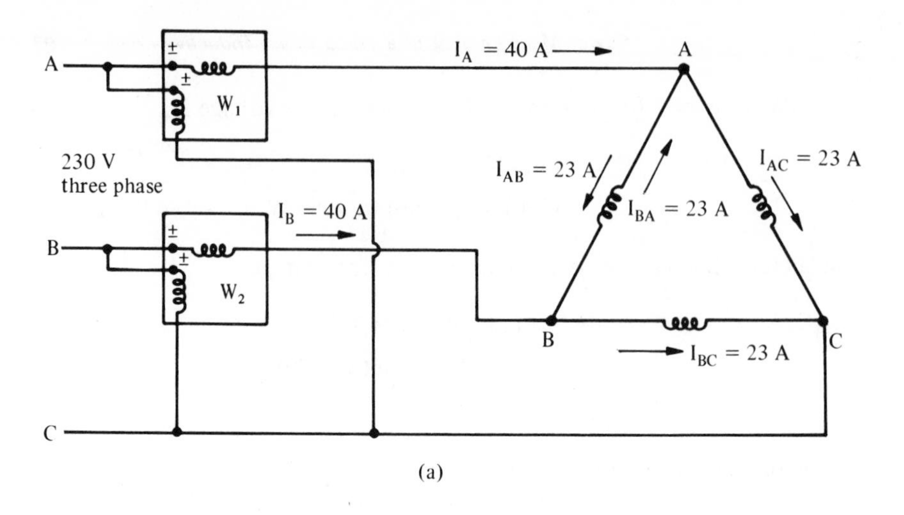

(a)

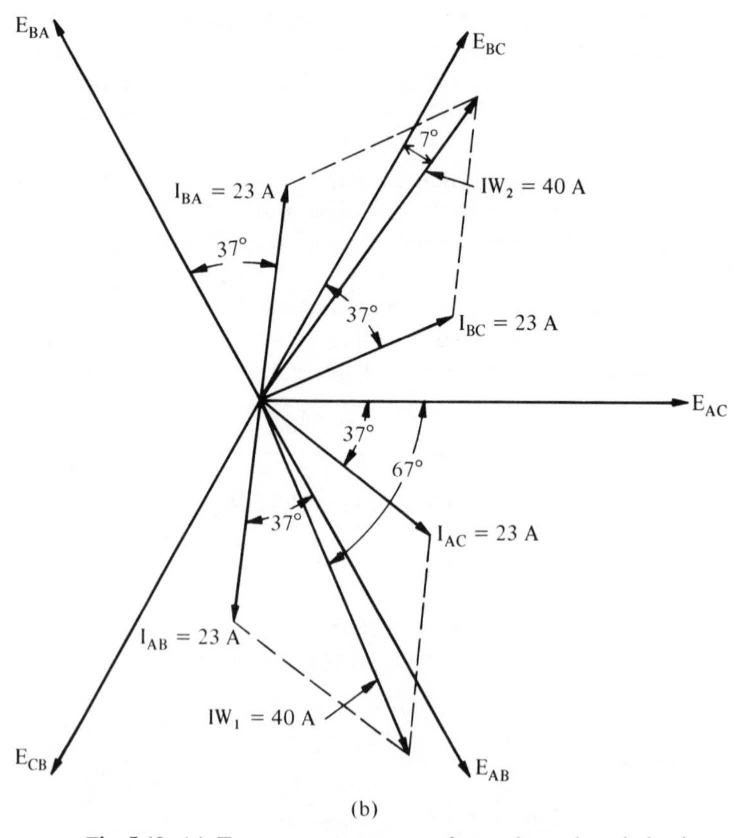

(b)

Fig. 5-19. (a) Two wattmeters measuring a three-phase inductive load with pf = 0.8. (b) Vectors showing wattmeter voltages and currents.

This leads to the following relationships for the two wattmeter readings, respectively:

$$W_1 = E_L I_L \cos(30 + \theta) \tag{5-5}$$

$$W_2 = E_L I_L \cos(30 - \theta) \tag{5-6}$$

in which θ (as always) is the angle between the *phase* voltage and the *phase* current.

Note from Eqs. (5-5) and (5-6) that both wattmeters read the same when $\theta = 0$. The difference between the wattmeter readings as shown above provides a measure of the power factor of the system. We can actually determine the power factor of the reactive three-phase load (if balanced) by obtaining a ratio of smaller wattmeter readings to the larger and by finding the power factor from the trigonometric curve of Fig. 5-20.

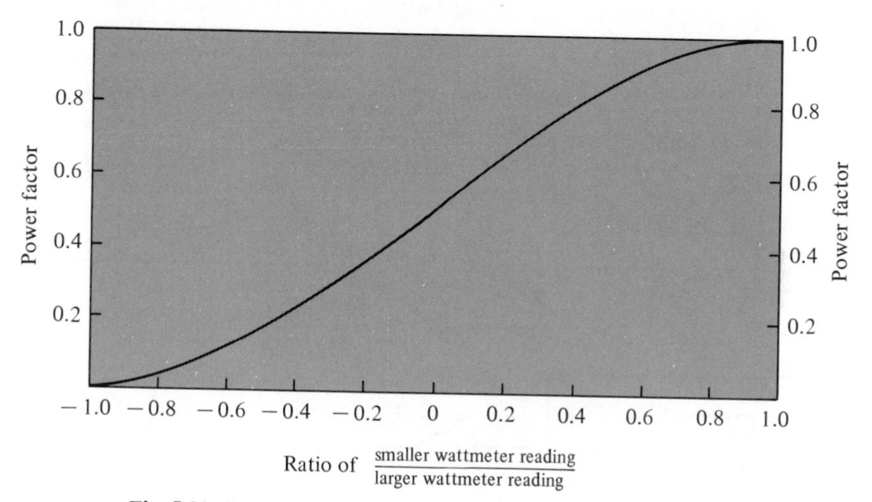

Fig. 5-20. Determination of power factor from wattmeter readings.

It should be noted from Fig. 5-20 that if one wattmeter reads 0, then the power factor is 0.5. Should the power factor be less than 0.5, the angle between one wattmeter current and voltage would be greater than 90°; hence, that particular reading would be considered negative and would have to be subtracted from the other. The amount to subtract may be found by reversing the current connections so the wattmeter indicates up scale rather than a reversed manner.

The power factor also can be computed from the relation

$$\tan\theta = \sqrt{3}\left(\frac{W_2 - W_1}{W_2 + W_1}\right) \tag{5-7}$$

where W_2 is always the larger reading wattmeter. Knowing W_2 and W_1 we can determine, respecturly, tan θ, θ, and cos θ. The last is the power factor of the balanced reactive three-phase load.

5-10 ***Power measurement of a three-phase, four-wire load.*** Sections 5-6 and 5-7 showed one method of connecting single-phase loads to a three-phase system. The neutral conductor makes certain that unbalanced loads do not cause unequal load voltages. When this system is balanced the two-wattmeter method of power measurement will properly measure the power. The meters must be connected according to Fig. 5-17, with no connections made to the neutral. However, this type of four-wire, three-phase load is often unbalanced and the two-wattmeter method will measure current in only two of the lines.

Examination of Fig. 5-21 shows that if only load *C* were operating (an extreme case of unbalance) no current would pass through either of the wattmeters. Note that load *C* would be supplied with current by means of the neutral and line *C*. In this case both wattmeters would indicate zero power.

To properly measure a three-phase, four-wire unbalanced load (always a possibility) three wattmeters must be used. Figure 5-22 shows how these connections are made. Note that the voltage coils are all connected to

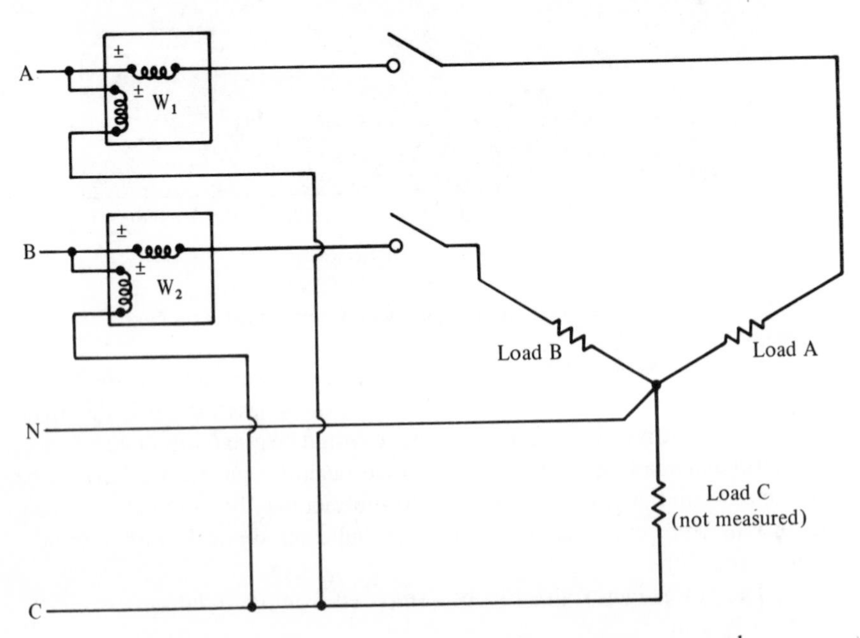

Fig. 5-21. Two-wattmeter method is unable to measure or record properly an unbalanced three-phase, four-wire load.

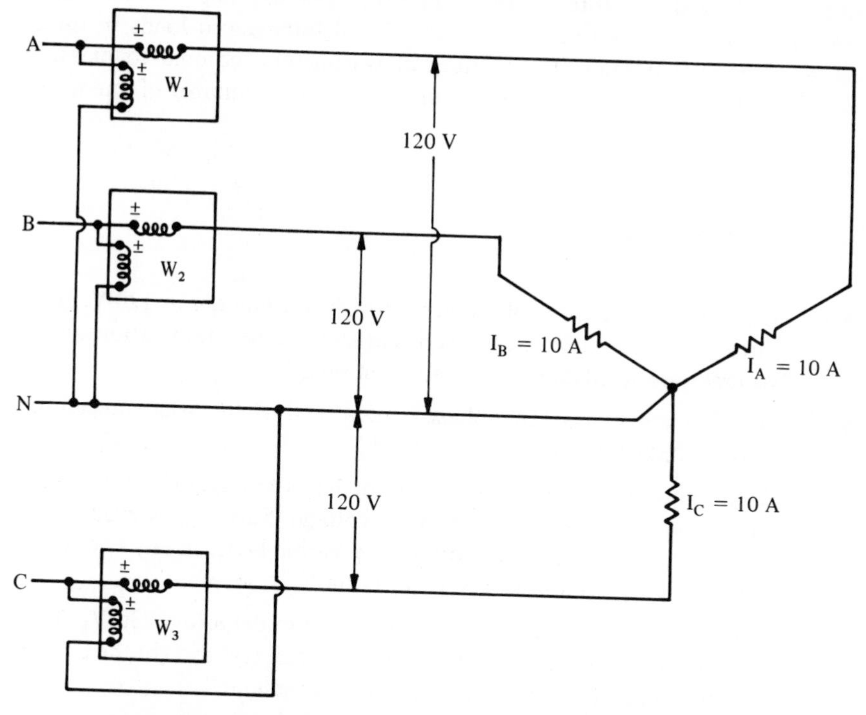

Fig. 5-22. Three-wattmeter method of measure or recording a three-phase, four-wire load.

the neutral. This arrangement permits each phase to be measured by a separate wattmeter, and the total power is the sum of their readings.

Figure 5-22 also shows each phase consisting of a resistive load of 10 A at 120 V. Each wattmeter will indicate (120)(10) = 1200 W. The total power is 1200 + 1200 + 1200 = 3600 W. This agrees with

$$\begin{aligned} P_T &= 3E_P I_P \cos\theta \\ &= (3)(120)(10)(1) = 3600 \text{ W} \end{aligned} \tag{5-3}$$

This particular load is balanced; therefore, we also may use Eq. 5-4 to calculate the total power. $I_L = I_P = 10$ A.

$$\begin{aligned} E_T &= (\sqrt{3})(E_P) = (1.732)(120) = 208 \text{ V} \\ P_T &= (\sqrt{3})(E_L I_L) \cos\theta \\ &= (\sqrt{3})(208)(10)(1) = 3600 \text{ W} \end{aligned} \tag{5-4}$$

The three-wattmeter method may be used to measure balanced and unbalanced loads in three-phase, four-wire systems. The two-wattmeter method may be used to measure balanced and unbalanced loads in three-phase, three-wire systems. The number of wattmeters required to measure unbalanced loads, therefore, is always one *less* than the number of line wires.

PROBLEMS

5-1 Each phase of a three-phase generator has a voltage of 120 V. If all three phases are connected in series in all possible combinations, what voltages are available across the open ends?

5-2 When connecting a three-phase generator in delta, what precaution must be taken before delta is closed?

5-3 Each phase of a three-phase generator has a voltage of 265 V. When connecting this source in wye, what voltage must be observed across any two lines? What other voltage is possible between any two lines? Is this still a three-phase wye connection?

5-4 Three resistors of 25 Ω each are connected in delta to a 230-V, three-phase supply. Calculate (a) the current in each resistor, (b) the current in each line, and (c) the total power of the system.

5-5 The current in each phase of a balanced delta, 230-V, three-phase load is 12 A, pf = 1.0. Calculate (a) the line current, (b) the power per phase, and (c) the total power.

5-6 A 35-kW resistive load is connected to a three-phase, 460-V supply so that the load is balanced. Calculate the current in each line.

5-7 The current in each line of a 230-V, three-phase system is 85 A. The pf of the load is 0.8. Calculate the power of this three-phase load.

5-8 Calculate the pf of a 25-kW, 230-V, three-phase balanced load; the line current is 75 A.

5-9 Calculate the power of a three-phase, delta-connected, 230-V load. The line current is 35 A, and the pf is 0.75.

5-10 A 15-kW, three-phase, 230-V load is equally divided and is connected in delta. However, one phase has been disconnected so that the operating load is actually only 10 kW. Determine the three line currents.

5-11 Three 10 Ω impedances, each with a pf of 0.8, are connected in delta to a 230-V, three-phase supply. Calculate (a) the phase current, (b) the line current, and (c) the total power.

5-12 A 4-, 7-, and 10-kW load are connected in delta to a 230-V, three-phase supply. Calculate (a) the total power, (b) each phase current, and (c) each line current.

5-13 Three 10-Ω resistors are connected in delta to a 230-V, three-phase supply. Calculate (a) the phase current, (b) the line current, and (c) the total power.

5-14 Three 10-Ω resistors are connected in wye to a 230-V, three-phase supply. Calculate (a) the phase voltage, (b) the phase current, (c) the line current, and (d) the total power (compare this with Prob. 5-13c).

5-15 Three 5-kW resistive, 120-V loads are connected in wye to a 208-V, four-wire, three-phase supply. Calculate (a) the line current and (b) the total power.

5-16 Three resistors of 10, 20, and 30 Ω, respectively, are connected in wye to a 120–208-V, four-wire, three-phase supply. Calculate (a) the current in each line, (b) the total power, and (c) the current in neutral.

5-17 A 120–208-V, three-phase, four-wire supply is supplying a lighting load equally divided between each line and neutral. Each phase consists of fifty 100-W lamps operating at 120 V. Calculate (a) the total power of the lighting load, (b) the current in each line, and (c) the current in neutral.

5-18 Three resistors of 10, 15, and 30 Ω each are connected in delta to a three-phase, 230-V supply. What is the total power of the system?

5-19 When properly connected, two wattmeters measuring power in a three-phase, 230-V balanced load indicate 5000 W each. Calculate (a) the total power of the system, (b) the power factor of the system, and (c) the current per line.

5-20 When properly connected, two wattmeters measuring power in a three-phase, three-wire, 480-V balanced system indicate 8000 and 4000 W. Calculate (a) the total power of the system, (b) the power factor of the system, and (c) the current per line, and (d) assuming a delta-connected load, what is the current in each phase?

5-21 A three-phase balanced load with a line current of 60 A is to be measured with a watt-hour meter to record the energy used. How many and what size conductors must be extended to the watt-hour meter location?

5-22 Three 277-V lighting loads, each rated at 15 kW with a pf of 0.8, are connected in wye to a 480–277-V, three-phase, four-wire supply.
(a) What is the current in each line?
(b) Calculate the total power using the three-phase power equation.

5-23 Three 277-V lighting loads—one of 15 kW, one of 12 kW, and the third of 5 kW—each with a pf of 0.8, are connected in wye to a 277–480-V, three-phase, four-wire supply. Calculate the current in each line.

5-24 Three resistors of 20, 30, and 50 Ω each are connected in delta to a 230-V, three-phase supply. Calculate (a) the current in each phase, (b) the total power of the system, and (c) the current in each line.

5-25 A 5000-, 3500-, and 2000-W load consisting of electric heaters are connected in delta to a 230-V, three-phase supply. Calculate (a) the current in each phase and (b) the total power in the system.

5-26 A 50-kW, 230-V, three-phase load consists of heating elements. The load is connected in delta and is balanced. What is the current per phase?

Chapter Six

Transformers

6-1 ***Transformer principles and construction.*** Transformers are a very important part of an electrical system. A transformer is an electromagnetic device designed to *transfer* electrical energy from one electric circuit to another. The voltages of the two circuits may be the same or may differ by any ratio. The transformer is a machine with a magnetic circuit and two or more electric circuits. It is a device that has no moving parts and therefore can operate at a very high efficiency, up to 98 or 99%. It requires no maintenance and is very reliable.

The magnetic circuit of a transformer is built of a laminated core of high permeability (see Sec. 4-3) steel sheets, forming a closed magnetic circuit. On this core is placed an *exciting winding*. When connected to an ac source, the resultant current will set up an alternating flux. This exciting winding is called the *primary* or input side, because it is connected to the power source and is the source of magnetization. Any other winding placed on this core will be *magnetically coupled* with the alternating flux. Voltages will be *induced* in these windings because of *electromagnetic induction*. A stationary conductor in a changing magnetic field can induce a voltage, as is the case when a conductor is physically moved through a constant magnetic field. This inductively coupled winding is called the *secondary* and will be the source of voltage for a load at the *output* terminals.

Each turn on each of the secondary windings induces a voltage. These turns in series will permit any desired voltage in the secondary circuit, dependent only on the number of turns, compared to the primary, which set up the flux.

The voltage ratio of the transformer can be expressed as

$$\frac{E_1}{E_2} = \frac{N_1}{N_2} \tag{6-1}$$

where E_1 and N_1 are voltage and turns of the primary winding and E_2 and N_2 are those of the secondary.

There can be no change of frequency in a transformer. The frequency of the flux changes follows the primary exciting current; therefore, any induced voltage must have the exact *same* frequency.

Figure 6-1 illustrates a transformer core with primary and secondary windings.

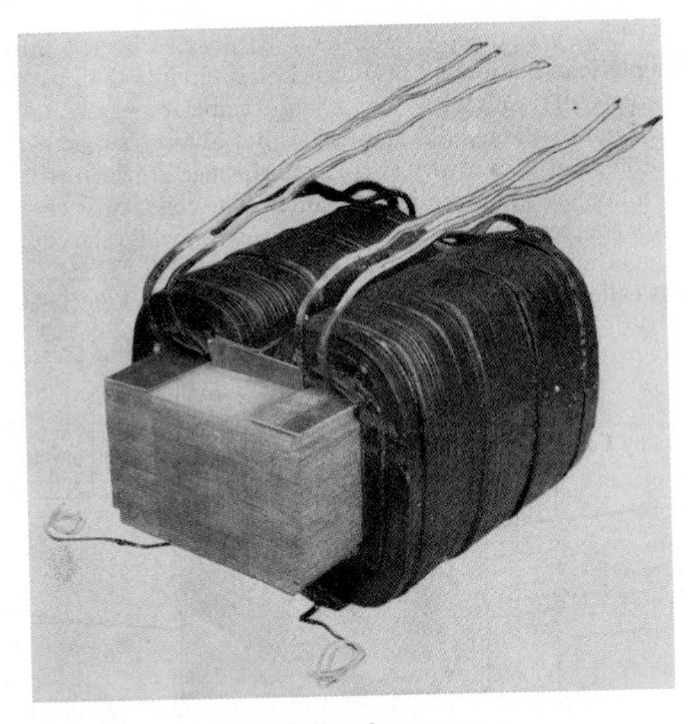

Fig. 6-1. Transformer core.

Example ***6-1.*** A transformer has two windings of 60 and 240 turns, respectively. Correct excitation is assured if 120 V at 60 Hz is applied to the 60-turn winding.

(a) Calculate the voltage available from the 240-turn winding.
(b) Calculate the volts per turn ratio on both sides.
(c) If the 60-turn winding has a 5% tap, calculate the primary voltage applied to 95% of the winding that will permit the same output voltage as (a).
(d) If the 240-turn winding has a 5% tap, what is voltage output when 95% of the turns are used? The primary is the same as a.

Solution

(a)
$$\frac{E_1}{E_2} = \frac{N_1}{N_2} \qquad \frac{120}{E_2} = \frac{60}{240} \tag{6-1}$$

$$E_2 = \frac{(240)(120)}{60} = 480 \text{ V}$$

(b)
$$\frac{480 \text{ V}}{240 \text{ t}} \quad \text{or} \quad \frac{120 \text{ V}}{60 \text{ t}} = 2 \text{ V per turn}$$

(c) (0.95)(60) = 57 turns as a primary and the same excitation will be assured if

$$\frac{2 \text{ V}}{\text{t}}(57 \text{ t}) = 114 \text{ V applied to the primary}$$

(d)
$$(0.95)(480) = 456 \text{ V}$$

6-2 ***Transformer magnetization.*** The assurance of the proper magnitude of flux in a transformer core is a complicated problem, affected by many variables in its design. Some of these are the core area and length, permeability of core material, physical placement of turns, circular mil area of the wire used, and desired magnitude of the excitation current—this is the primary current when no load is connected to the secondary.

To obtain the proper excitation current in any winding used as a primary, we observe one simple rule: Connect *any* winding to the *voltage* and *frequency* for which it has been designed. The transformer nameplate provides this information either directly or by inference.

The primary winding not only carries the excitation current but has a voltage induced in it as does the secondary winding, since their turns are linked by the same flux. This self-induced primary voltage must *oppose* the applied voltage. Therefore, the primary excitation current that flows is a result of the *difference* between these two primary voltages, the applied line voltage and induced primary voltage. In a well-designed transformer with little magnetic leakage, this difference is very slight (less than 1% of the applied voltage), resulting in a *small* excitation current compared to the full load primary current.

When a voltage less than rated voltage is applied to the primary winding, the flux is below the designed level. Induced voltages tend to follow Eq. (6-1) but will decrease sharply when the transformer supplies current to a load. If voltages above rated voltage are applied to the primary, the excitation current increases, the magnetic core saturates (flux cannot increase), and the induced voltage cannot increase. This causes a greater difference between induced and applied voltage in the primary, causing a rapid increase in primary current that overheats the transformer. Figure 6-2 shows a graph

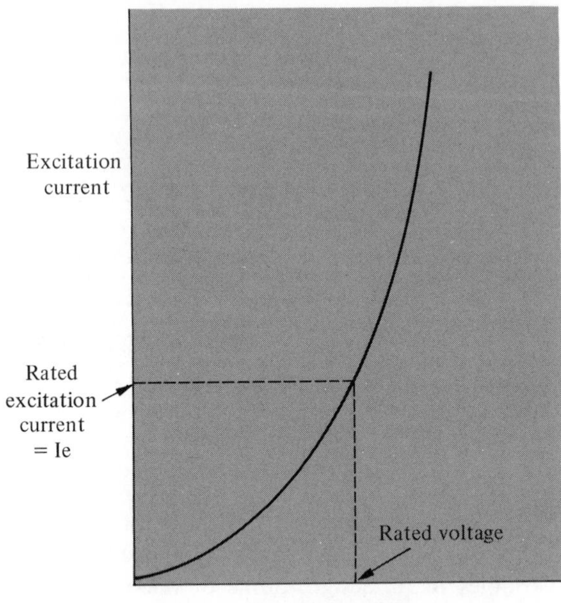

Fig. 6-2. Graph of excitation current vs. applied voltage in typical transformer primary.

of excitation current changes as primary voltage is changed. At rated voltage the primary current sets up the proper flux and, for the above reasons, rated voltage should be applied.

Frequency changes cause the opposite response. A lower frequency lowers the flux linkages per unit of time, causing a lower induced voltage and a higher excitation current with considerable overheating. A higher frequency increases the flux linkages, increases induced voltage, and causes a lower excitation current with poor voltage regulation.

6-3 ***Voltage–current ratios with load.*** When a load is connected to the secondary winding of a transformer, a secondary current flows. This tends to lower the magnetizing flux slightly. According to Lenz's law any current caused to flow by an induced voltage (secondary) will flow in such a direction as to oppose the flux that caused the induced voltage. This demagnetizing action slightly decreases the induced voltage or countervoltage in the primary. Since the net difference between the primary applied and induced voltages has now increased, more current can flow in the primary to compensate for the increased secondary load current.

For proper transformer operation two conditions must be satisfied at all times:

1. The flux in the core must remain almost constant.
2. The ampere turns due to load current in the secondary must be equal and opposite to those in the primary.

The relationship between these two statements is shown in Fig. 6-3. The windings of the transformer of Ex. 6-1 are shown with an assumed excitation (no load) current of 2 A in Fig. 6-3a. We shall identify this excitation current as I_e. In Fig. 6-3b the vector diagram shows the induced voltage E_2, with the excitation current lagging behind the applied voltage by nearly 90°. At a different scale, the excitation mmf of (60 t)(2 A) = 120 At is shown in Fig. 6-3c.

In Fig. 6-3d a resistive load of 5 A is connected to the secondary. But I_1N_1 must equal I_2N_2, as shown in Fig. 6-3e. Note that the original 120 At is the *only* mmf acting in the core. But

$$I_2N_2 = (5)(240) = 1200 \text{ At}$$

For I_1N_1 to equal 1200 At,

$$N_1 = 60 \text{ t}, \quad \text{therefore} \quad I_1 = \frac{1200 \text{ At}}{60 \text{ t}} = 20 \text{ A}$$

Figure 6-3f shows the secondary current and the addition of the two primary currents, 2 A excitation, and 20 A load current. Note that the 2 A, because of its angle, does not materially add to the 20 A necessary for the load. Therefore, in most instances, we may neglect the excitation current when calculating primary and secondary load currents.

As voltage and turns in any winding are proportional, this leads to the following relationship:

$$\frac{E_1}{E_2} = \frac{I_2}{I_1} \tag{6-2}$$

or

$$E_1I_1 = E_2I_2 \tag{6-2a}$$

where E_1 and I_1 are primary and E_2 and I_2 are secondary, or load, voltages and currents.

We should note here that with respect to the magnetic circuit and producing flux, I_1 and I_2 are in *opposite* directions. This justifies statement 1 in Sec. 6-3. Load changes do *not* materially change the flux in the core.

Equation (6-2a) states that the power transferred to the secondary is

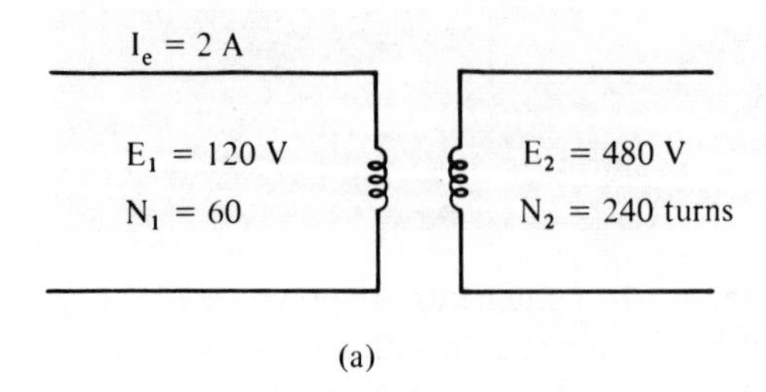

(a)

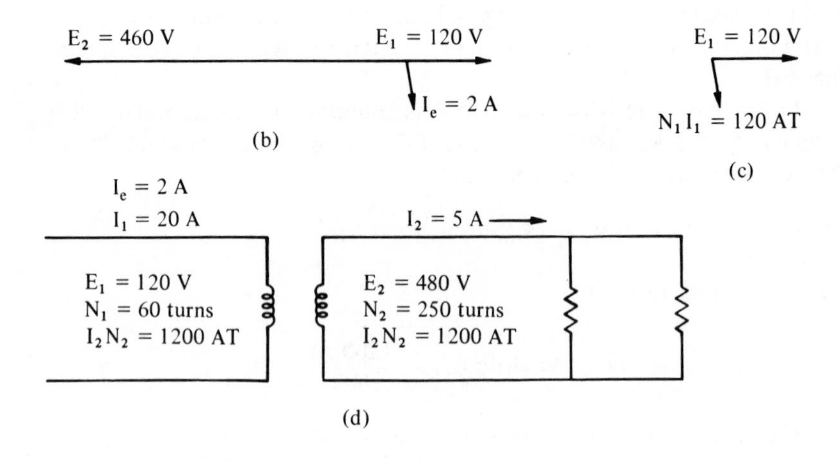

(d)

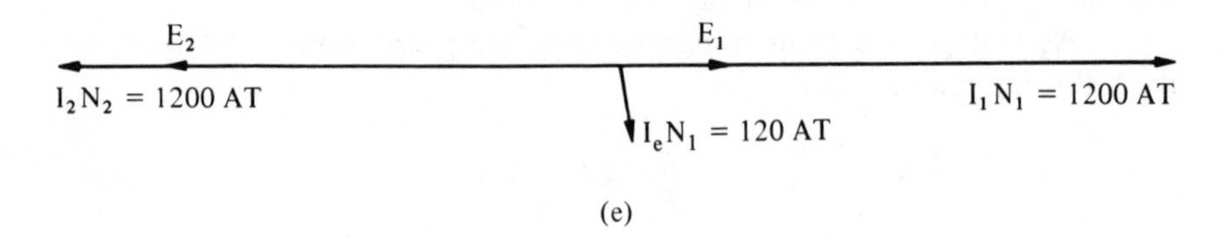

(e)

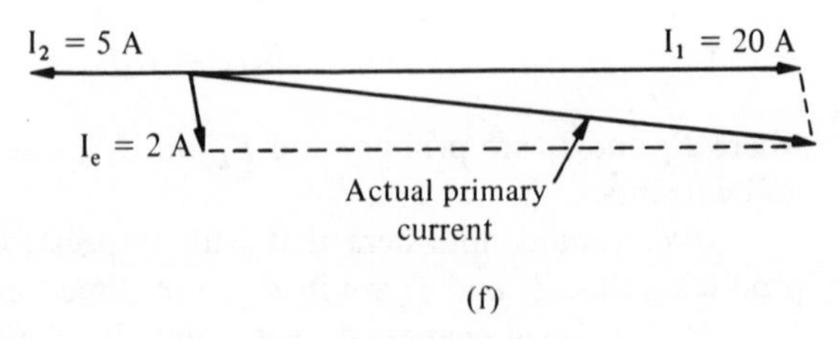

(f)

Fig. 6-3. Magnetomotive forces and currents with and without load on secondary.

the same as the power drawn by the primary from the ac supply. Since the transformer is a highly efficient device, for most purposes and calculations Eq. (6-2a) may be used.

Example **6-2.** A 550-V, 110-V transformer has a 35-A load. Calculate (a) the primary current and (b) the volt-ampere rating of the transformer.

Solution

(a)

$$\frac{E_1}{E_2} = \frac{I_2}{I_1} \tag{6-2}$$

$$I_1 = \frac{E_2 I_2}{E_1} = \frac{(110)(35)}{550}$$

$$= 7 \text{ A}$$

(b)

$$\text{VA} = E_1 I_1 = E_2 I_2$$

$$= (550)(7) = (110)(35)$$

$$= 3850 \text{ VA} = 3.85 \text{ kVA}$$

6-4 ***kVA and current ratings of windings.*** All electrical equipment must have a current rating, indicated either directly or indirectly on the device itself. This rating shows the maximum current that the particular device can safely carry without overheating and limits the equipment to a safe specified temperature rise.

Transformers, therefore, must be rated in kilovolt-amperes. If rated in kilowatts and if the load were operated at a power factor less than unity, it would require more current to operate the load as the power factor is decreased. When the rating is given in kilovolt-amperes, a definite voltage and current limit has been assigned.

The rating of a transformer is intended to indicate the safe limit of load that can be connected to its secondary circuit, which of course could be either its high or low voltage side, depending on whether the particular application is to step up or step down the voltage of the system [see Eq. (6-2a)].

This kilovolt-ampere load rating also applies to the primary since it must take from the power supply the same power that the secondary delivers to a load—if the small transformer losses are neglected, as in Eq. (6-2a). If each section, or high and low voltage circuits, contain only one coil, then the current rating of either coil is obtained by multiplying the kilovolt-ampere rating by 1000 to obtain the volt-amperes and then dividing by the voltage rating of that coil, as shown in Ex. 6-3.

Example 6-3. What is the current rating of each winding of a 25-kVA, 2400/240-V distribution transformer?

Solution

The current rating of the high-voltage winding is

$$I = \frac{(\text{kVA})(1000)}{E} = \frac{(25)(1000)}{2400} = 10.3 \text{ A}$$

the current rating of the low-voltage winding is

$$I = \frac{(25)(1000)}{240} = 103 \text{ A}$$

If either section of the transformer contains more than one winding, then *each* (load or excitation) winding must account for a proportionate share of the kilovolt-ampere rating. The simplest procedure to find the current rating of each coil in either high- or low-voltage sections is to divide the transformer's rating by the number of coils in one section, then to proceed as in Ex. 6-3. The procedure is given in Ex. 6-4.

Ordinarily, multicoiled transformer coils in each section have the same voltage rating, so they may operate in parallel.

Example 6-4. Calculate the current rating of each coil of a 50-kVA, 2300–4600/115–230-V distribution transformer that has two coils for each section.

Solution

Each winding would be rated at 25 kVA. Therefore, the current rating of each 2300-V winding is

$$I = \frac{(\text{kVA})(1000)}{E} = \frac{(25)(1000)}{2300} = 10.8 \text{ A}$$

The current rating of each 115-V winding is

$$I = \frac{(25)(1000)}{115} = 216 \text{ A}$$

6-5 ***Winding polarity and connections.*** To permit transformers to be more versatile, many have more than one winding for both circuits. To properly identify the various windings, the letter *H* is used to identify the *high*-voltage windings and the letter X to identify the *low*-voltage ones.

H_1 and H_2 would be used if only one high-voltage winding were used, H_3 and H_4 for the second, and H_5 and H_6, etc. for subsequent high-voltage windings. The same system is used with X or low-voltage windings. The significance of the numbers is important. All *even* numbers signify the *same* instantaneous polarities of the induced voltages. Odd numbers then will have the opposite polarity. This is similar to + and − signs on batteries. Small transformers may identify the same end or polarity of separate windings by a dot.

When windings are used as primary or exciting windings, the correct number of turns per volt of applied voltage for a given transformer must be observed to assure proper excitation.

Windings to be used as secondary or load circuits must follow the same observance of polarities. If we wish the instantaneous ac voltages of two windings to *add*, we connect odd- and even-numbered terminals together. To be connected in parallel the two voltage sources must be connected with the even-numbered terminals together, and the same is true for the odd numbered terminals.

Figure 6-4 shows how a 1200–2400/120–240-V transformer would be connected to a 1200 or 2400-V supply. Possible secondary voltages are 120 or 240 V. It should be noted that if the common terminals of Fig. 6-4b or d are extended, a three-wire, single-phase, 120–240-V supply is now available.

If each 1200-V winding had 2400 turns, then each 120-V winding must have 240 turns. This means 2 V/turn. It may be observed that each primary arrangement of Fig. 6-4 has the same series of turns per volt of applied voltage. The H_3, H_4 winding in Fig. 6-4a and b makes no excitation contribution as the turns are parallel. It does, however, permit the transformer to provide more load current and must be connected to utilize the full transformer kilovolt-ampere rating.

In the event of unknown terminal markings of transformer windings, they can be easily identified. The high-voltage section will be obvious because smaller wires are necessary compared to the higher current ratings of the low-voltage section. Individual windings can be detected with an ohmmeter or a continuity tester. Figure 6-5 shows this procedure for a 230–460/115–230-V transformer.

1. Connect any winding to the proper voltage according to the nameplate information, as in Fig. 6-5a.
2. Arbitrarily assign H_1 and H_2 to this winding (or X_1 and X_2 if it is a low-voltage winding).
3. Check the other three voltages; they should be as indicated in Fig. 6-5a.
4. Connect one lead of the other 230-V winding to H_2, as in Fig. 6-5b.

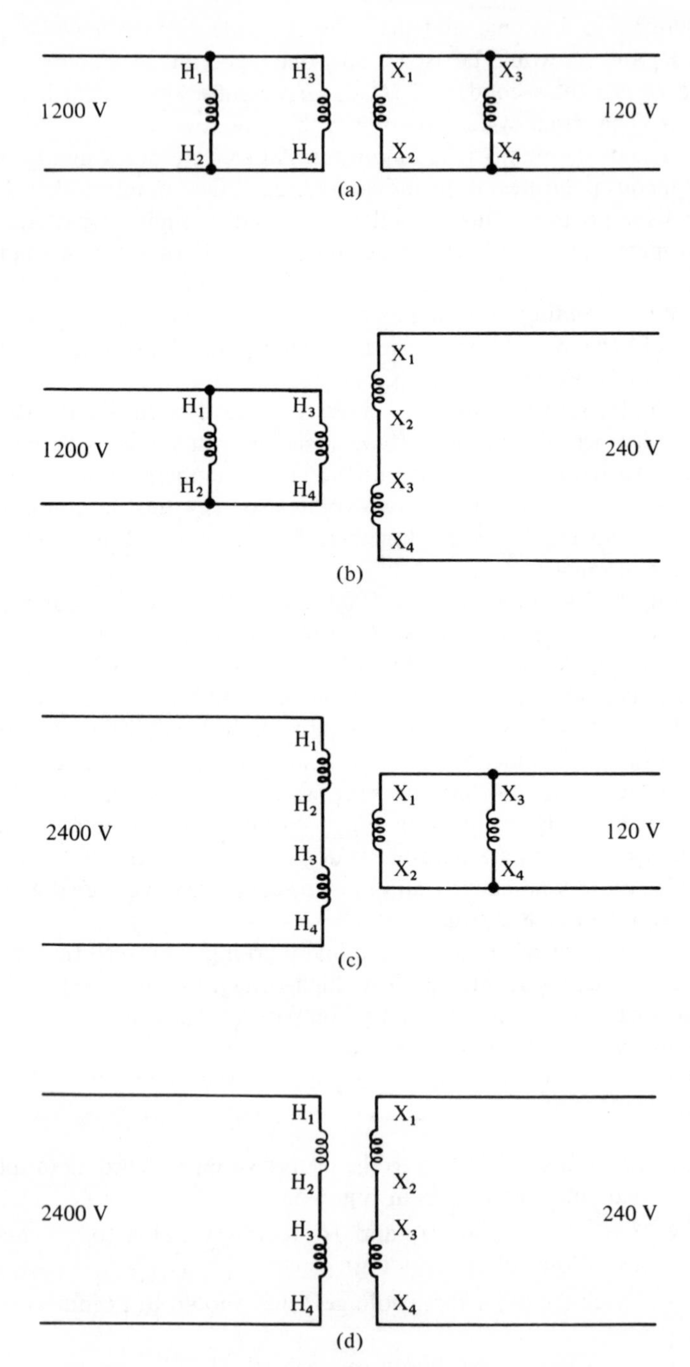

Fig. 6-4. Connections of transformer with four windings.

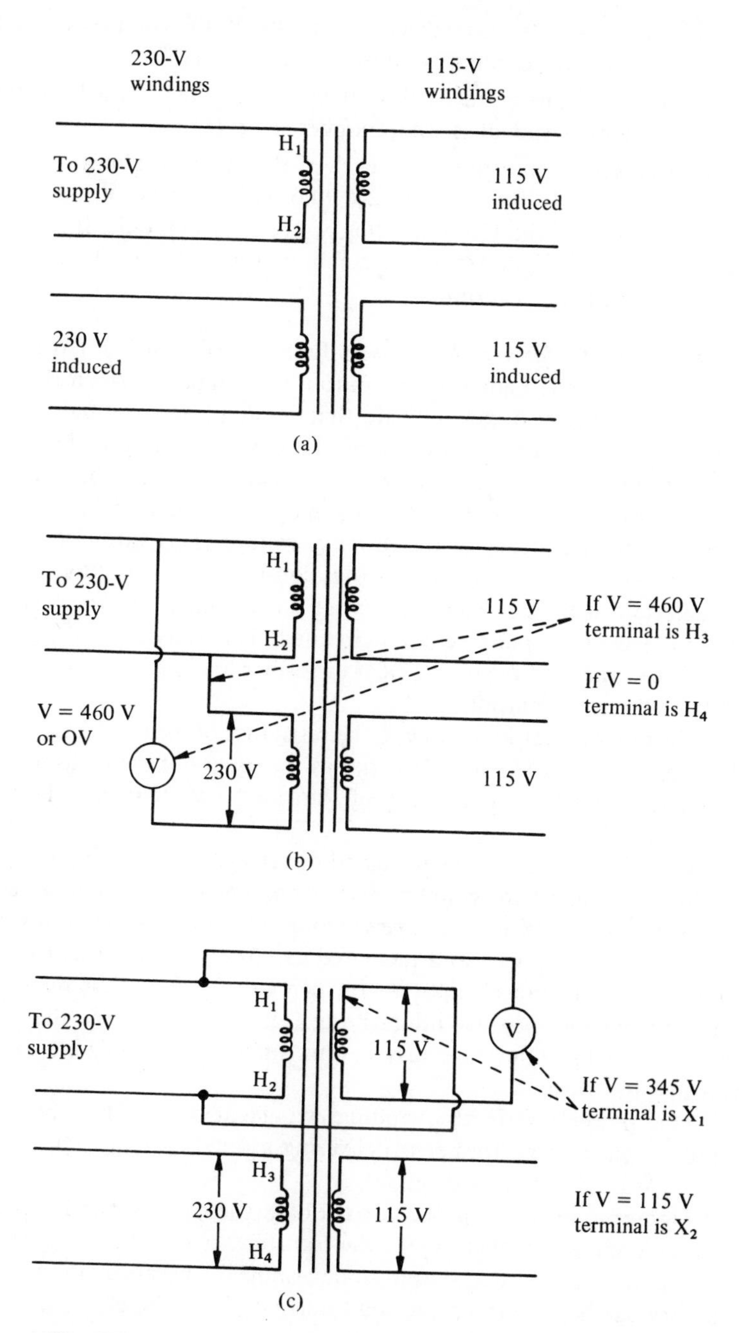

Fig. 6-5. Procedure to determine polarity identification of transformer windings.

5. If the voltmeter as connected reads 460 V, the voltages are additive and the top terminal is odd-numbered or H_3.
6. If the voltmeter reads 0 V, the voltages are subtractive and the common terminal is even-numbered or H_4.
7. Repeat the procedure with each of the 115-V windings. If the voltmeter reads the sum of 230 and 115 V, or 345 V, the common terminal of the 115-V winding is X_1, as in Fig. 6-5c. If it indicates 230 − 115 V, or 115 V, the common terminal is X_2. X_3 and X_4 can be similiarly determined.

6-6 ***Per cent impedance and efficiency.*** Equation (6-2) states that the input volt-amperes are equal to the output volt-amperes. When the load is resistive this also can indicate that input power is equal to output power. This is not exactly accurate since all devices have some losses.

A transformer has two sources of power loss. A small amount of power is necessary to force the flux in the magnetic core material to alternate according to the frequency of the primary voltage. This variation gives rise to eddy current and hysteresis losses that constitute the *core* loss. It can be easily determined by measuring the input power when the core has proper excitation. Secondary must be open circuited. This is a constant power loss that must be expended regardless of transformer loading. It is usually about 1–1.5% of the rated output.

The other power loss is due to the resistance of the windings, identified as the copper loss, or I^2R loss. It varies as the square of the load current. At full load it would be about 1–2% of the rated output in a well-designed transformer.

The copper loss can be evaluated directly by a short-circuit test. The low-voltage windings are connected in series and are short-circuited: No output is possible. The high-voltage windings are also connected in series. A very *low* voltage is applied and increased carefully until rated current flows in all windings. The power input is now being dissipated in the resistance of all the windings as they are all carrying rated current. Since the primary voltage is very low, the core loss in this situation is negligible and only copper loss is measured.

This power loss in the windings would also be lost as heat in the windings when the transformer is delivering a rated current to a load. This is true regardless of the load pf.

Figure 6-6 shows the procedure to obtain these copper and core losses. We can determine the efficiency of the transformer by adding the core loss and copper loss to the rated output to determine the necessary input. Copper losses vary as the square of the load current (I^2R), therefore, at one-half rated kilovolt-ampere load the copper loss would be one-quarter of the

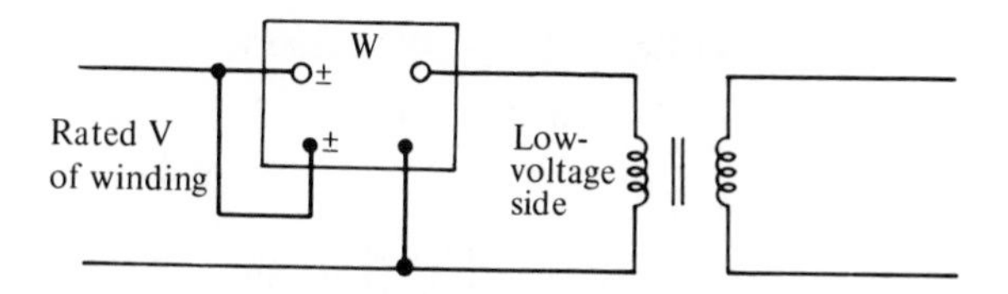

(a) Open-Circuit Test to Determine Core Loss in Watts

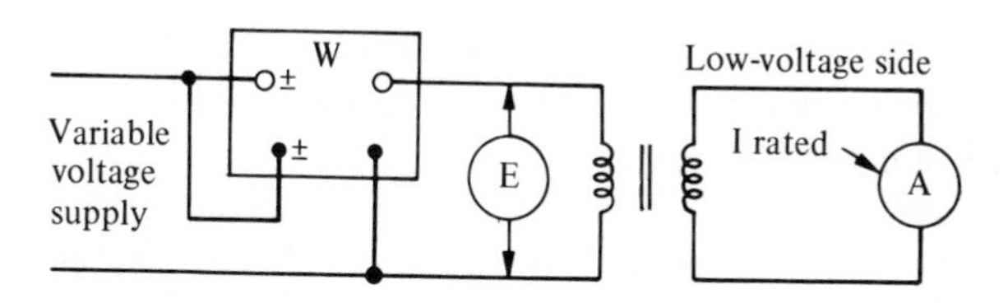

(b) Short-Circuit Test to Determine Copper Loss in Watts

Fig. 6-6. Determination of transformer losses.

full load value. The core loss is constant regardless of load, since it varies only with the rated voltage.

Example **6-5.** A 25-kVA, 550–220-V transformer has a core loss of 250 W, and its copper loss at rated kilovolt-ampere load is 350 W.

(a) Calculate the efficiency of the transformer at rated kilovolt-amperes, unity pf load.

(b) Calculate the efficiency at rated kilovolt-ampere load with 0.7 pf.

Solution

(a)

$$\text{Output power} = 25{,}000 \text{ W}$$

$$\text{Total losses} = 250 + 350 = 600 \text{ W}$$

$$\text{Input power} = \text{Output} + \text{losses} = 25{,}000 + 600 = 25{,}600 \text{ W}$$

$$\text{Efficiency} = \left(\frac{\text{output}}{\text{input}}\right)(100) = \left(\frac{25{,}000}{25{,}600}\right)(100) = 97.5\%$$

(b)

$$\text{Output power} = EI(\text{pf}) = (25{,}000)(0.7) = 17{,}500 \text{ W}$$

$$\text{Losses} = 250 + 350 = 600 \text{ W}$$

$$\text{Input power} = \text{output} + \text{losses} = 17{,}500 + 600 = 18{,}100 \text{ W}$$

$$\text{Efficiency} = \left(\frac{17{,}500}{18{,}100}\right)(100) = 96.5\%$$

From Ex. 6-5 it is obvious that very little error is introduced in transformer calculations when using Eq. 6-2, where output kilovolt-amperes are assumed to equal the input.

A second important measure of a transformer's performance (in addition to kilovolt-amperes) is its per cent impedance. This parameter can be understood from Fig. 6-6b. This measure involves the voltage necessary to force rated current to flow through the short-circuited transformer, as compared to rated voltage.

$$\text{Per cent } Z = \left(\frac{E_{sc}}{E_{rated}}\right)(100) \tag{6-3}$$

Per cent Z is a measure of the transformer's ability to maintain a constant output voltage throughout the entire range of load. Calculation of such *voltage regulation* is somewhat involved, since it varies with the load and pf. On the whole, however, the voltage change is small.

Per cent impedance becomes important and must be considered when *separate* transformers are connected in parallel. They must have the same per cent impedance to enable them to share the load evenly. The transformer with the lowest per cent Z would try to maintain a slightly higher voltage and therefore tend to supply most of the load. The per cent impedance of a transformer is usually given as part of the nameplate information.

Example **6-6.** In Fig. 6-6b the transformer is rated at 550/220 V. With the low-voltage winding shorted, it requires 25 V across the 550-V winding to permit rated current through both windings of the transformer. Calculate the per cent impedance.

$$\text{Per cent } Z = \frac{E_{sc}}{E_{rated}}(100) = \frac{25}{550}(100) = 4.45\% \tag{6-3}$$

6-7 ***Transformers in three-phase systems.*** In many large buildings it becomes necessary to transform a three-phase feeder to one at a lower voltage. Each phase of a three-phase system must keep its identity, and its time difference with respect to the other two phases, from the generator to the motor where the generator's rotation is reproduced. A separate primary and secondary, therefore, must be provided for each phase. This is possible with a three-phase transformer having a three-legged core, as shown in Fig. 6-7. A disadvantage of this arrangement is the necessity of replacing the entire unit, should one winding become defective. Usually (three or two) *separate* single-phase transformers are used.

Four possible ways of connecting transformer windings are primaries in *wye* or *delta* and secondaries in wye or delta. Several considerations dictate the use of each configuration. Wye connections, however, require more careful attention. If the possibility of unbalanced loads exists, a wye-connected primary requires a neutral back to the source to assure equal voltage across each primary winding. Similiarly, a wye-connected secondary requires a neutral to the load to maintain equal load voltages. If loads are balanced, any configuration with primary and secondary connections is permissible. However, harmonics in the magnetic circuits make the wye-wye connection undesirable except in very high-voltage transmission lines.

When connecting the windings of the three primary and three secon-

Fig. 6-7. Three-phase transformer core. *Courtesy of G.E. Company.*

dary sections, we follow the same procedures as those presented in Figs. 5-5 and 5-11. In these diagrams phases *A*, *B*, and *C* represent the secondary windings. If we follow the polarity markings of the windings as developed in Sec. 6-5 in both primary and secondary circuits, no testing of the output voltages is necessary. This assumes, of course, that the polarity markings are

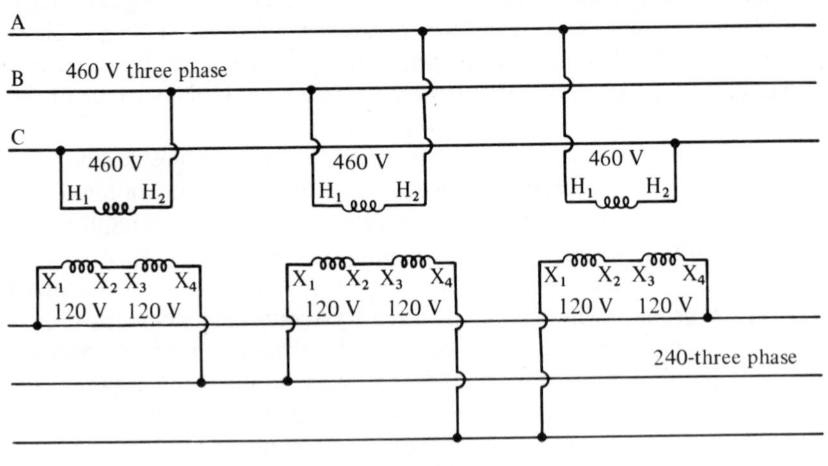

(a) Delta-to-Delta Connection

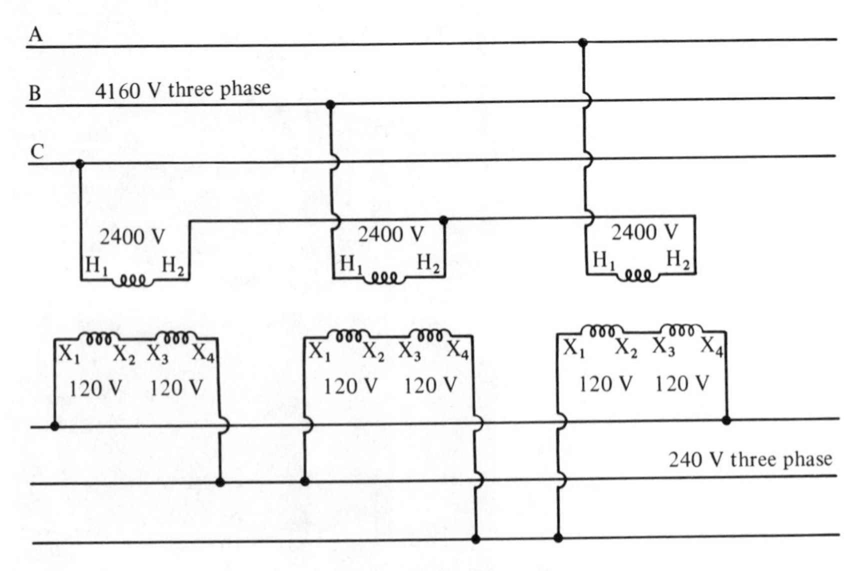

(b) Wye-to-Delta Connection

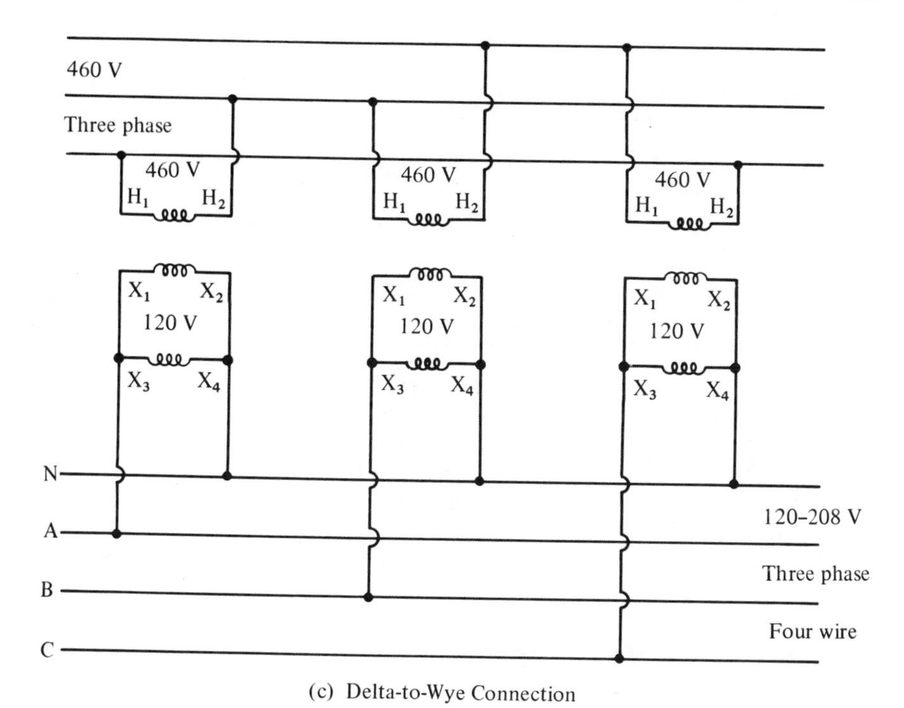

(c) Delta-to-Wye Connection

Fig. 6-8. Three-phase connections of transformers.

correct and, like all human endeavors, are subject to error. Before closing any delta-connected supply, caution dictates a check to assure that the voltage across the three series-connected phases is zero.

Figure 6-8 shows the proper connections of windings to assure correct three-phase output. The diagrams are shown in line rather than the geometric wye and delta. The identity of each transformer winding is more easily followed. The voltages used for transformer windings, input and output circuits, are for typical applications. Note in wye connections the $\sqrt{3}$ relationship between line voltage and phase (winding) voltage. The wye-wye connection is not shown because it is infrequently used in interior wiring systems.

Example 6-7. A 50-kW, 120-V, unity pf load is supplied from a 460-V, three-phase line. Draw a diagram of this system with proper transformer connections using a delta-wye configuration. Transformers are rated at 460/120–240 V. Calculate (a) the minimum transformer rating, (b) the current in the primary and the secondary of each transformer, and (c) the line current in the line and load circuits.

Solution

Refer to Fig. 6-9. (a) If the load is balanced each transformer provides one third of the load at unity pf: 50 kW = 50 kVA.

$$\frac{50 \text{ kVA}}{3 \text{ transformers}} = 16.67 \text{ kVA} \quad \text{rating of each transformer}$$

(b)
$$I_1 = \frac{\text{VA}}{\text{V}} = \frac{16{,}670}{460} = 36.2 \text{ A} \quad \text{(460-V winding)}$$

$$I_2 = \frac{16{,}670}{120} = 139 \text{ A} \quad \text{(120-V winding)}$$

(c) The line current in a 460-V line is

$$I_L = \frac{P_T}{(\sqrt{3})(E_L)(\text{pf})} = \frac{50{,}000}{(1.732)(460)(1)} = 62.7 \text{ A}$$

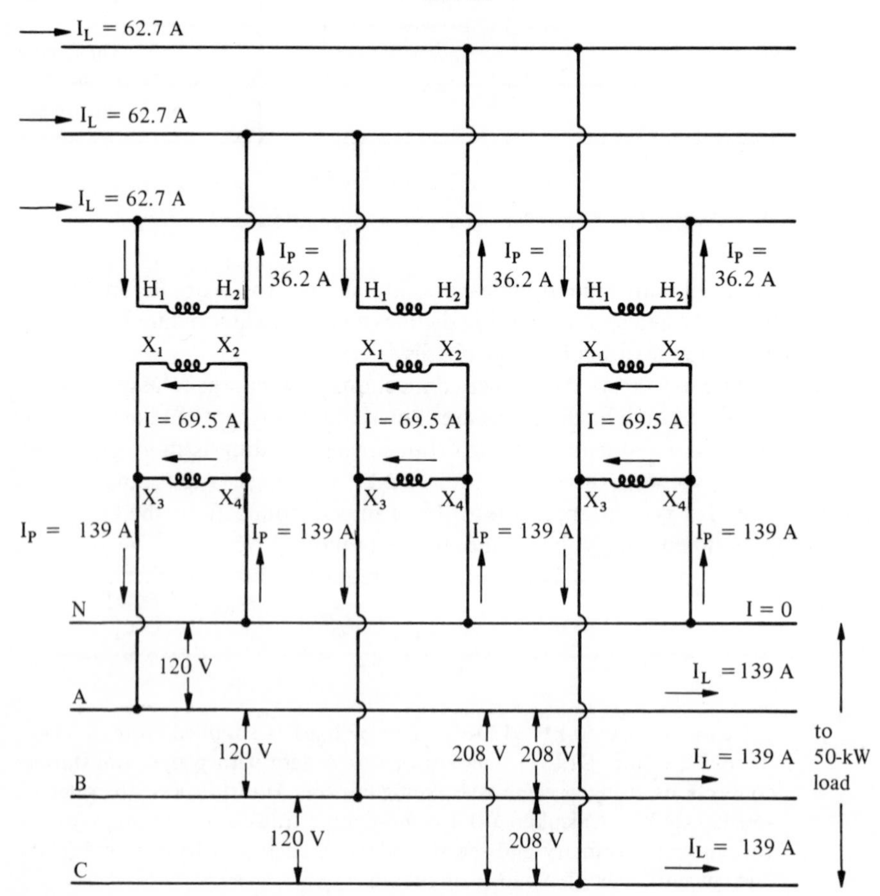

Fig. 6-9. Solution to Ex. 6-7.

Note that this agrees with

$$I_L = I_p\sqrt{3} = (36.2)(1.732) = 62.7\text{ A} \qquad (5\text{-}2)$$

The line current in a 120/208-V line is

$$E_L = E_p\sqrt{3} = (120)(1.732) = 208\text{ V} \qquad (5\text{-}5)$$

$$I_L = \frac{50{,}000}{(1.732)(208)(1)} = 139\text{ A}$$

This agrees with the transformer secondary (phase) current. In wye, phase and line are in series: $I_L = I_p$.

Example ***6-8.*** Repeat the calculations of Ex. 6-7. The load is 50 kW, balanced but inductive, with 0.7 pf.

Solution

(a) The minimum transformer rating is

$$\text{Total kVA of load} = \frac{\text{kW}}{\text{pf}} \qquad (4\text{-}6)$$

$$\frac{50{,}000}{0.7} = 71.5\text{ kVA} \quad \text{(total)}$$

$$\frac{71.5}{3} = 23.85\text{ kVA} \quad \text{(each transformer)}$$

(b) The current in the primary and the secondary at each transformer is

$$I_1 = \frac{\text{VA}}{\text{V}} = \frac{23{,}850}{460} = 51.8\text{ A} \quad \text{(460-V winding)}$$

$$I_2 = \frac{23{,}850}{120} = 198\text{ A} \quad \text{(120-V winding)}$$

(c) The line current in a 460-V line is

$$I_L = \frac{P_t}{(\sqrt{3})(E_L)(\text{pf})} \qquad (5\text{-}4)$$

$$\frac{50{,}000}{(1.732)(460)(0.7)} = 88.5\text{ A}$$

The line current in a 120/208-V line is

$$I_L = \frac{50{,}000}{(1.732)(208)(0.7)} = 198\text{ A}$$

As is always the case, a lower power factor load requires more current and larger equipment. The unity pf load required 16.67-kVA transformers, but the 0.7-pf load increased their minimum rating to 23.85 kVA.

6-8 ***Transformers connected in open delta.*** One advantage of connecting transformers in a delta-to-delta configuration is the ability to operate the system with only two transformers. A three-phase supply is possible with only two supplies if they are displaced by 120°.

Refer to Fig. 5-4a. Extend a wire from the junction of *DB*. Draw the vector representing E_{EA} opposite to the vector E_{AE}. We now have three voltages; E_{AD}, E_{BE}, and E_{EA}, all 120° apart. Generators are neither built nor used with only two windings because of the resulting unsymmetrical stator. Such a connection is known as the *open-delta* or *V-V* connection.

The open delta is a very practical transformer arrangement for temporary or emergency conditions. It is not a perfect arrangement, however, for a number of reasons. The maximum kilowatt load on each transformer cannot exceed 86.6% of its kilovolt-ampere rating when load pf is unity. The current in one transformer will lag its voltage by 30°; in the other the current will lead by 30°. Therefore, each transformer operates with a pf of 0.866 when the load pf is 1.

This also causes the voltage regulation (change in the voltage with the load) to differ slightly in each transformer (from 2 to 5%), depending on the pf of the load.

The advantage of the open-delta system is that if one transformer in a delta-delta system is defective, it can simply be disconnected and the open delta results. The permissible kilovolt-ampere load, however, is not 66.7% of the original transformer bank but is 86.6% of 66.7, or 57.7%.

Figure 6-10a shows two transformers connected in open delta. Note that the connections are identical to the delta, except that the third transformer is omitted. A three-phase, resistive, balanced load of 20.1 A per phase is connected in delta to the transformers.

$$\text{Line current} = \sqrt{3}(I_p)\ (1.732)(20.1) = 34.8\text{ A}$$

$$\text{Total power} = E_L I_L \sqrt{3}\cos\theta = (230)(34.8)(1.732)(1) = 13{,}850\text{ W}$$

In Fig. 6-10b, I_{AB} and I_{AC} are shown on the vector diagram in phase with their respective voltages E_{AB} and E_{AC}. They add to form the current in line I_{CA} and I_{CB} and are also added vectorially to become the line current I_c.

Note that I_c is the same current supplied by transformer 2 *leading* its voltage E_{CB} by 30°. I_a is the same current supplied by transformer 1 and is *lagging* behind its voltage E_{AB} by 30°. Therefore, each transformer has a pf of

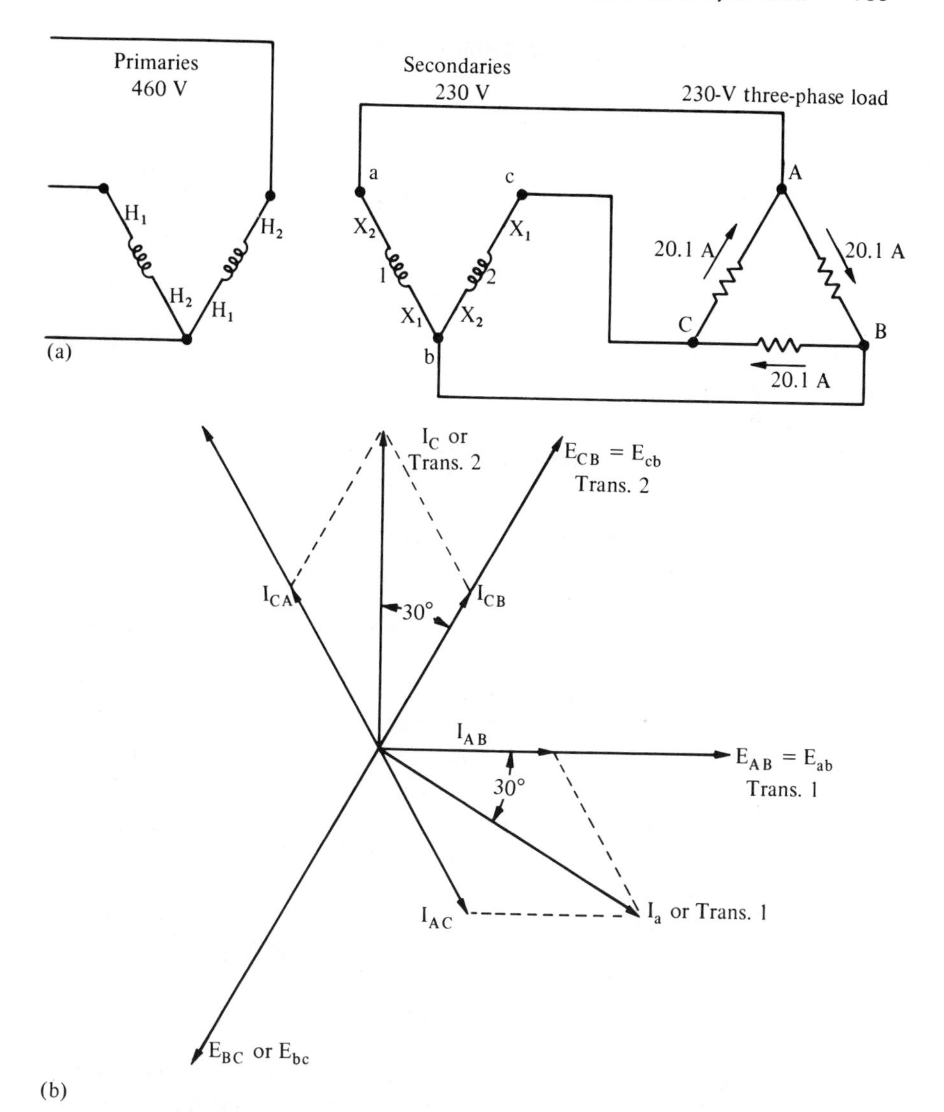

Fig. 6-10. (a) Transformers in open delta supplying a three-phase, balanced-resistive load. (b) Vector relationships.

cos 30° or 0.866. The power delivered by each transformer is

$$P = EI \cos 30^\circ \tag{4-5}$$

$$= (230)(34.8)(0.866) = 6925 \text{ W}$$

The power delivered by both transformers is

$$P = (6925)(2) = 13{,}850 \text{ W} \quad \text{(the same as the load)}$$

The kilovolt-ampere rating of each transformer is

$$\text{kVA} = \frac{EI}{1000} = \frac{(230)(34.8)}{1000} = 8 \text{ kVA}$$

The maximum power that three 8-kVA transformers can deliver to a unity pf load, if connected in delta, is

$$(8)(3) = 24 \text{ kW}$$

The power delivered by two transformers in open delta is 13.85 kW, and the power ratio is

$$\text{Ratio} = \frac{13.85}{24} = 0.577$$

Two transformers in open delta, therefore, can supply only 57.7% of the power of three transformers connected in delta.

6-9 *Scott or T connection to obtain two phase.*

Except for isolated small power plants, all power in the United States is generated and distributed as three phase. Consequently, all polyphase machines and equipment are manufactured to utilize three phases. The electrical industry has not always been so universal. Many early areas were developed using two-phase power supplies. When utility companies changed to three phase, all two-phase equipment became obsolete.

A two-phase supply provides two voltages displaced by 90°. Any polyphase system may be transformed to another polyphase system with appropriate transformer connections. A unique arrangement using two transformers can convert a three-phase supply to two phase. This arrangement is called a Scott connection. It permits individual facilities requiring two phase to continue in operation if desired or if it should prove to be economically practical.

The secondary transformer windings that comprise a two-phase supply must have slight modifications. One, called the *main* transformer, must have a center tap. The other, called the *teaser*, must have 86.6% of its turns tapped.

Figure 6-11a shows the connections of the two transformers. A 460-V, three-phase supply is transformed to a 230-V, two-phase supply. The vector diagram in Fig. 6-11b shows that the voltages E_{FG} and E_{HJ} are displaced by 90°, the requisite to obtain a two-phase supply.

The three-phase supply voltages are shown in the same configuration as Fig. 5-5, 120° apart. The primary of the main transformer is center tapped: The two voltages E_{bd} and E_{dc} add to comprise the voltage E_{BC} and, therefore, are shown as its respective parts.

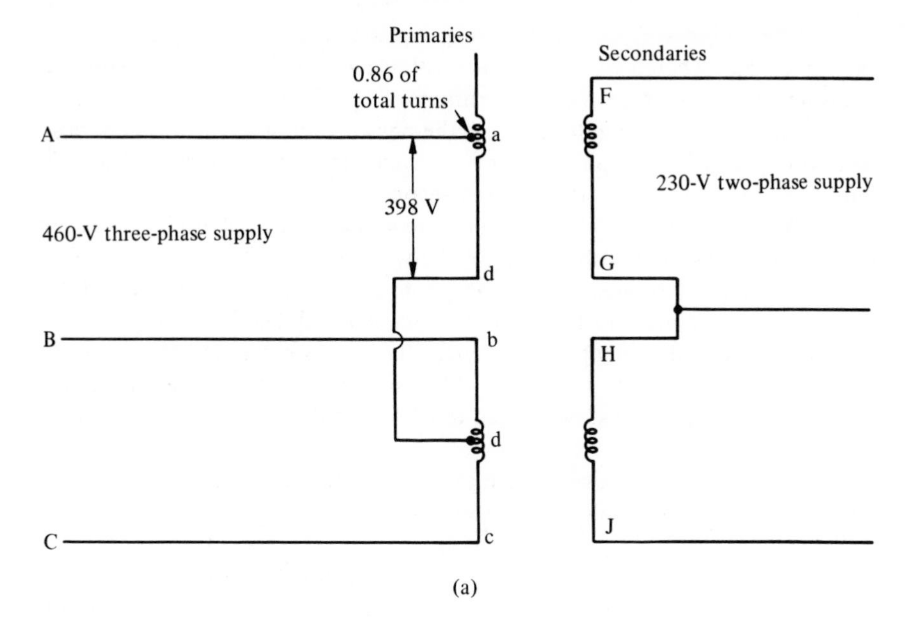

(a)

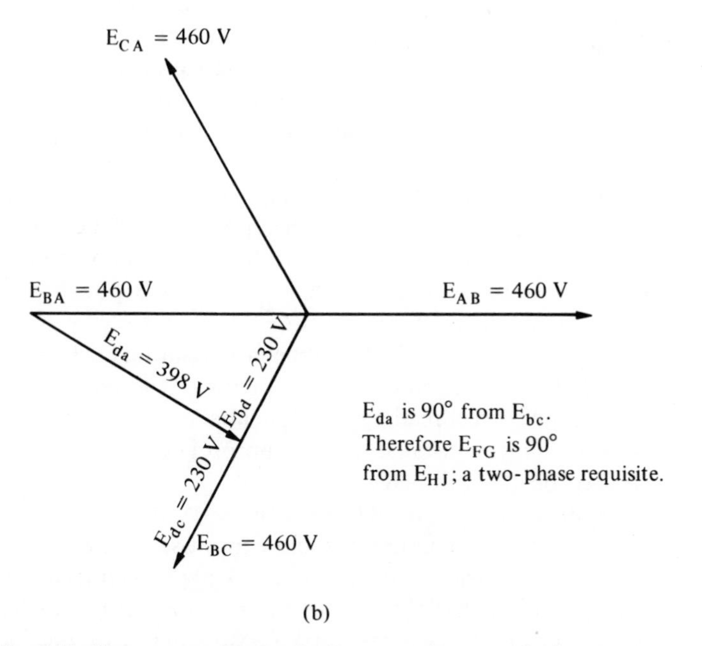

(b)

Fig. 6-11. Scott connection to obtain a two-phase supply from a three-phase line.

The voltages E_{bd} plus E_{da} must be equal to the voltage represented by the vector E_{BA}. Therefore, the primary voltage E_{da} of the teaser transformer is displaced by 90° from E_{BC}, the primary of the main transformer.

The secondary windings follow the phase differences of the primaries; therefore, a 230-V, two-phase system is available. Do *not* misinterpret this output as an open delta, three-phase secondary.

With a resistive load the current and voltage are in phase in the teaser transformer. However, only 86.6% of the winding is used. Note that 86.6% of rated voltage is applied to the primary to assure proper excitation. In the main transformer primary the current in one half will lead its voltage by 30° and in the other half will lag behind by 30°. This transformer will then operate at 0.866 pf. Because of these two different 0.866 factors, the two-phase kilovolt-ampere load must be limited to 86.6% of the combined transformer kilovolt-ampere rating.

The Scott connection also may be used to transform two phase to three phase, but this is seldom necessary.

6-10 ***Instrument transformers.*** Instrument transformers are designed primarily for the measurement of voltage, current, power, energy, or vars. The measurement of high voltages and currents is simplified by the use of instrument transformers. The *potential transformer* is used to measure high-voltage circuits with a low-scale voltmeter. Ordinarily, circuits up to 600 V can be measured *directly* because voltmeters, wattmeters, and watt-hour meters can have potential coils and scales that will operate satisfactorily up to this voltage. However, the higher voltage meters are expensive. Potential circuits used in metering also are safer when high voltages are reduced by means of potential transformers. As in any transformer, each winding must be designed to operate approximately at its *rated* voltage to assure proper magnetization, and precautions should be taken to assure no overvoltage. As the only load on the secondary of a potential transformer is the high-impedance voltmeter, the ratings of the potential transformer need be only in the order of a few watts.

Figure 6-12 shows a 150-V voltmeter measuring a 2300-V supply. The meter will read 115 V; a multiplication factor of the potential transformer ratio will give the correct reading. The potential transformer has a ratio of 20 to 1. On a permanent installation, an appropriately calibrated meter scale could be used on the voltmeter.

Ammeters rated at over 100 A become large and expensive to manufacture because of internal connections and current-sensing elements. Also, the necessity of bringing large conductors to the desired meter location introduces connection problems that the *current transformer* can eliminate.

Some current transformers are constructed with only a solid bar (one turn) as a primary that is connected in series with the load to be mea

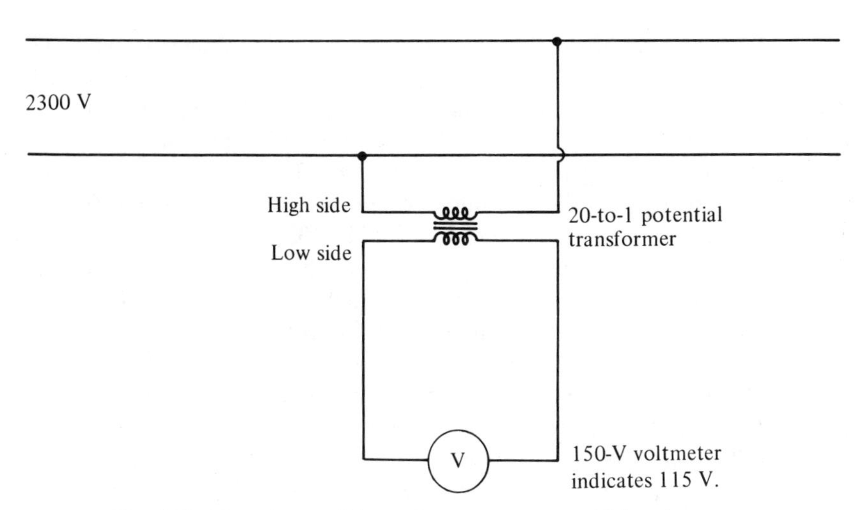

Fig. 6-12. Use of potential transformer to measure high voltage.

Fig. 6-13. Current transformers. *Courtesy of General Electric Company.*

sured. A more simplified current transformer is the *doughnut* type. The conductor passing through the center magnetizes the core because of the magnetic lines around the single (one-turn) conductor. Figure 6-13 illustrates both types. The secondary has many turns of small wire and is connected directly to a low scale, generally a 5-A ammeter. The ratio of turns between the two circuits depends on how high a current is to be anticipated. If 500 A is the maximum current, then the ratio would be 500 to 5. Actually this would be 100 to 1, but since 5-A ammeters are used, the ratio when referred to 5 would indicate the actual current that the transformer could safely carry in the primary and still keep the ammeter reading within scale limits.

A current transformer is also a small device, since very little power is consumed. The low-impedance ammeter when connected across the secondary practically constitutes a short circuit. Precaution must be taken to keep this winding short-circuited with a switch, provided for the purpose, if the ammeter is not connected. This must be done to prevent a high-induced voltage in the secondary because of a primary load current that is *not* dependent on the load current in the secondary. This violates the second condition in Sec. 6-3 and causes the primary to act as an inductance, which creates an objectionable voltage drop in the primary circuit. Since the current transformer

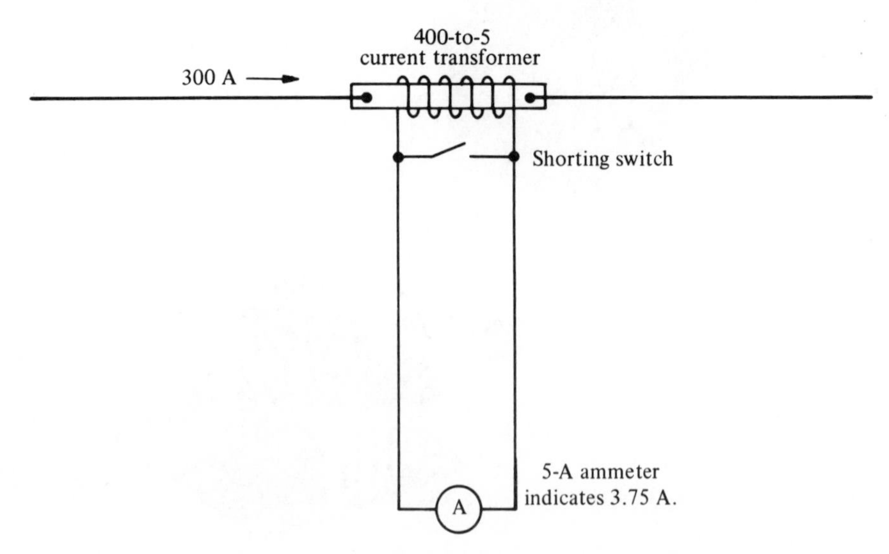

Fig. 6-14. Small ammeter measuring large current by use of current transformer.

is a step-up voltage transformer, the open-circuit voltage is very high and daugerous.

Figure 6-14 shows a 5-A ammeter measuring a large current by use of a 400-to-5 current transformer. This one ammeter could be used in one leg of a three-phase line to measure the three-phase current if the load were balanced. Motor loads are always balanced on three-phase circuits.

The relationship between line and ammeter currents and the transformer ratio can be shown by

$$\frac{I_L}{I_A} = \frac{400}{5} \tag{6-4}$$

where I_L is the current to be measured and I_A is the ammeter current measured in amperes. This right side of the equation is the current transformer ratio.

Example ***6-9.*** A 200-to-5 current transformer is used with a 5-A ammeter to measure a line current of 145 A. If the ammeter has its original 5-A scale, what will it indicate?

Solution

$$\frac{I_L}{I_A} = \frac{200}{5} = \frac{40}{1} \tag{6-4}$$

$$\frac{145}{I_A} = \frac{200}{5} = \frac{40}{1}$$

$$I_A = \frac{145}{40} = 3.625 \text{ A}$$

Example ***6-10.*** A 350-kW, 2300-V, three-phase resistive load is measured with two small wattmeters. Two 100/5-A current transformers and two 2300/115-V potential transformers make this possible.

(a) Draw a sketch of the circuit.

(b) Calculate the reading on each wattmeter.

Solution

(a) See Fig. 6-15. (b) The current in each line is

$$I = \frac{350{,}000}{2300\sqrt{3}} = 87.9 \text{ A}$$

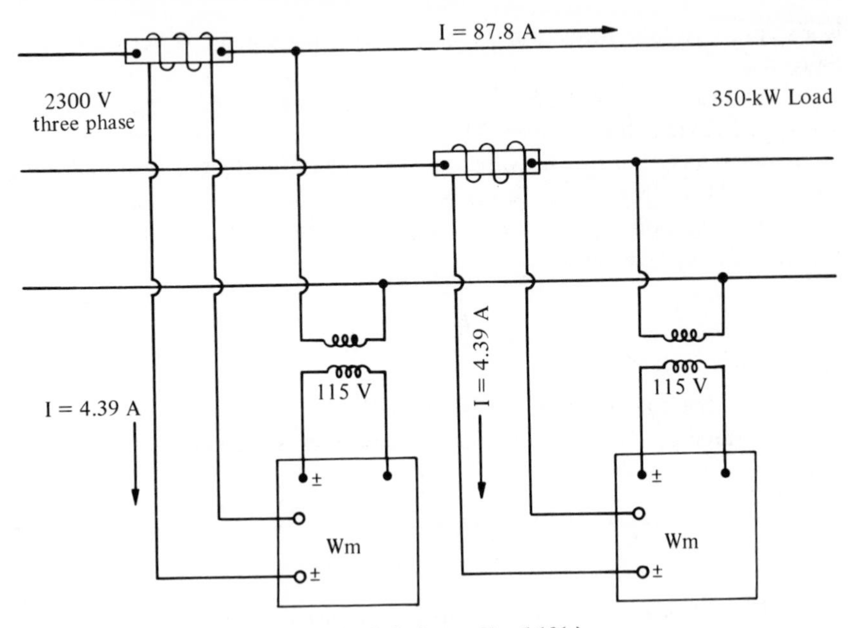

Fig. 6-15. Solution to Ex. 6-10(a).

The current to each wattmeter is

$$\frac{I_L}{I_A} = \frac{100}{5} = \frac{20}{1}$$

$$\frac{87.9}{I_A} = \frac{100}{5} = \frac{20}{1}$$

$$I_A = \frac{87.9}{20} = 4.39 \text{ A}$$

The voltage across the wattmeters is 115 V. Each wattmeter indicates (pf = 1.0).

$$W = EI \cos 30^\circ = (115)(4.39)(0.866) = 437 \text{ W} \qquad (5\text{-}5)$$

To check with the original load,

$$\begin{aligned} P &= (437 \times 2)\left(\frac{2300}{115}\right)\left(\frac{100}{5}\right) \\ &= (437 \times 2)(20)(20) \\ &= 350{,}000 \text{ W} \end{aligned}$$

6-11 ***Autotransformers.*** When the desired voltage ratio between two circuits is less than two to one, it is often expedient to use an autotransformer. An autotransformer is one that has the primary and secondary circuits electrically connected. This is in contrast to the electrically isolated windings in the conventional transformer circuits. For safety reasons high-voltage circuits must be insulated from those of lower voltages.

The interconnection of primary and secondary circuits permits some of the power required by the load to be conducted from the supply to the load. This is also in contrast to the conventional transformer that must transform all the power of the load. Figure 6-16 shows a one-line diagram illustrating how the amount of power that must be transformed will be determined by the voltage ratio between supply and load circuits. The necessary kilovolt-ampere transformer rating for a particular load can then be determined.

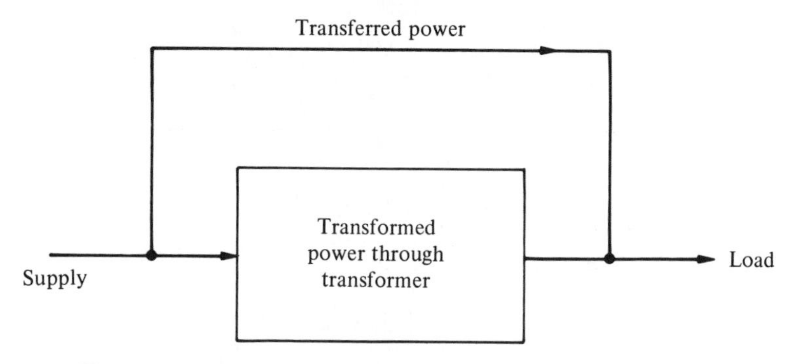

Fig. 6-16. Power flow diagram showing autotransformer principle.

To calculate autotransformer circuit and performance problems we must follow a few basic rules:

1. Select a conventional transformer that has the correct combination of windings to match the supply to the desired load voltage.
2. Observing polarity indentifications, make the necessary connections.
3. Calculate the current to load I_2 based on the load kilovolt-amperes.
4. Calculate the current from the supply I_1 based on $E_1 I_1 = E_2 I_2$, where E_1 and I_1 are the supply circuit voltage and current and E_2 and I_2 are those of the load. These values are not necessarily those of the transformer.
5. Draw a diagram of the circuit, and determine the current in any transformer winding. Multiply by the voltage of that winding; dividing by 1000 will give the kilovolt-ampere rating of the transformer necessary for that particular load.

Example **6-11.** It is desired to operate a 50-kVA, single-phase, 550-V load from a 440-V supply.

(a) Draw a sketch of the transformer showing all the voltages and polarity identifications.

(b) Calculate the kilovolt-ampere rating at the transformer.

Solution

(a) See Fig. 6-17. A 440-V winding is necessary to connect to the 440-V supply for proper excitation. As 110 V must be added to this, a 110-V winding must be connected in series, odd to even. Therefore, X_1 is connected to H_2. E_{CD} is 550 V. If the 110-V winding were reversed, E_{CO} would be (440 − 110) or 330 V.

(b)
$$I_2 = \frac{50{,}000}{550} = 90.7 \text{ A}$$

$$I_1 = \frac{50{,}000}{440} = 113.4 \text{ A}$$

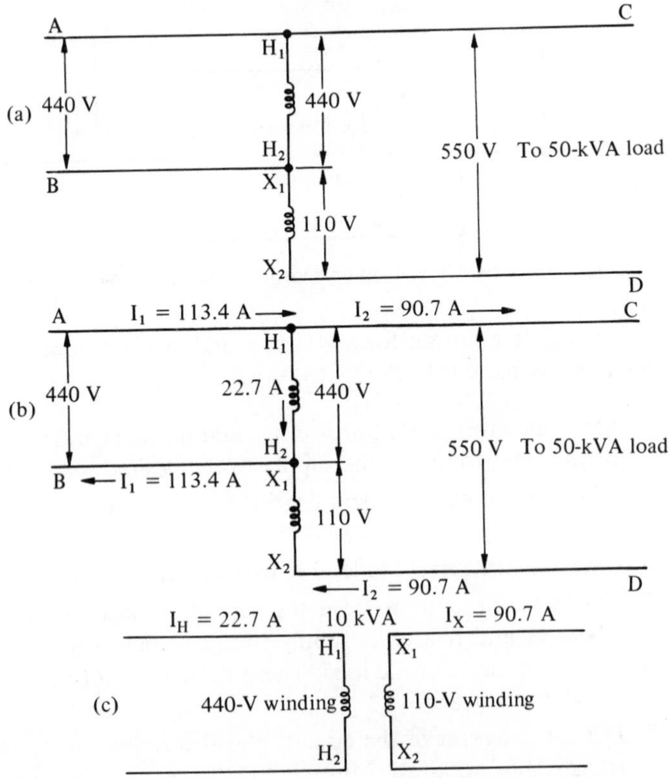

Fig. 6.17 (a) Solution to Ex. 6-11, part (a). (b) Solution to part (b). (c) Voltage, current, and kVA rating of transformer.

The current from H_1 to H_2 is

$$113.4 - 90.7 = 22.7 \text{ A}$$

For the $H_1 - H_2$ winding,

$$\text{kVA} = \frac{(440)(22.7)}{1000} = 10 \text{ kVA}$$

For the $X_1 - X_2$ winding, the kilovolt-ampere rating of the transformer is

$$\text{kVA} = \frac{(110)(90.7)}{1000} = 10 \text{ kVA}$$

Figure 6-17b shows that the solution to Ex. 6-11 must be a conventional 440–110-V transformer rated at only 10 kVA. Only one-fifth of the load need be transformed. Four-fifths is transferred or conducted from the supply to the load.

This relationship between transformer rating and voltage ratio can be shown by

$$\text{Transformer kVA} = \frac{E_L}{E_H + E_L} \quad \text{(load kVA)} \tag{6-5}$$

where E_L is the low-voltage winding and E_H is the high-voltage winding of the transformer needed to obtain the required voltages.

In Ex. 6-11, therefore, using Eq. (6-5),

$$\begin{aligned} \text{Transformer kVA} &= \left(\frac{110}{440 + 110}\right)(50 \text{ kVA}) \qquad (6\text{-}5) \\ &= 10 \text{ kVA} \end{aligned}$$

PROBLEMS

6-1 A 2400/240-V transformer has 1000 turns on the high-voltage winding. How many turns are necessary on low-voltage winding?

6-2 A 480/120–240-V transformer has 240 turns on the 480-V winding. Calculate the number of turns on each low-voltage winding.

6-3 A 1200/240-V transformer has the proper excitation if 2 V is applied to each turn. Calculate the number of turns on each winding.

6-4 A 2300/120–240-V transformer has 48 turns on each 120-V winding. The high-voltage winding is tapped at four places, each tap consisting of five turns. Calculate all possible primary voltages that will maintain the same 120–240-V output.

6-5 A 2400/240-V transformer is rated at 20 kVA. Calculate the current rating of each winding.

6-6 A 2400/120–240-V transformer is rated at 25 kVA.
(a) Calculate the current rating of the high-voltage winding.
(b) If windings are connected for a 120-V load, what is the maximum current the transformer can deliver to the load?
(c) If the load operates at 240 V, what is the maximum current the transformer can deliver to the load?
(d) When the transformer is supplying a 240-V load at 0.8 pf, calculate the maximum power and current to the load within the transformer's kilovolt-ampere rating.

6-7 A 1200–2400/120–240-V transformer is rated at 10 kVA. Sketch all possible ways this transformer may be used. Calculate the primary and secondary currents for each configuration.

6-8 A 25-kVA, 600/120–240-V transformer has two low-voltage windings.
(a) Draw a diagram showing the proper connections for a 120–240-V, three-wire secondary system.
(b) Determine the rated high side current.
(c) Determine the rated low side current per coil.
(d) Determine the rated low side current for a 120-V load.
(e) Determine the rated low side current for a 240-V load.

6-9 A 1200/240-V transformer has taps brought out of the low-voltage winding to obtain 220–240 V in 5-V increments. Draw a diagram of this transformer showing all the taps. Express all taps in per cent of rated voltage of winding.

6-10 A 2-kVA transformer has two 10-V windings and two 20-V windings. It has a 220-V winding ordinarily to be used as a primary. Sketch all possible output voltages, maximum currents, and kilovolt-amperes of maximum load. Show the proper polarity identification.

6-11 Testing of a 50-kVA, 1200/240-V transformer shows the following data:

Open circuit test: 1200 V, 5 A, 800 W
Short-circuit test: low side shorted, readings taken on high-voltage side: 30 V, 41.7 A, 700 W

(a) Determine the excitation current.
(b) Calculate the per cent impedance.
(c) What is the cost per month, 24 hr per day, of keeping the trans-

former connected but without load? The energy cost is 2 cents/kWh.

(d) Calculate the efficiency of the transformer at rated load, unity pf.

(e) Calculate the efficiency at rated kilovolt-amperes load, pf = 0.6.

6-12 A 25-kVA, 480/120–240-V transformer is tested, and the core loss is found to be 400 W. The copper loss at the rated load is 700 W.

(a) Calculate the efficiency at $\frac{1}{2}$ and the rated kilovolt-ampere load with unity pf.

(b) Calculate the efficiency at $\frac{1}{2}$ and the rated kilovolt-ampere load with 0.7 pf.

6-13 A transformer has four coils, two rated at 120 V and two at 600 V. The 600-V windings are connected in series to a 1200-V supply. With no load on the 120-V windings the input to 600-V windings is 200 W and 3 A at 1200 V.

(a) Determine the power and the current if the 600-V windings are connected in parallel to a 600-V supply.

(b) Determine the power and the current if one 600-V winding is connected to the 600-V supply.

6-14 A group of 240-V, three-phase motors are supplied by three transformers from a 2400-V supply. The maximum load is 20 kW at 0.75 pf. Use a delta-delta transformer bank.

(a) Calculate the kilovolt-ampere rating of the transformers.

(b) Sketch one transformer showing all voltages and currents at the rating obtained in (a).

(c) Calculate the load that may remain connected if one transformer is removed and operation is open delta.

6-15 A 60-kW, unity pf, 120-V load is supplied by three transformers from a 480-V supply.

(a) Sketch this system using a delta-wye connection. Show the polarity identification of all windings and calculate the currents in each winding.

(b) Calculate the kilovolt-ampere rating of each transformer based on currents of (a).

6-16 A three-phase industrial load is computed to be 200 kVA but is anticipated to increase to 325 kVA in a short time. If an open delta transformer connection is to be used temporarily for the 200-kVA load, what rating of transformers could be installed? The same size transformer is to be added when the load becomes 325 kVA to complete the delta.

6-17 A 375-A load is measured with a 5-A ammeter with a 400–5-A current transformer. What does the ammeter indicate?

6-18 A 690-A load is measured with a 5-A ammeter with an 800–5-A current transformer. What does the ammeter indicate?

6-19 A 13,800-V line is measured with a 150-V voltmeter in conjunction with a 100-to-21 potential transformer. What does the voltmeter indicate?

6-20 A 500-kW, single-phase, 1100-V load with 1.0 pf is measured with a small wattmeter. A 10-to-1 potential and an 800-to-5 current transformer are used.
(a) What does the wattmeter indicate?
(b) The wattmeter must be multiplied by what factor to determine the power of the load?

6-21 A 250-kW, 460-V, unity pf, balanced three-phase load is measured with two small wattmeters.
(a) What ratio current transformers should be used?
(b) What do the wattmeters indicate?

6-22 A 250-kW load, 460-V, 0.8 pf, balanced three-phase load is measured with two small wattmeters.
(a) What ratio current transformers should be used?
(b) What do the wattmeters indicate?

6-23 A 10-kVA, 2300/230-V transformer is connected as an autotransformer to raise a 2300-V line to 2530 V.
(a) What is the maximum kilovolt-ampere load at 2530 V?
(b) What part of this load is transformed?
(c) What part is transferred from the supply to the load?

6-24 A 550-V, 40-kW, 0.8 pf, single-phase load is to be supplied from a 440-V supply.
(a) Sketch an autotransformer circuit showing the winding voltages to fulfill these requirements.
(b) Determine the kilovolt-ampere rating of the transformer.
(c) Calculate the input line current at 440 V and the load current at 550 V.
(d) Calculate the current in each winding of the transformer.

6-25 A 110-V line must be increased to 118 V.
(a) What voltages must an autotransformer provide?
(b) What size transformer is necessary for a 50-kW, 1.0 pf load?
(c) What size transformer is necessary for a 50-kW, 0.75 pf load?

6-26 A 90-V load is to be connected to a 115-V supply by the use of an autotransformer.
(a) Sketch the circuit showing all voltages.
(b) Calculate the transformer kilovolt-amperes if a 90-V load requires 85 A.

Chapter Seven

Electrical Machines

7-1 ***Electrical machine generalizations.*** Electrical machines are defined as electromagnetic devices designed to convert mechanical energy into electrical energy or electrical energy into mechanical energy.

An electric generator driven by a prime mover, such as a steam turbine, diesel engine, or waterfall, is an energy source of fairly constant voltage. As electrical loads are connected to that voltage, current flows and power is transferred from the electric generator to the electrical load. The prime mover is the source of that power, and hence the energy conversion is completed from prime mover to load.

The voltage is generated by rotating a group of conductors, or coils, in a magnetic field or vice versa. This action produces an alternating voltage in the form of a sine wave. Usually in an alternator the magnetic field is rotated and, therefore, the electrical load is connected to the stationary coils. In a dc generator the coils are rotated and the field is stationary. The electrical load is connected to the coils through a set of brushes that slide on a commutator. This commutator permits the current to flow in only one direction and, therefore, this machine is a dc source.

The winding structure of either machine wherein the voltage is generated is called the *armature*. The magnetic poles and exciting windings are called the *field*.

Figure 7-1 illustrates the frame and magnetic poles of a six-pole dc machine. Figure 7-2 shows an armature of a large dc generator.

Electric motors recreate the mechanical energy and power put into the prime movers that drive the generators. Here the electrical energy is converted back to mechanical energy. The connecting wires between the generator and motor instead of a shaft, belts, and pulleys attest to the

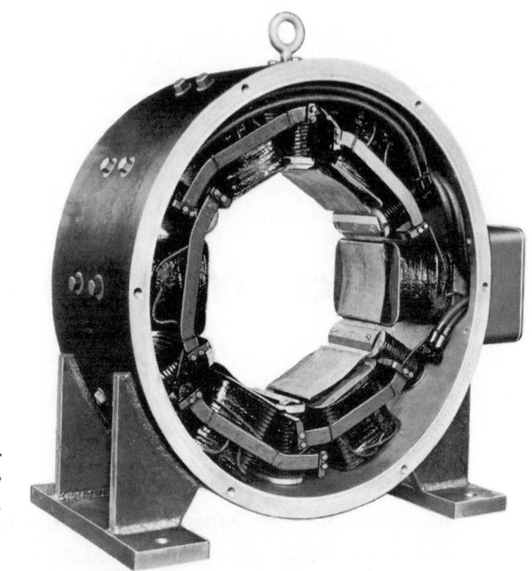

Fig. 7-1. Direct-current dc machine frame and poles. *Courtesy of Fairbanks Morse Motors, Manufactured by Colt Industries' Motor and Generator Operation.*

Fig. 7-2. Direct-current armature. *Courtesy of Fairbanks Morse Motors, Manufactured by Colt Industries' Motor and Generator Operation.*

versatility of electric power in transferring mechanical energy over vast distances.

7-2 ***Direct-current motor principles.*** *Torque* is produced in a dc motor by the interaction of the two circuits, armature and field. This introduces a fourth and final basic fact of electricity (the first three are summarized in Sec. 4-3). When a conductor carrying a current is placed in a magnetic field, there is a force acting on that conductor. Figure 7-3 illustrates how the strengthened flux on one side of the conductor develops the torque that moves the conductor and rotates the armature. The amount of torque needed to turn the mechanical load on the motor shaft is developed from the two sources of flux, the current in the field winding, and the current in the armature conductors. Our discussion of the three types of dc motors will concentrate on how these two currents are changed, either by changes in

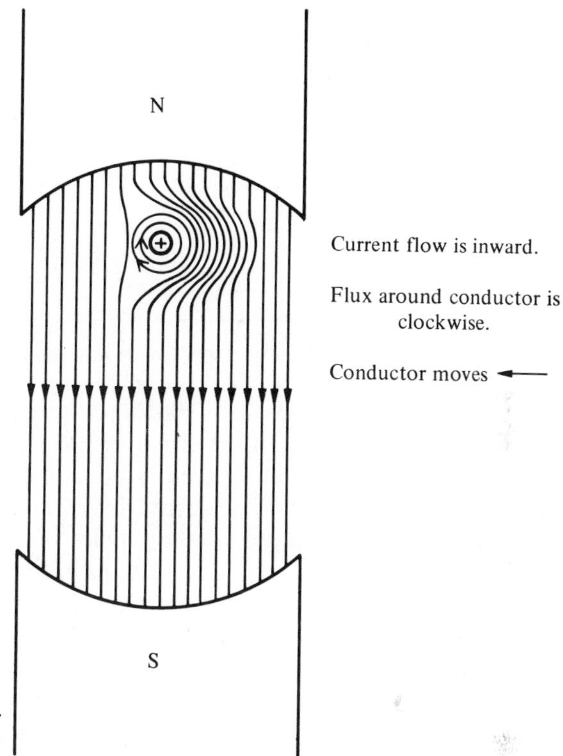

Fig. 7-3. Current-carrying conductor in a magnetic field.

mechanical load or by controllers external to the motor. It should be noted that dc generators and motors are identical in construction.

7-3 ***Shunt motor characteristics.*** By designing a field winding with many turns, a small current can set up a strong magnetic field (Sec. 4-3).

Therefore, the resistance of the field windings are relatively high and are connected directly across the line voltage for which the motor is designed. Field windings are only necessary to produce flux in an electromagnetic circuit. The windings would not be necessary if the poles were made of permanent magnetic materials. Such materials would not be practical since the flux could not be changed, and one of the advantages of dc motors would be lost—that of speed control.

The armature windings must carry large currents to produce the necessary torque and therefore, must be designed with low resistance. In large motors this may be less than 0.25 Ω.

The shunt motor is a dc motor whose field is shunted across its armature, with both connected to the dc line as in Fig. 7-4.

The difficulty with this arrangement is in starting the motor. The low armature resistance causes a very high starting current if the armature winding

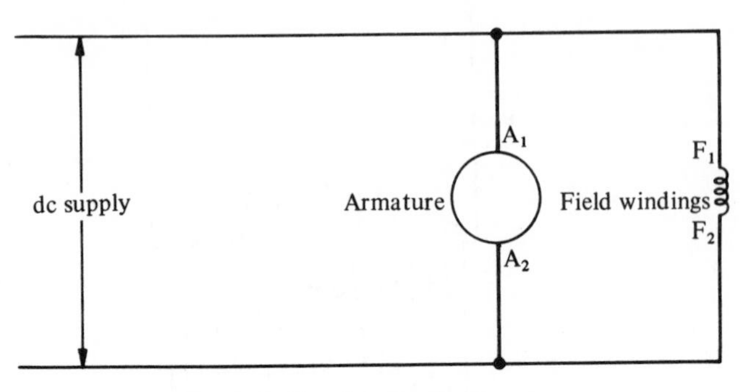

Fig. 7-4. Circuits of a shunt motor.

is directly across the line. This starting current must be limited by connecting a resistance in series with the armature. Because of commutator limitations, the starting current in all dc motors should be limited to 150–175% of rated full load current. Excessive currents (over 200%) can seriously damage a commutator and its armature winding. Once the armature starts to rotate the armature conductors, which now are passing through a magnetic field, will induce a (counter) voltage due to generator action. This voltage called *counter emf* (cemf) will oppose the applied voltage, limiting the armature current, and the resistance may be removed. This cemf is directly proportional to the speed of the armature. To assure smooth acceleration the starting resistance is removed in sections, as the cemf increases with speed.

Figure 7-5 shows a diagram of a shunt motor connected to a three-point *manual starter*. It is called three-point because it has three external connections. It is also identified as a *no field release* starter. A magnetic coil that holds the handle in the running position (with all resistance removed) is in series with the field winding. Any break in the field circuit will cause the cemf to go nearly to zero, resulting in the same inrush of current as if the starter resistance were not used when starting. This is a very dangerous situation that will cause the motor speed to increase beyond safe limits. The no field release will cause the handle to fall back to the off position, shutting down the motor. Although our description is for a manual starter, the same functions can be built into an automatic dc motor starter. This requires only push buttons for operation.

Once started, with the armature across the line, the armature current will vary only with a change in mechanical load.

$$I_A = \frac{E - \text{cemf}}{R_A} \tag{7-1}$$

where I_A is the armature current in amperes, E is the line voltage in volts, cemf is countervoltage generated by rotating the armature in volts, and R_A is the armature resistance in ohms.

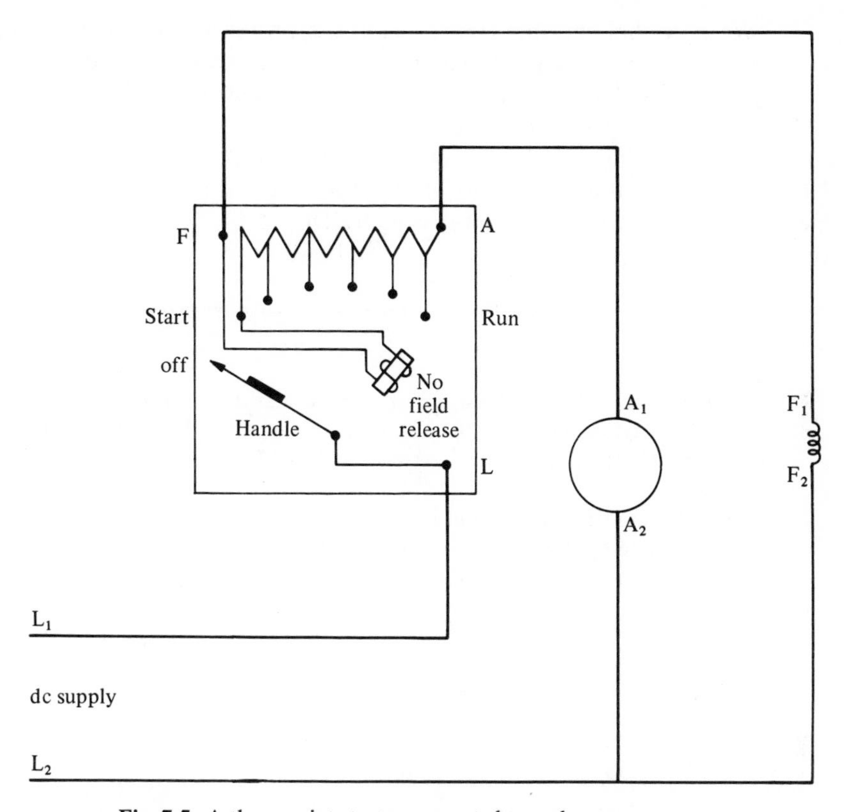

Fig. 7-5. A three-point starter connected to a shunt motor.

The only variables in Eq. (7-1) are I_A and cemf. Since cemf will vary with the speed of the motor, a small decrease in speed because of increased mechanical load will decrease the cemf. This will increase the denominator on the right side of the equation, which means that the armature current must increase. This will provide the additional torque necessary to turn the increased mechanical load.

Although actual measurement of cemf is impossible unless the motor is operated as a generator, we can substitute in the equation the variables that affect cemf. These are speed S and flux ϕ. Rewriting the equation, we get

$$I_A = \frac{E - \phi S}{R_A}$$

In terms of speeds,

$$S = \frac{E - I_A R_A}{\phi} \tag{7-2}$$

We note that the speed of the motor is proportional to the dc line voltage and inversely proportional to the flux. If the field windings are directly across the line and the line voltage is constant, the variables in Eq. (7-2) are speed and armature current.

When making calculations of electrical quantities in electric motors, 1 horsepower (hp) is equivalent to 746 W. The output is usually measured in mechanical units. One housepower is equivalent to lifting 33,000 ft-lb/min or 550 ft-lb/sec.

Example 7-1. A 230-V shunt motor with no load has a speed of 2000 rpm with its armature current at 6 A. At a full load of 10 hp the armature current is 36 A. The field current is constant at 2 A. The armature resistance is 0.4 Ω.

(a) Calculate the speed at full load.
(b) Calculate the total motor current at full load.
(c) Calculate the efficiency of the motor at full load.

Solution

(a)

$$S = \frac{V - I_A R_A}{\phi} \tag{7-1}$$

when $I_A = 6$ A,

$$2000 = \frac{230 - 6(0.4)}{\phi}$$

when $I_A = 36$ A,

$$S = \frac{230 - 36(0.4)}{\phi}$$

Divide one equation by the other (ϕ is constant and cancels out in division).

$$\frac{2000}{S} = \frac{230 - 6(0.4)}{230 - 36(0.4)}$$

$$= \frac{227.6}{215.6}$$

$$S = \frac{(215.6)(2000)}{227.6} = 1890 \text{ rpm}$$

(b) Total current = 36 A + 2 A = 38 A

(c)

$$\text{Power input} = EI$$
$$= (230\text{ V})(38\text{ A}) = 8750\text{ W}$$
$$\text{Power output} = (10\text{ hp})(746\text{ W/hp}) = 7460\text{ W}$$
$$\text{Efficiency} = \left(\frac{\text{output}}{\text{input}}\right)(100) = \left(\frac{7460}{8750}\text{W}\right)(100)$$
$$= 85.5\%$$

It is noted from Ex. 7-1 that the speed of the motor changes only about 5% throughout the range of the load, from no load to full load. For this reason we can describe the shunt motor as virtually a constant speed motor throughout its entire range of load.

7-4 ***Compound motor characteristics.*** A compound dynamo is one with two (or more) separate field windings. In addition to the shunt field it has a series field winding with a few turns that are connected in series with the armature. The series field will then carry the same current as the armature and will contribute flux according to the load or armature current. These turns are placed on the same magnetic circuit as the shunt field. The relative direction of the current in the two windings will determine if the flux contributions of each field winding are additive or subtractive. If the mmfs are in the *same* direction, the machine is *cumulative* compounded. If they are *opposite*, it is identified as *differential* compounded. At rated load the series field flux contribution is $\frac{1}{4}$ to $\frac{1}{3}$ of the shunt field.

The compound dynamo is used more frequently as a generator than as a motor. As a generator it can maintain a constant voltage as load current changes, if it is cumulative compounded. A differential compound motor will tend to produce a rising speed with load increases as compared to the shunt motor whose speed drops slightly. However, it will have poor torque characteristics at heavy loads and an extremely unstable speed characteristic. For this reason it is never used. A cumulative compound motor will have better torque but poorer drooping speed characteristics than the shunt motor. Most dc motors are shunt motors and series motors (Sec. 7-5). Nearly all dc dynamos operated as generators are compounded. The same dynamo will perform either function: as a motor or as a generator.

The starting problem of the compound motor is the same as that of the shunt motor. A resistance starter must be used.

Figure 7-6 shows the relationship between armature current, speed, and torque in the typical shunt and cumulative compound motor.

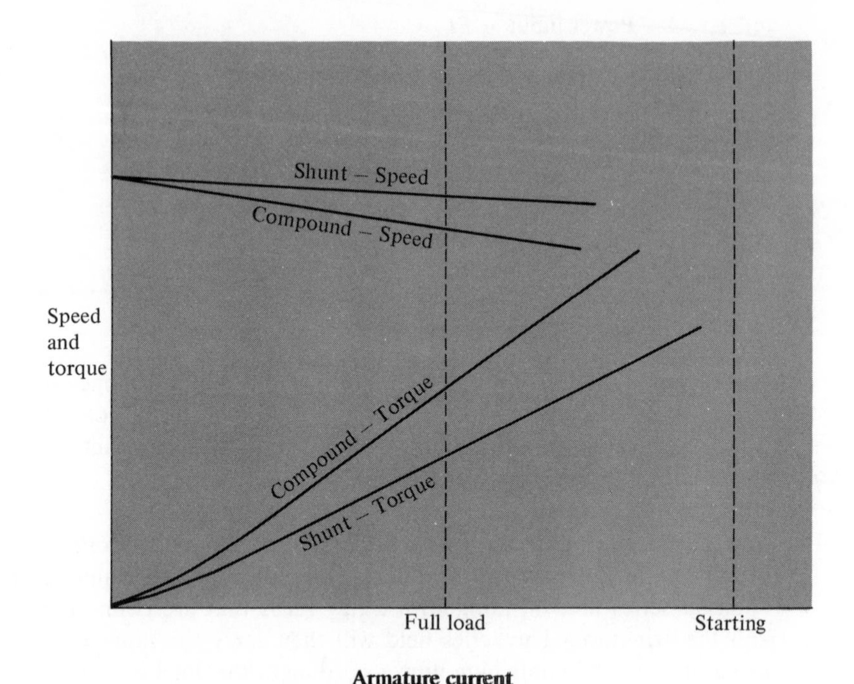

Fig. 7-6. Direct-current motor characteristics.

7-5 ***Series motor characteristics.*** A series motor has its field winding connected in series with the armature. Its only source of magnetism is the armature current. Since the armature current depends on the load, the flux will be very low at small loads, causing this type of motor to have a very high no load speed. This can be a hazard for large motors if the speed becomes excessive.

Series motors are usually used in applications where the load is always mechanically connected to its shaft to prevent any runaway speed. Very high starting torque can be produced because of high starting current. Locomotion (traction) vehicles using electric motors are an excellent application for series motors. The starting current is controlled by means of a resistance starter. Figure 7-7 shows its characteristics in terms of armature current.

7-6 ***Methods of speed control of dc motors.*** The use of dc motors permits wide ranges of speed. We should note that if rated voltage is applied to both field and armature circuits, the motor will operate at rated nameplate speed. Examination of Eq. (7-2) shows that if the armature voltage is reduced, but the flux is maintained with rated voltage on the field, the speed will be reduced in proportion to the reduced voltage.

Several methods of obtaining this variable voltage dc supply are

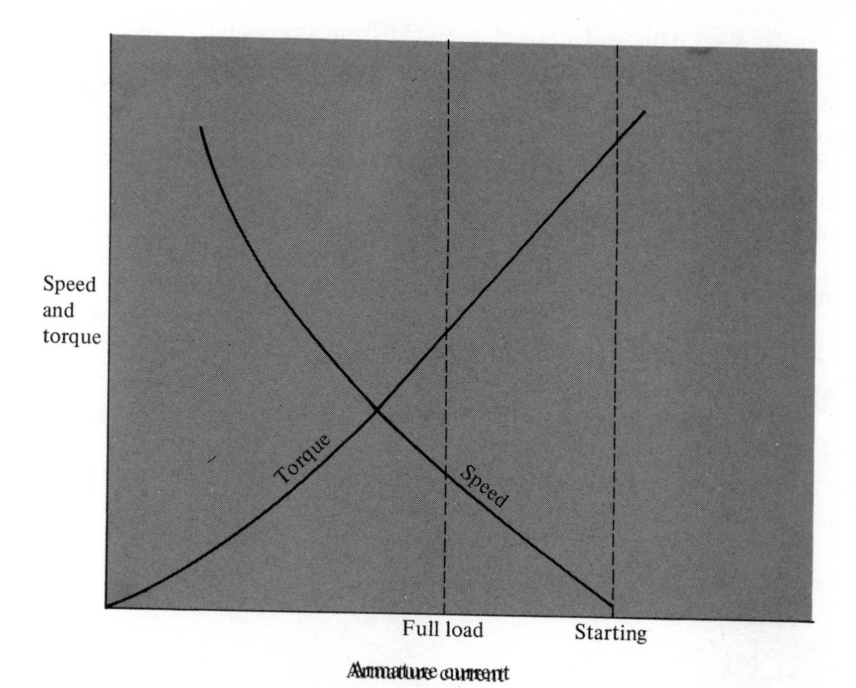

Fig. 7-7. Series motor characteristics.

possible. Resistance units are one possibility, but much power is wasted and a different value of resistance is needed for each desired speed. Also, speed regulation is poor as load is changed.

The Ward–Leonard system of speed control uses a separate dc generator, usually driven by an ac motor. By controlling the small field current, we control the output voltage of the generator applied to the armature of the variable speed motor. This is obviously a very expensive system, but it gives a wide range of speed control and is used for very large dc motors.

An excellent method for medium size and smaller dc motors is the use of silicon controlled rectifiers (SCR). These are solid state devices that can convert ac to dc very efficiently and at the same time control the dc voltage. Silicon controlled rectifiers are available for currents of hundreds of amperes and can be used to control medium size dc motors by armature voltage control from rated speed to zero.

To *increase* the speed of the dc motor *above* rated speed is quite simple. Equation (7-2) shows that a reduction of flux increases speed. The shunt field circuits of dc machines are designed with many turns. As a result the field current is usually from 3 to 6% of rated armature current. Reducing this current with a small rheostat is the only control needed to increase speed. The resistance of the rheostat should be of about the same magnitude as the field winding. This will prevent the field from becoming so weak that the speed will be beyond the centrifugal limit of the armature. At excessive speeds armature conductors can be thrown out of their slots and the motor severely damaged.

Direct current motors are easily reversed. Examination of Fig. 7-3 shows that reversing either the armature or the field connections changes the direction of flux set up by the poles with respect to the armature. The north pole will be south and vice versa with respect to the armature. This will cause the conductor to move to the right. If the armature current is reversed instead of the field current, the same will be true: The conductor will move to the right. The lines of flux around the conductor then will be counterclockwise. Figure 7-8 shows the connections to reverse direction.

When changing shunt field connections of a compound motor, the series field also must be changed. This will assure the same compounding. This is one reason why most reversals are accomplished by reversing armature connections.

To determine if a compound motor is connected cumulative or differential, a simple test is conclusive.

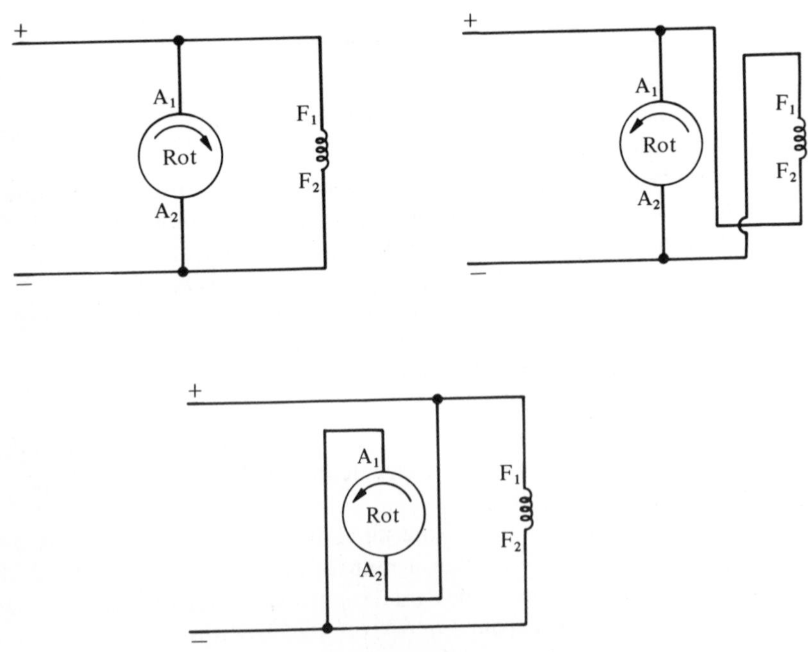

Fig. 7-8. Connection changes to reverse a shunt motor.

1. Operate the motor as a shunt motor with the series field disconnected.
2. Observe the direction of rotation.
3. Operate as a series motor with the shunt field disconnected.
4. Observe the rotation but do not permit the motor to accelerate. If it has the same rotation, the fields are additive; replace the shunt connections as in 1. The motor will be cumulative compounded.

7-7 ***Three-phase induction motor principles.*** The squirrel cage induction motor is by far the most commonly used type of ac motor. It is economical to build, has good operating characteristics, and requires virtually no maintenance. Small fractional horsepower sizes usually are wound for single phase, but over 1 hp, three-phase operation is more practical. The three-phase windings produce more uniform torque. Single-phase motors are used mostly in residences where a three-phase power supply is usually not available or necessary.

The three-phase squirrel cage induction motor has two circuits, a stator and a rotor. The stator consists of three (phase) windings placed symmetrically in a slotted, laminated sheet, steel magnetic circuit. These three (phase) windings must be identical as to turns and wire size and may be connected to the three-phase source in either delta or wye. Figure 7-9 illustrates the stator of a large three-phase induction motor.

The rotor circuit is constructed by imbedding a group of aluminum,

Fig. 7-9. Stator of squirrel cage motor. *Courtesy of U. S. Electric Motors. Division of Emerson Electric Co.*

copper, or brass bars in another magnetic circuit consisting of an iron rotor. This iron rotor is free to rotate in the stator. The conducting rotor bars are short-circuited at the ends, which gives the appearance of a squirrel cage if removed from the iron. The squirrel cage is usually poured in a molten state into voids in the iron magnetic circuit. Currents are *induced* in the bars by transformer action from the stator. For this reason the stator is sometimes called the primary and the rotor the secondary. The secondary of the squirrel cage motor is not an insulated winding with a finite number of turns but simply a very low-resistance circuit. Figure 7-10 shows the rotor in a cutaway view of an induction motor.

Another type of three-phase motor is the *wound rotor*, in which the rotor is an insulated, three-phase, wye-connected winding. Three slip rings permit the ends of this winding to be connected to an external circuit.

The *synchronous motor* has electromagnetic poles as a rotor, requiring dc for excitation. The stator of the synchronous motor is identical to the induction and wound rotor motors and operates with the same principle.

Figure 7-11a shows a simplified version of the three-phase stator windings, used on these polyphase motors, with only two slots filled with conductors. Each phase can produce a north and south pole, depending on the current direction. This is called a two-pole winding (two poles per phase). Figure 7-11b shows the three currents that will flow in each of the three wye-connected stator windings. As this is a 60-Hz power supply each sine wave requires $\frac{1}{60}$ sec to complete its cycle. The time axis has been marked off into six periods of time, each representing 60° of rotation or $\frac{1}{360}$ sec. The arrows in Fig. 7-11b, representing I_A, I_B, and I_C, are 120° apart and show instantaneous currents but are continously changing in magnitude and direction. Note from Fig. 7-11a that their sum at any instant is zero. Also, for any instant—time 1, for example—the currents in phases A and C are positive, while in

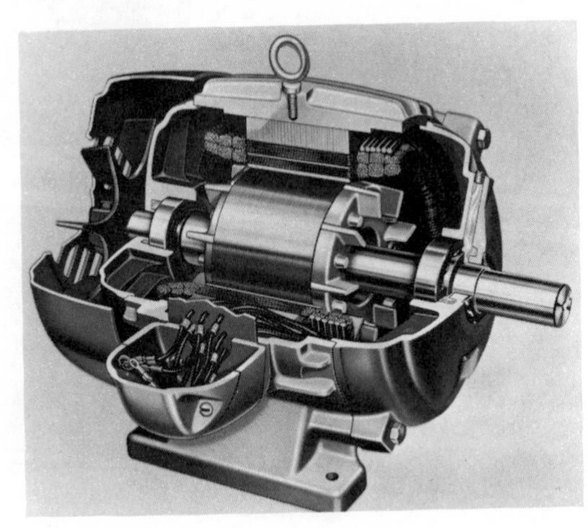

Fig. 7-10. Rotor in cutaway view. *Courtesy of U. S. Electric Motors. Division of Emerson Electric Co.*

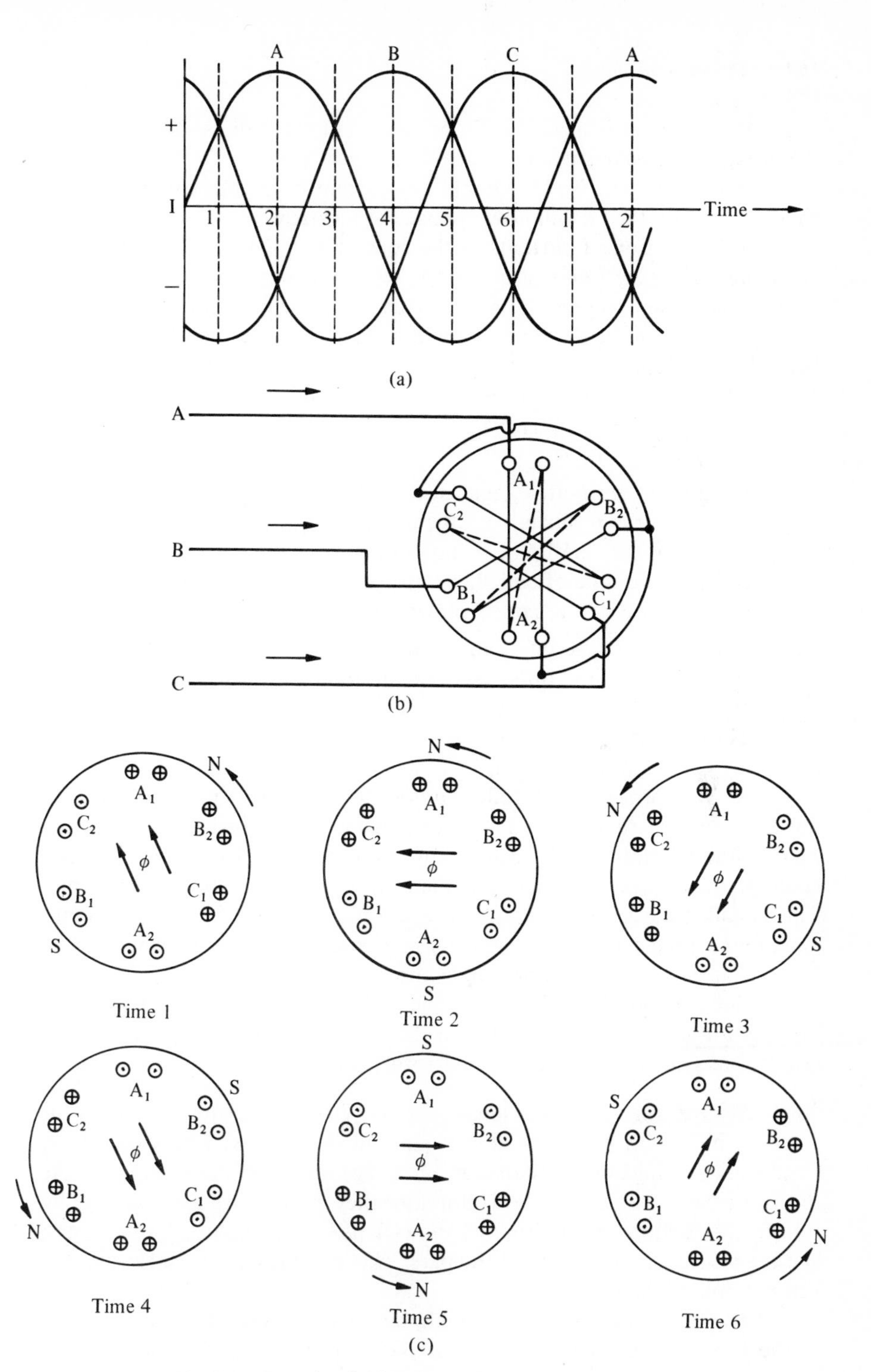

Fig. 7-11. Rotating field in three-phase two-pole stator.

phase B the current is negative. To be consistent we shall assume that when the instantaneous currents are positive, as indicated in Fig. 7-11a, they agree with the arrow in Fig. 7-11b. Should any current be in the negative portion of the sine wave, then it will flow opposite to its arrow.

The direction of currents in the coils of Fig. 7-11c is indicated by a dot representing the point of an arrow, when the current flows toward the reader. The + sign, representing the tail end of an arrow, indicates a current flowing away from the reader. We are looking at the stator conductors parallel to their position.

At time 1 the current direction in the six coils are as follows:

A_1 Current is positive, with the arrow of Fig. 7-11b, and flows in, as in Fig. 7-11c.
B_1 Current is negative, against the arrow, and flows out.
C_1 Current is positive, with the arrow, and flows in.
A_2 Opposite side of coil A_1 flows out.
B_2 Opposite side of coil B_1 flows in.
C_2 Opposite side of coil C_1 flows out.

The cumulative result of these currents flowing in the same direction in the top three coils, and opposite in the bottom coils, result in a north pole on the right side of the stator and a south pole at the left at time 1.

The two parallel arrows in the rotor position indicate the flux direction instant by instant at 60° intervals.

As this procedure is followed for the next five time intervals, the north pole will actually rotate counterclockwise and return to the starting point, although nothing physical on the stator has actually moved. This has taken place in $\frac{1}{60}$ sec. The speed of the rotating field, therefore, is 3600 rpm. This is called the *synchronous speed* of the motor. This effect can be simulated by rotating a permanent magnet in a circle under a table while a steel ball on top proves that "something is moving."

The method of developing torque in the rotor and causing it to follow the rotating field differs somewhat in the three types of three-phase motors (induction, synchronous, and wound rotor).

7-8 ***Squirrel cage induction motor characteristics.*** The squirrel cage rotor is built with a very small air gap and is equipped with close-fitting ball bearings rather than sleeve bearings that must have clearance. As the rotor bars are cut by the stator flux they have a voltage induced by transformer action. A current will flow in the short-circuited rotor bars, causing a magnetic flux around the bars. This develops a torque causing the rotor to follow the rotating field.

The starting current will be high, but of short duration, and decreases as the rotor approaches the speed of the rotating field. The rotor cannot reach this synchronous speed. If it could, no flux would cut the rotor, there would be no induction or rotor current or torque, and the motor would slow down. The actual rotor speed slips behind the rotating field sufficiently so it can induce enough current to produce the torque needed to satisfy the demands of the mechanical load.

This inability to keep up with the synchronous speed is an important measure of an induction motor's performance and is called *slip*. It may be measured in revolutions per minute. It also is expressed in per cent of synchronous speed. When written as a decimal its symbol is s.

$$\text{Per cent slip} = \left(\frac{N_{syn} - N_{rotor}}{N_{syn}}\right)(100) \tag{7-3}$$

and slip,

$$s = \frac{N_{syn} - N_{rotor}}{N_{syn}} \tag{7-4}$$

$$N_{slip} = (s)(N_{syn}) \tag{7-4a}$$

$$N_{rotor} = N_{syn} - N_{slip} \tag{7-5}$$

where N is measured in revolutions per minute.

The windings of Fig. 7-11 are wound for two poles per phase. If each phase were wound for four poles the north pole on the stator would make only half a revolution in the same period of time. This gives us a relationship between synchronous speed and the number of poles in each phase winding. Of course frequency also affects the rotating field speed, but with 60 Hz practically universal we may consider it a constant.

We can say that the frequency of a generated voltage is equal to the number of pairs of poles a coil cuts per second. This is

$$f = \left(\frac{P}{2}\right)\left(\frac{N_{syn}}{60}\right)$$

or

$$N_{syn} = \frac{(120)(f)}{P} \tag{7-6}$$

where f is the frequency in hertz; P is the number of poles per phase on the stator; and N_{syn} is revolutions per minute of the stator rotating field.

Therefore, the rotating field speed N_{syn} is in exact synchronism with the generator at the source. The number of poles on the stator winding determines the motor speed.

Table 7-1 lists the possible synchronous speeds for various magnetic pole configurations.

Table 7-1. Synchronous speeds of ac motors

Poles per Phase	N_{syn} at 60 Hz	N_{syn} at 50 Hz
2	3600	3000
4	1800	1500
6	1200	1000
8	900	750
10	720	600
12	600	500
14	514	428.6
16	450	375

Example **7-2.** A 60-Hz, 240-V induction motor with its stator wound for four poles per phase operates with a slip of 1% at no load. Calculate the rotor speed at no load.

Solution

$$N_{syn} = \frac{(120)(f)}{P} \tag{7-6}$$

$$= \frac{(120)(60)}{4} = 1800 \text{ rpm}$$

$$N_{slip} = (s)(N_{syn}) \tag{7-4a}$$

$$= (0.01)(1800)$$

$$= 18 \text{ rpm}$$

$$N_{rotor} = N_{syn} - N_{slip} \tag{7-5}$$

$$= 1800 - 18$$

$$= 1782 \text{ rpm}$$

As load is added to the motor of Ex. 7-2 more current and power must be supplied to the stator (or primary). As mechanical load is increased the motor slows down slightly. This increases the slip, induces more current in the rotor secondary, and by transformer action an increase in current and power is drawn by the stator from the power supply.

An induction motor with no load draws only a magnetizing current. The power input is small, only to provide for rotational losses. This causes the motor to operate with a very low pf. At half load the pf is still quite low, about 0.5, and this is one of the principal reasons for low pf loads in industry.

Induction motors should be selected accurately, according to the expected horsepower loads insofar as possible.

The nameplate of an induction motor lists the voltage and frequency of the power supply to which it must be connected. Full load input current, output horsepower, and rated speed also are necessary data that are included. The power factor and efficiency are not listed on the nameplate; however, at full load both are usually between 80 and 90%.

Example 7-3. When the motor of Ex. 7-2 delivers rated horsepower output, the slip has increased four times to 4% ($s = 0.04$). The stator current is 25 A. Assuming pf = 0.85 and the efficiency is 85, calculate: (a) the power input, the slip in revolutions per minute, (c) the rotor speed, and (d) the output in horsepower.

Solution

(a)

$$P_T = (\sqrt{3})(EI \cos \theta) \qquad (5\text{-}4)$$
$$= (1.732)(240)(25)(0.85)$$
$$= 8850 \text{ W}$$

(b)

$$N_{\text{slip}} = (s)(N_{\text{syn}}) \qquad (7\text{-}4\text{a})$$
$$= (0.04)(1800)$$
$$= 72 \text{ rpm}$$

(c)

$$N_{\text{rotor}} = N_{\text{syn}} - N_{\text{slip}} \qquad (7\text{-}5)$$
$$= 1800 - 72$$
$$= 1728 \text{ rpm}$$

(d)

$$\text{Efficiency} = \frac{\text{output}}{\text{input}}$$
$$\text{Output} = (\text{input})(\text{efficiency})$$
$$= (8850 \text{ W})(0.80)$$
$$= 7550 \text{ W}$$
$$\text{hp} = \frac{7550}{746} = 10.1 \text{ hp}$$

This is a typical 10-hp motor. Note that the speed changed from 1782 to 1728 rpm from no load to full load. This is good speed regulation, and the induction motor may be called a "constant speed motor."

7-9 ***Wound rotor induction motor characteristics.*** The wound rotor motor has an insulated winding with the same pole configuration as the stator. By means of three slip rings on the rotor shaft, the windings can be connected to three wye-connected external resistors.

When the motor is started, these resistances cause the input current to the stator to be more in phase with the line voltage. The starting pf will be higher than most squirrel cage motors, thereby developing a higher starting torque and a lower starting current. As the motor accelerates, the resistances may be removed and the rotor shorted at the brushes. When short-circuited the rotor performs the same as the squirrel cage rotor.

The same resistors (if rated for continuous duty) may be used as a

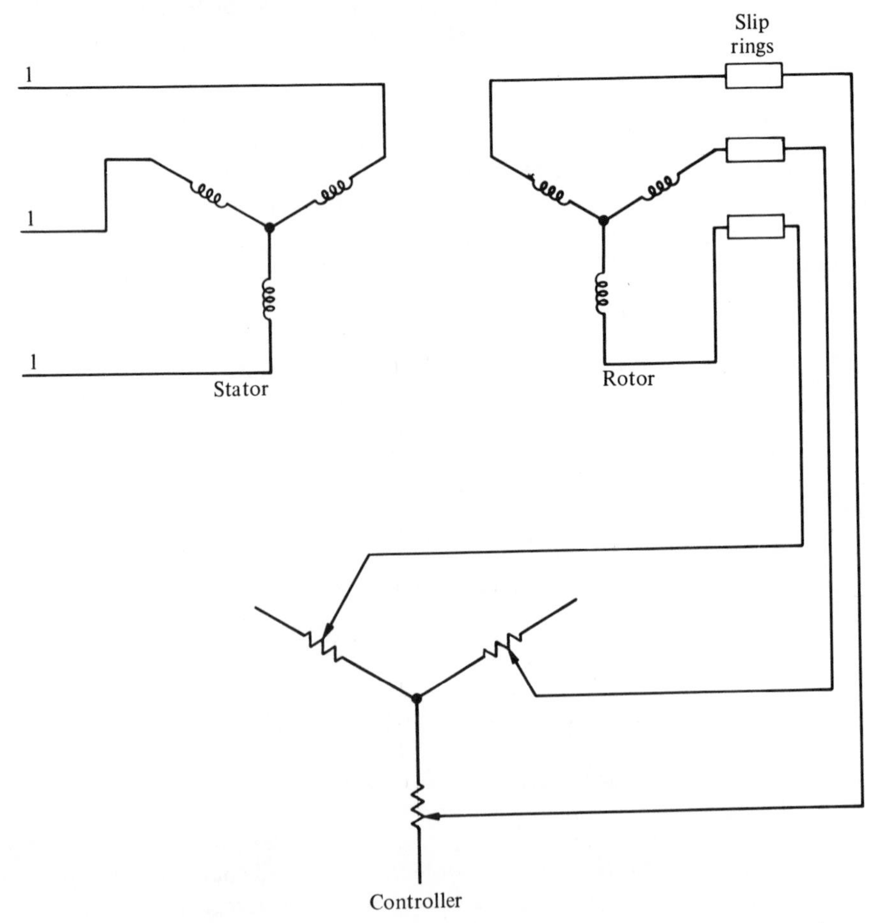

Fig. 7-12. Wound rotor induction motor connected to starter and speed controller.

variable speed control. Resistance in the rotor circuit will increase slip for a particular value of torque. This results in a decrease in speed.

Figure 7-12 shows a wiring diagram of a wound rotor motor connected to its controlling resistors.

Figure 7-13 illustrates a rotor from a large wound rotor motor. Note the brush-holders for connections to an external circuit.

Wound rotor motors are used where very high starting torques are necessary, such as for compressors, or where speed control is required below synchronous speed. Because of I^2R losses in resistors, the efficiency is decreased compared to an ordinary polyphase induction motor.

Fig. 7-13. Rotor from wound rotor motor. *Courtesy of Fairbanks Morse Motors, Manufactured by Colt Industries' Motor and Generator Operation.*

7-10 *Synchronous motor characteristics.* The synchronous motor, as its name implies, operates at synchronous speed ($s = 0$). The stator has the same winding and resultant rotating field as the induction motor. The difference is in the rotor. Instead of depending on slip to set up a magnetic field, a permanent or electromagnetic field structure is used as the rotor.

The south pole of the rotor will be attracted to the north pole of the stator rotating field, and they will stay in absolute synchronism once they are synchronized. Should excessive load pull the motor out of synchronism, the stator draws a high current and operation becomes unstable and must be shut down.

Small, single-phase, synchronous motors such as those used in electric clocks and timing devices, and small polyphase (nonexcited) synchronous motors use permanent magnets in the field or rotor structure.

The electromagnetic poles of most polyphase synchronous motors require a dc source that places some limit on the versatility of this type of motor. This machine is identical to that of a three-phase generator, except that if specifically designed as a motor, the rotor poles also have a few squirrel cage bars to assist in starting. When close to synchronous speed the dc is applied and the rotor synchronizes with the stator. A polyphase synchronous motor must not be started with the dc field energized.

One of the assets of this type motor, however, is the ability to vary the field current. This will vary the induced voltage in the stator windings. For a fixed load this will vary the power factor of the motor. With reduced field current (underexcited) the stator current will lag the line voltage. With increased field current (overexcited) the current will lead the voltage. The phase angle can be shifted from a leading to a lagging stator current by changing the field current. For some value of field current and load the pf can be adjusted to unity. This would be minimum stator current for that particular horsepower load.

Synchronous motors are built from the smallest clock motors to those of thousands of horsepower. Large motors can be adjusted to draw a leading current, thereby correcting the pf of a load with low lagging pf—perhaps because of underloaded induction motors. A large synchronous motor driving a fixed load such as a ventilation or a refrigeration system can improve the power factor of the entire plant substantially with the proper field current adjustment. If desired, it can be operated with *no* mechanical load on its shaft, when adjusted to draw a leading current, and it is called a *synchronous condenser* where operated this way. Its electrical characteristics are now the

Fig. 7-14. Rotor of synchronous motor. *Courtesy General Electric Company.*

same as those of a capacitor, except of course it is a dynamic and not a static device.

Figure 7-14 shows the rotor of a six-pole synchronous motor. Note the field windings and the squirrel cage bars imbedded in the face of the poles. The small winding on the shaft is its *exciter* (a dc shunt generator). This provides the dc for the magnetic poles.

Synchronous motors must be rated in kilovolt-amperes in addition to horsepower. Caution must be taken not to exceed its current rating when used to correct pf. As the pf of a motor is adjusted to a lower value, the line current must increase if mechanical load is constant.

Example 7-4. The load in a plant is 200 kW at 460 V and is three phase with a pf of 0.6 lagging. A synchronous motor with no mechanical load (synchronous condenser) is to be operated with the correct field excitation to correct the pf to 0.85 lagging. Assume the pf of the motor to be zero. Calculate (a) the load current before correction, (b) the load current after correction, (c) the current to the motor, and (d) the minimum kilovolt-ampere rating of the motor.

Solution

The following is also shown graphically in Fig. 7-15:

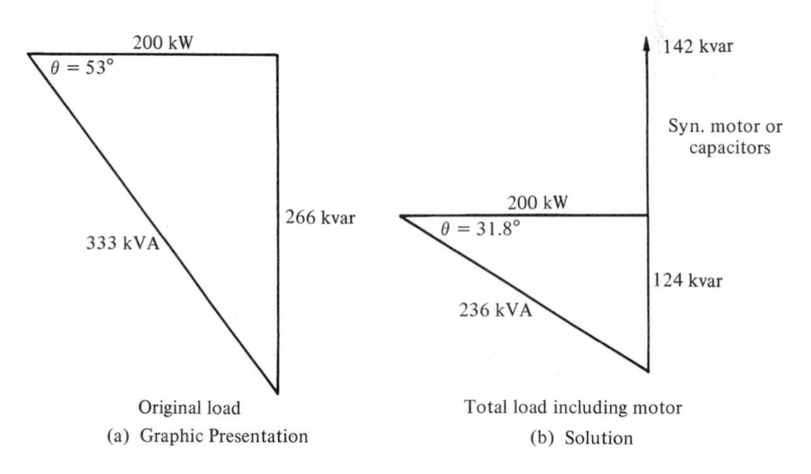

Figure 7-15

(a)
$$I = \frac{P}{(\sqrt{3})(E \cos \theta)} \tag{5-4}$$

$$= \frac{200{,}000}{(1.732)(460)(0.6)} = 418 \text{ A}$$

(b) The original load is

$$\text{kW} = 200 \qquad \cos\theta = 0.6 \qquad \theta = 53^\circ \qquad \sin\theta = 0.8$$

$$\text{kVA} = \frac{200}{\cos\theta} = \frac{200}{0.6} = 333 \text{ kVA}$$

$$\text{kvars} = \text{kVA} \sin\theta = (333)(0.8) = 266 \text{ kvars}$$

With the motor added the pf must be 0.85.

$$\text{kW} = 200 \qquad \cos\theta = 0.85 \qquad \theta = 31.8^\circ \qquad \sin\theta = 0.525$$

$$\text{kVA} = \frac{200}{0.85} = 236 \text{ kVA}$$

$$\text{kvars} = (236)(0.525) = 124 \text{ kvars}$$

To reduce lagging kilovolt-amperes reactive from 266 to 124, the motor must contribute 266 − 124 = 142 leading kvars. The load current after correction is

$$I = \frac{P}{\sqrt{3}\,(E\cos\theta)} = \frac{200{,}000}{(1.732)(460)(0.85)} \tag{5-4}$$
$$= 295 \text{ A}$$

(c) The current to the motor is

$$I = \frac{\text{kvars}}{\sqrt{3}\,E} = \frac{142{,}000}{(1.732)(460)} = 178 \text{ A}$$

(d) The electrical load of the motor is 142 kvars, the power is zero, and the pf = 0; therefore, its load may be considered to be 142 kVA at pf = 0.

Note the decrease in the current from 295 to 178 A—a much better overall installation. The static capacitors of course can duplicate this pf correction. Their rating would be 142 kvars at 460 V.

Example 7-5. The load in a plant is 200 kW at 460 V and is three phase with a low pf of 0.6 lagging. This is to be corrected to 0.85 lagging. A synchronous motor with a mechanical load that causes an input of 40 kW is added to the system.

(a) Calculate the kilovolt-ampere rating of the motor.
(b) If the motor is 90% efficient, what is its mechanical load in horsepower?
(c) Calculate the current before correction.
(d) Calculate the power and the current of the system after correction.

Solution

The following is also shown graphically in Fig. 7-16:

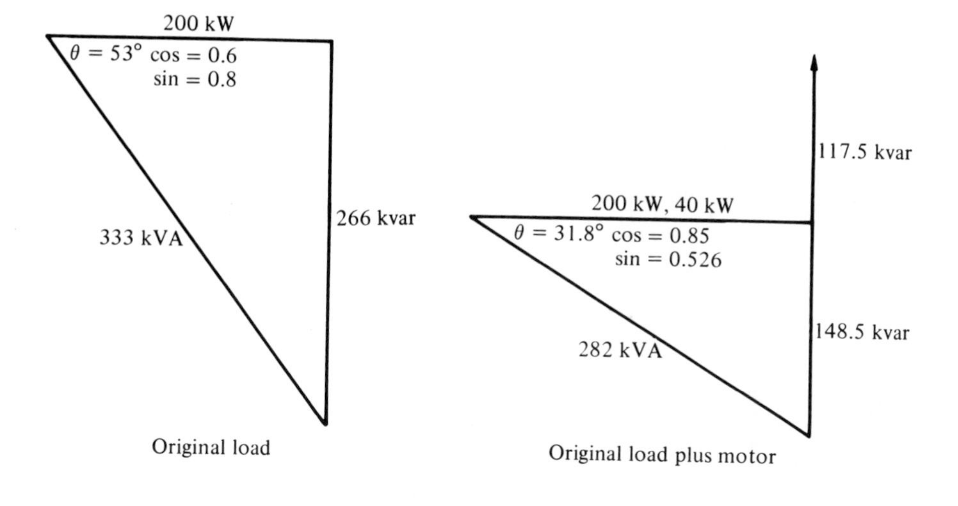

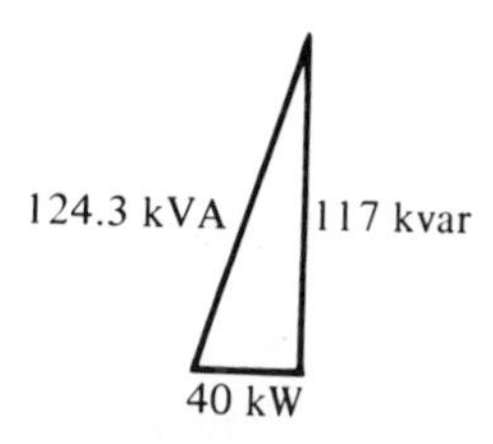

Fig. 7-16. Graphical solution to Ex. 7-5.

(a) The original load is

$$\text{kW} = 200 \qquad \cos\theta = 0.6 \qquad \theta = 53^\circ \qquad \sin\theta = 0.8$$

$$\text{kVA} = \frac{200}{\cos\theta} = \frac{200}{0.6} = 333 \text{ kVA}$$

$$\text{kvars} = \text{kVA} \sin\theta = (333)(0.8) = 266 \text{ kvars}$$

The original load plus the synchronous motor are

$$\text{kW} = 200 + 40 = 240 \text{ kW}$$

$$\cos\theta = 0.85 \qquad \theta = 31.8^\circ \qquad \sin\theta = 0.526$$

$$\text{kVA} = \frac{\text{kW}}{\cos\theta} = \frac{240}{0.85} = 282 \text{ kVA}$$

$$\text{kvars} = 282 \sin\theta = (282)(0.526) = 148.5 \text{ kvars}$$

To decrease 266 kvars to 148.5 kvars, the motor must contribute 266 − 148.5 = 117.5 *leading* kvars.

$$\text{kVA of motor} = \sqrt{40^2 + 117.5^2} = 124.3 \text{ kVA}$$

$$\text{Efficiency} = \frac{\text{output}}{\text{input}} \qquad \text{output} = (\text{input})(\text{efficiency})$$

$$\text{Output} = (40)(0.9) = 36 \text{ kW}$$

$$\text{hp output} = \frac{36{,}000}{746} = 48.4 \text{ hp}$$

(c)
$$I = \frac{\text{VA}}{\sqrt{3}\,E} = \frac{333{,}000}{\sqrt{3}\,(460)} = 418 \text{ A}$$

(d)
$$I = \frac{282{,}000}{\sqrt{3}\,(460)} = 354 \text{ A}$$

From the results of (c) and (d) it is noted that the addition of the synchronous motor with an added load of 40 kW can decrease the total current 15.3%.

This assists in making a more economical electrical installation. If a 48.4-hp load were provided by an induction motor, none of this saving could be realized.

7-11 ***Single-phase motors.*** With only a single-phase supply available, refer to Fig. 7-11b and assume that only the current of phase A is available.

In Fig. 7-11c and b only the coils A_1 and A_2 are on the stator. At time 2, when the current is positive and maximum, the north and south poles of the magnetic field will be in the same position, as indicated in Fig. 13-4c.

One half cycle (180°), or $\frac{1}{120}$ sec later at time 5, the north and south poles will have changed places. At time 1, after a total elapsed time of $\frac{1}{60}$ sec, the north and south poles are back at the same place as time 1.

The pole changes are not instantaneous but follow the smooth transition of the sine wave of current. This also can be considered to be a rotating field on the stator. But does it rotate clockwise (cw) or counterclockwise (ccw)? It should be obvious that it has no direction. The next impulse is as far away in a cw direction as ccw. If a squirrel cage rotor could receive an initial push in either direction, the rotor would induce a current as did the three-phase rotor and would then continue to turn in that direction as a result of the developed torque.

We can develop a starting "push" with another stator winding identified as a starting winding. The starting winding must have less inductance and more resistance, and when both are connected in parallel their currents will be displaced by the angle θ. The magnetic poles resulting from these two currents also will be displaced by the same angle θ if each winding is wound for two poles. Refer to Fig. 7-18.

The direction of rotation of the rotor will depend on the relative direction of the two currents. To reverse rotation, the starting winding terminals must be interchanged.

The starting winding can contribute very little torque when the rotor is at full speed. A centrifugal mechanism on the rotor shaft opens a switch on the end plate of the motor and opens the starting winding circuit.

This type of motor is called a *split-phase induction motor* because the one-phase voltage available is made to split the current into two separate circuits for starting purposes.

Torque can be improved by connecting a capacitor in series with the starting winding. This will "split" the phase even more. A larger angle between the two currents will increase the displacement of the stator poles.

Figure 7-17 illustrates a split phase motor with its starting capacitor visable on top. Figure 7-18 shows the vectors of starting currents and wiring diagrams of both types of split phase motors.

Split-phase motors can perform most small motor applications up to about 2 hp. Over 2 hp they become expensive and the centrifugal switch is always a maintenance problem. For motors over 2 hp, a three-phase motor is more practical.

Other types of single-phase motors are *repulsion induction*, *universal*, *shaded pole*, and *synchronous*. Each has its own particular application.

The repulsion induction motor has a very high starting torque. The rotor is like that of a dc motor with the coils connected to a commutator, although rotor currents are induced. On start a set of brushes short-circuit a few of the coils; during acceleration a centrifugal ring short-circuits all coils on the rotor and it operates like a squirrel cage motor.

The universal motor is similar to the dc series motor with a very high

Fig. 7-17. Single-phase split-phase motor. *Courtesy of General Electric Company.*

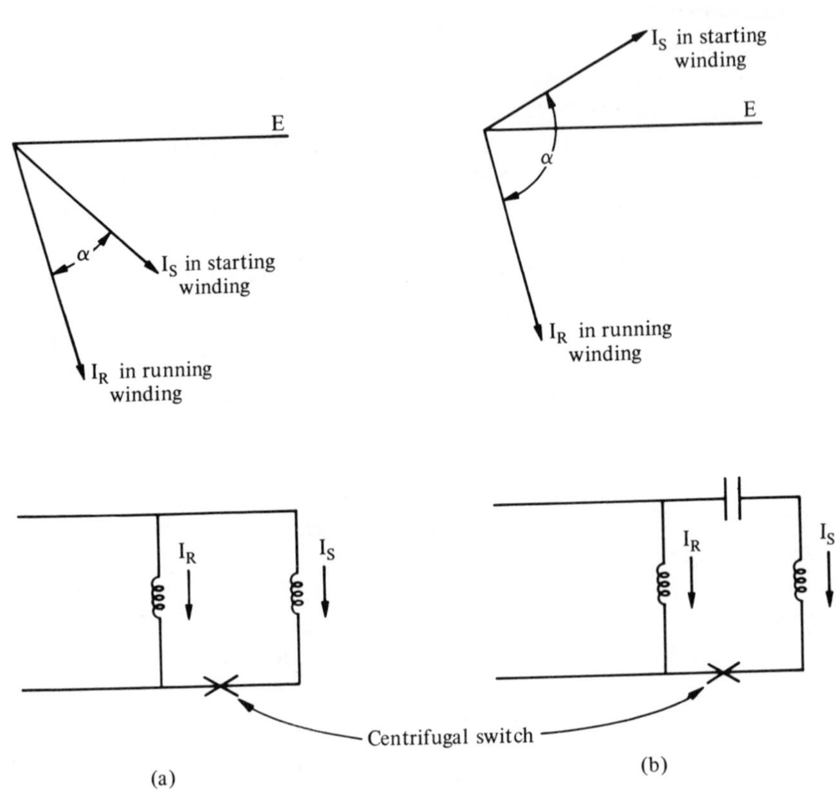

Fig. 7-18. Current vectors and diagrams. (a) Split-phase motor. (b) Capacitor-start motor.

no load speed. It is usually built in small sizes for domestic appliances such as vacuum cleaners and mixers.

Shaded pole motors have very little starting torque. The stator has salient poles rather than distributed ones. One side of the pole has a slot cut. Around this part of the pole a current is induced in a copper ring that sets up a small pole out of phase with the main pole. This type of motor cannot be reversed (although there are types of reversible shaded pole motors). It is used in small sizes for driving loads such as fans since only small starting torque is necessary. Figure 7-19 shows this pole arrangement. The rotor is the typical squirrel cage; therefore, speed is a function of line frequency and the number of stator poles.

The single-phase synchronous motor is a miniaturized version of the motor described in Sec. 7-10. It is used for electric clocks or any timing mechanism that requires absolute constant speed. The single-phase stator sets the speed, and small permanent magnets permit the rotor to turn at

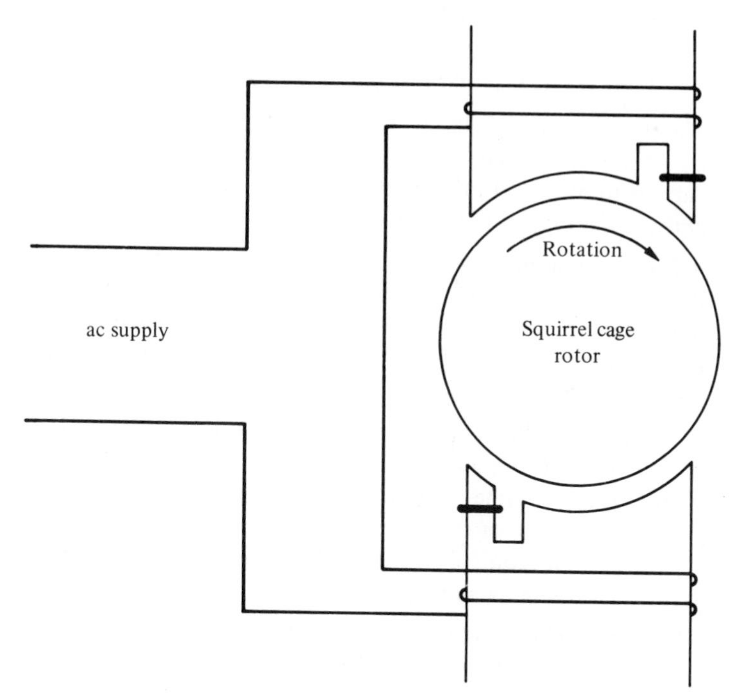

Fig. 7-19. Typical shaded-pole single-phase motor.

synchronous speed. A small shading coil starts the rotor. Rotation cannot be reversed, but a clock running backward would be only wishful thinking.

PROBLEMS

7-1 The armature of a 230-V shunt motor requires 32 A when delivering a rated horsepower load. The field current is 2.5 A. Calculate (a) the motor current and (b) the total power input at rated load.

7-2 The armature of a shunt motor carries 40 A at rated load. Calculate the starting resistance necessary to limit the starting current to 150% of full load current. The resistance of the armature is 0.5 Ω. The supply voltage is 240 V.

7-3 The field circuit of a 230-V shunt motor takes 2 A at rated voltage. A rheostat is to vary the field current from 1.2 to 2 A. What should be its maximum resistance?

7-4 A 230-V motor operates at 2000 rpm. Its speed-voltage characteristic is linear.

(a) What is the speed of the motor if the armature voltage is changed to 170 V with a constant torque load?

(b) What voltage should be applied to a shunt field, 170 or 230 V?

7-5 The armature and field current to a 230-V shunt motor are 23 and 2 A, respectively. The motor is 80% efficient. Calculate (a) the total current to motor, (b) the input power, (c) the output power in watts, and (d) the output power in horsepower.

7-6 A 230-V shunt motor operates at 2400 rpm when the armature current is 3 A at no output. The armature resistance is 0.5 Ω. With an added load, the armature current increases to 20 A. Calculate (a) the speed if the flux is assumed to be constant and (b) the speed if the flux is reduced by 5%.

7-7 A three-phase, 460-V ac motor is driving a dc generator. It is desired to operate a 20 hp, 240-V dc shunt motor from this dc supply. Each of the three machines is 80% efficient. Calculate (a) the horsepower rating of the ac motor, (b) the input power in watts to the ac motor, (c) the current required by the dc motor, and (d) the current required by the ac motor. Assume pf to be 0.85.

7-8 What is synchronous speed for an induction motor having (a) two poles, (b) ten poles, (c) six poles at 50 Hz, and (d) four poles at 25 Hz?

7-9 A two-pole induction motor has a slip of 0.03 at rated horsepower load. What is its full load speed?

7-10 A six-pole induction motor with no load on its shaft operates at 1188 rpm. Calculate (a) the slip in revolutions per minute, (b) the per cent slip, and (c) the speed if the slip increases four times with added load.

7-11 A 240-V, three-phase induction motor at no load takes 1000 W and 12 A. Calculate (a) the power factor of the motor and (b) the power factor if the increased load requires the motor to take 6650 W and 20 A.

7-12 A large induction motor is started with its windings connected in delta. The line current at start is 300 A. What reduction in starting current would be possible if the motor were started in a wye configuration?

7-13 (a) What input power is necessary for a 240-V, three-phase motor to maintain a 25-hp output? The power factor is 0.8, and the efficiency is 85%.

(b) Calculate the line current.

7-14 Calculate the change in slip of an induction motor if the speed changes from 1790 rpm at no load to 1750 rpm at full load.

7-15 At rated output of 25 hp, a three-phase, 230-V induction motor requires an input of 69 A and 22 kW. Calculate (a) the efficiency and (b) the power factor.

7-16 A four-speed induction motor has two sets of windings that may be interconnected so that the motor can operate with two, four, or eight poles.

(a) What speeds are possible if the slip at full load in each instance is 3%?

(b) Are any intermediate speeds possible? Explain.

7-17 At what speed does a 60-Hz, 18-pole synchronous motor operate?

7-18 A large synchronous motor used for ship propulsion may be connected for 16 or 32 poles. The frequency of the system can be varied from 20 to 70 Hz. Calculate the speed range of the motor.

7-19 A 460-V, three-phase synchronous motor rated at 150 A is to be used to correct pf, operating as a synchronous condenser with no load on its shaft.

(a) What is the maximum leading kilovolt-amperes reactive the motor can receive from the line?

(b) If the mechanical load is added so that the input power is 50 kW, how many leading kilovolt-amperes reactive can the motor contribute to correct the pf?

7-20 It is necessary to vary the field current of a synchronous motor from 3 to 10 A for pf control. Its field resistance is 23 Ω.

(a) What dc voltage is necessary?

(b) What is the maximum resistance of the controlling rheostat?

7-21 A 200-kW load has a lagging pf of 0.7. This is to be corrected to 0.85 by adding a synchronous motor.

(a) What is the minimum kilovolt-ampere rating of the motor if it is operating with no load?

(b) What is the minimum kilovolt-ampere rating of the motor if the motor has a 50-hp load and is 85% efficient?

Chapter Eight

The Electrical System

8-1 ***From generating station to substation.*** This chapter briefly traces the complete electrical system from the generating station to the final outlet. Explanations are in very general terms until the power enters the building. Thereafter, the details of the equipment used and the terminology are more specific.

Modern generating stations can generate large amounts of electric power. Some steam turbine units have nameplate ratings as high as 500,000 kW (500 megawatts). Coal, oil, gas, or nuclear energy are fuels used to generate steam pressures over 2000 lb-ft/in.^{2}. These stations must be located near large quantities of water. Large modern units are very efficient; they can produce 1 kWh of electrical energy from 0.7 lb of coal. One kilowatt-hour operates a household toaster for about 1 hr. It would be very difficult to concentrate the heat from 0.7 lb of coal to operate a toaster by any other method.

All generators are three phase and operate at the maximum possible speed to generate hertz, 3600 rpm. This is an efficient speed for steam turbines. Voltages range from 12,000 to 22,000 V. Several of these units would be incorporated in one station; two or more can be in parallel when load demands are high.

Of interest in a visit to a large generating station is the vast amount of mechanical equipment, pumps, boilers, etc. necessary to convert the fuel to steam. When the energy is converted to electricity we see three copper bars several inches wide extending from the generator, where the power flow is now in a most convenient form. Figure 8-1 illustrates a 400,000-kW steam turbine generator.

Natural waterfalls and man-made dams are excellent sources of power to turn generators. These hydro plants are usually smaller in capacity than steam plants. This depends on their proximity to concentrated loads in large cities.

Fig. 8-1. A 400-MW turbine generator. *Courtesy of Boston Edison Company.*

After leaving the generator, the voltage is raised to a suitable level for transmission to various substations. This transmission voltage will vary from 66,000 to 110,000 V. Longer distances, distances greater than 100 mi, may use voltages as high as 345,000 V. A generally accepted voltage gradient for transmission is about 1000 V/mi.

In concentrated areas, as in New York City, power at the generated voltage may be conducted directly to buildings. No point in Manhattan is more than 5 mi from a power plant.

8-2 ***Distribution of power.*** Substations are located close to smaller cities that have no generating station. The high efficiency of very large generating stations, the availability of large amounts of cooling water, and efficient high-voltage transmission are important factors in choosing the location of a generating station.

The function of the substation is to step down the high transmission voltage, with transformers, to an appropriate voltage for distribution. This is usually an outdoor installation; a small building may be necessary to provide for relay and monitoring devices associated with the numerous lower voltage circuits that originate at the substation. Figure 8-2 illustrates a 66,000-V substation, and Fig. 8-3 is an installation with underground lines. Individual lower voltage circuits originate at these substations.

The distribution voltage may vary from 2300 to 13,800 V depending on the distances that the circuits must extend. All these circuits must be

Fig. 8-2. A 66,000-V substation.

Fig. 8-3. A substation with underground lines.

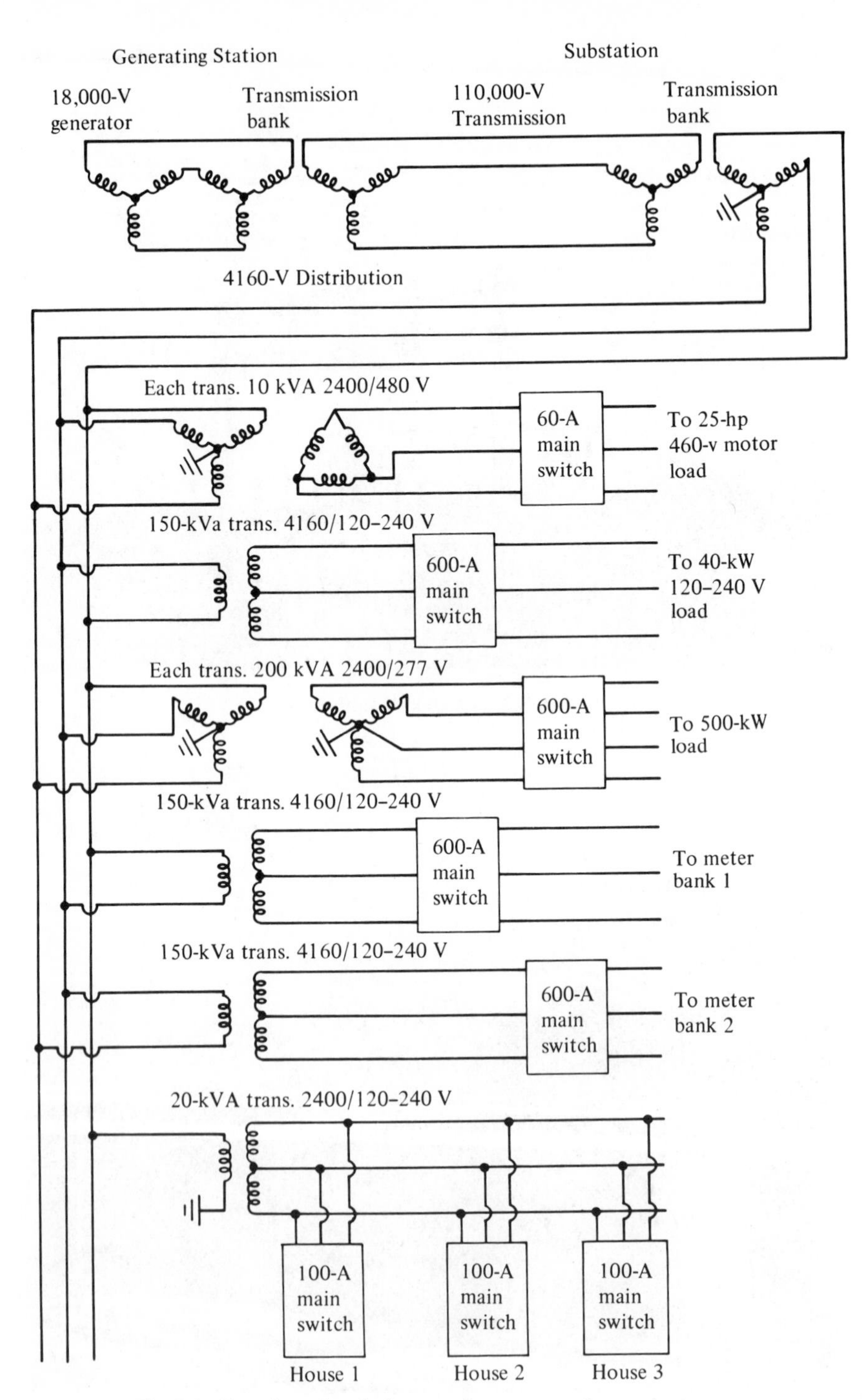

Fig. 8-4. Complete electrical system from power plant to customer.

protected with very fast-acting circuit breakers to protect transformers in the event of short circuits in the distribution lines.

Each individual customer is served from these distribution lines. The voltage is once again stepped down to provide the necessary voltages and phases for each particular load. Figure 8-4 shows a diagram of this system from the generator to each customer's main switch. Four different occupancies and their associated electrical characteristics are discussed (see below). The distribution voltage for this system is 2400–4160 V, three phase, and wye-connected secondaries. The neutral is usually a grounded cable that supports the three line conductors.

(1) A store has a 40-kW, single-phase, 120–240-V load and a group of three-phase motors with a total rating of 25 hp at 240 V. This would require a 50-kVA, 2400/120–240-V transformer and three 10-kVA, 2400/240-V transformers. They would be located as close to the building as possible.

(2) A large manufacturing plant has a total load of 500 kW. Some of it requires 120–240 V; most of the lighting operates at 277 V. Three-phase power requires 480 V. The best arrangement would be three 200-kVA transformers in a vault inside or adjacent to the building. This would end the utility company's responsibility. The 115–230-V loads would be served by 480/120–240-V transformers throughout the building.

Fig. 8-5. Three-phase transformer bank. *Courtesy of Square D Company.*

Figure 8-5 illustrates a typical three-phase transformer bank. Primary high-voltage lines are from an underground distribution system. The transformers shown are of historical significance; they are the power supply for Independence Hall in Philadelphia.

(3) An apartment building has a maximum load of 300 kW, 120–240 V, and single phase. This building might best be served by two 150-kVA, 4160/120–240-V transformers at different locations. This would enable shorter conductors to serve each apartment. The transformers could be in a vault inside the building or outside if distribution were from overhead lines.

(4) A group of single-family houses in the suburbs would be served by a single 2300/115–230-V transformer. Small transformers are used because 115–230 V should not extend over 400 or 500 ft. One transformer would serve only 4 or 5 houses.

8-3 ***Service entrance equipment.*** When distribution is overhead, the utility company usually secures the service drop conductors to the building.

All design and construction beyond this point is the responsibility of the owner, and the NEC regulations now apply. If distribution is from an underground system, the underground service installation is usually made by the utility company. However, local conditions may vary.

The service entrance equipment consists of necessary conductors, metering equipment, main disconnecting means, and suitable overcurrent devices for the various groupings of electrical loads throughout the buildings. Connection to a grounding electrode is made at this point.

The main disconnecting means are in a readily accessible location as close as possible to the point of entrance of the service conductors. Therefore, the service conductors are relatively short in length. This is a design advantage in addition to safety because these conductors are not protected against overcurrent at their source of power but only at the end.

As we follow the system through the four installations of Sec. 8-2 let us first define some of the terminology that shall be used throughout the following chapters.

SERVICE ENTRANCE CONDUCTORS. A set of conductors that connect the utility company's transformer or distribution system to the service equipment of the building to be supplied.

FEEDER. A set of conductors that supply a group of branch circuit overcurrent devices.

BRANCH CIRCUIT. A set of conductors that extend beyond the last overcurrent device in the system. A branch circuit usually supplies a small portion of the total load.

SWITCHBOARD. A large frame assembly that contains a number of

Fig. 8-6. Switchboard. *Courtesy of Square D Company.*

Fig. 8-7. Panel board with replaceable circuit breakers. *Courtesy of Square D Company.*

overcurrent devices for feeders. It may also contain instruments and may be accessible from both front and rear. Figure 8-6 illustrates a typical switchboard assembly.

PANEL BOARD. A smaller assembly of overcurrent devices, contained in a cabinet and accessible only from the front. The overcurrent devices, up to a maximum of 42, may be for either branch circuits or feeders. Figure 8-7 illustrates a panel board assembly equipped with replaceable circuit breakers for single- and three-phase circuits.

The service equipment necessary for each of the four typical installations of Sec. 8-2 is as follows:

(1) The store would require a 200-A, 120–240-V, single-phase service. A 60-A, 240-V, three-phase service would be of sufficient capacity for the present 25-hp load. The possibility of an increase in three-phase motors would justify the installation of a 100-A, three-phase service. Separate metering for each service is necessary.

(2) This plant would require only a 480-V, three-phase service of at

Fig. 8-8. Meters and breakers for an apartment building. *Courtesy of Square D Company.*

Fig. 8-9. Residential panel board.
Courtesy of Square D Company.

least 600 A capacity. A switchboard complete with main circuit breaker, current transformers for a kilowatt-hour meter, and individual circuit breakers for feeders could be contained in one complete fabricated assembly.

(3) Each service to this apartment building must be rated at least 600 A. Troughs or ducts would then carry power to banks of meters and circuit breakers for each apartment. Such an installation is illustrated in Fig. 8-8.

(4) A single residence with 10 kW or more initial load requires a 100-A service. If electric heating is installed, 200 A is necessary. Kilowatt-hour meters are usually installed on the outside to facilitate reading by the utility company. The service equipment would be one assembly of various capacity circuit breakers or fuse-holders for the several branch circuit ratings necessary in the modern electric home. Figure 8-9 illustrates a residential panel board complete with a 150-A main circuit breaker.

8-4 ***Feeders and subfeeders.*** The electrical loads in large buildings are supplied from many panel boards located throughout the building. Some are at considerable distances from the service entrance equipment. Each panel board is supplied with a separate feeder.

When many feeders are necessary, a second switchboard may be installed some distance from the service entrance. This switchboard is supplied with a feeder, and subfeeders supply the panel boards.

We shall investigate the feeders as needed in each of the four examples of Sec. 8-2:

(1) Two or more panel boards may be necessary to properly distribute the 40-kW, single-phase load. Each panel board would be supplied with a separate feeder.

A three-phase feeder would supply power to a three-phase panel board, which provides separate overcurrent devices for each motor.

(2) Three-wire, three-phase feeders are necessary for the motor loads, but four-wire, three-phase feeders are necessary for the 277-V lighting loads. Lighting and power loads would be supplied from separate panel boards and hence separate feeders.

The 120–240-V, single-phase loads can be supplied with a 480/120–240-V transformer. This requires a two-wire, 480-V feeder as a primary. The transformers should be located as close as possible to the 120-V loads. This enables small conductors to supply large 120-V loads.

(3) Each apartment must have a separate feeder supplying a paneboard in the apartment. This is for convenience and economy. It would be more economical to install one feeder than several branch circuits to a common area for all panel boards.

(4) A single-family residence would not require any feeders since the main switch and branch circuit overcurrent protection is usually in the same enclosure.

Should the branch circuit panel board be located in another area of the residence a feeder then would be necessary.

8-5 ***Branch circuits.*** The branch circuit principle is to assure the proper overcurrent protection for small and large devices. Branch circuits are rated according to the overcurrent devices protecting the circuit. For lighting, heating, and appliance loads, the ratings are 15, 20, 30, 40, and 50 A.

Fuse-protected branch circuits are supplied from panel boards. Switchboards are usually too large to provide the smaller overcurrent protection.

The proper fuse- or circuit-breaker-protected branch circuits for the four buildings of Sec. 8-2 could be allocated as follows:

(1) The store lighting would be connected to 15- or 20-A branch circuits. Circuits can carry only 80% of their capacity and must be designed very carefully.

Each motor must be connected to a three-phase circuit with a capacity of at least 125% of the motor nameplate rating.

(2) The single-phase, 277-V lighting units could be connected to 15- or 20-A branch circuits connected between each line wire and neutral.

If heavy duty mogul sockets with lamps of more than 300 W were used, any branch circuit from 15 to 50 A may be used.

The 120–240-V loads would be supplied with branch circuits appropriate to their current ratings. Provisions for three-phase motor circuits would be similar to (1).

(3) An apartment could be smaller in area than a single-family residence, but it would function as a family unit. Two or more 20-A circuits for appliances, separate circuits for specific appliances, and at least one 15-A circuit per 500 ft^2 of floor area for lighting are some of the more important branch circuit requirements. Figure 8-10 illustrates a typical apartment branch circuit panel board.

(4) The branch circuit requirements in a single-family residence would be similar to the apartments of (3) above, but more circuits might be necessary.

Fig. 8-10. Panel board for apartment. *Courtesy of Square D Company.*

PROBLEMS

8-1 A turbo generator is wye-connected and is rated at 10,400 V and 3200 A/phase.

(a) Calculate the line-to-line voltage.

(b) Calculate the kilovolt-ampere rating.

8-2 A generating station has three units as described in Prob. 8-1. The voltage is raised to 110,000 V for transmission. Calculate the total current at 110,000 V.

8-3 A substation receives 110,000 V and supplies distribution lines at 13,800 V. If transformers are connected wye-wye, calculate the primary and the secondary voltage ratings of the substation transformers.

8-4 When 13,800 V is used for three-phase distribution, how much current must be provided for (a) each 100 kW of load at unity pf and (b) each 100 kW of load at 0.8 pf.

8-5 Investigate the power source for your own home: the distance to "your" transformer and its kilovolt-ampere rating; the distribution voltage; the distance to the substation; the transmission voltage; and the distance to the nearest generating station. These might all be determined with a little research in the local office of your utility company. Perhaps a visit to its public relations office also might be beneficial.

Chapter Nine

Conductors and Raceways

9-1 ***Purpose and scope of the National Electric Code.*** The only "invitation" the flow of an electric current needs is a potential difference between two points. This flow is practically instantaneous, and very elaborate precautions must be taken to see that this current is permitted to flow only where and when it is safe to do so. A short circuit may be defined as one that has zero resistance. According to Ohm's law, the current in a short circuit is infinite or unlimited. This calls for the proper safeguards in all wiring systems so that this uncontrolled current (in the event of an accidental short circuit) does not become a fire hazard.

One of the most common causes of fires in buildings results from the improper use of electricity or from defective equipment. Usually the person who uses electrical devices knows little of the hazards involved. Because of this, electrical equipment must be properly installed and sized.

To assure the standardization and proper installation of electrical equipment the National Fire Protection Association (NFPA) has developed a complete set of rules to set minimum standards for electrical installations within buildings. The NFPA also determines manufacturing standards for many items, so that a fuse made in Oregon will fit a switch made in Connecticut. The minimum standards are incorporated in a book called the *National Electrical Code* (NEC). Each section of the NEC is supervised by a panel of experts in various areas who make revisions from time to time so that a revised code is published about every 3 years.

The adoption and enforcement of the NEC is usually left to the discretion of local authorities rather than being left to the state or national level authorities. Local wiring inspectors may differ in their interpretation of some provisions of the code since NEC standards are *minimum* requirements. In the absence of any local authority, the NEC prevails, usually under the authority of the electric utility company.

The NEC is not an instructional manual, but it sets an official minimum standard in the determination of many electrical design problems.

9-2 ***Conductor insulation.*** We defined conductors in Chapter 1. They perform the very important function of conducting electric current from the source to the load.

For many years natural rubber was used to insulate building wire conductors, but age and heat caused such insulation to dry out, to crack, and to become brittle. Since 1945 better-quality rubber and thermoplastic materials have been developed that not only permit thinner insulation on wires but also withstand temperature and remain good insulators indefinitely.

The amount and type of insulation on conductors is determined by the voltage that will exist between them. The type of insulation used is decided by the conditions under which the conductors must operate with regard to heat, moisture, or other conditions that might have a deleterious effect on insulation.

The building wiring systems that operate at less than 600 V may use the same conductors, since the first classification the NEC permits is 600 V. For higher voltages there are five classifications of maximum voltages: 1000, 2000, 3000, 4000, and 5000 V. The NEC requires these conductors to be marked with the first two numbers of this classification. No identification mark means 600 V; 30 indicates approval for 3000 V. The higher the voltage, the more the insulation necessary for NEC approval.

To properly identify the various types of insulation, the NEC has established a system of letters that indicate their characteristics. Some of the most common types are shown in Table 9-1.

Table 9-1. NEC letter system for identifying types of insulation

Trade Name	Type Letter	Maximum Operating Temperature (°F)	Application
Rubber	R	140	Dry locations
Heat-resistant rubber	RH	167	Dry locations
Heat-resistant rubber	RHH	194	Dry locations
Moisture- and heat-resistant rubber	RHW	167	Dry and wet locations
Thermoplastic	T	140	Dry locations
Moisture- and heat-resistant thermoplastic	THW	167	Dry and wet locations

From these identification letters it may be noted that the letters R and T refer to the material, rubber or plastic; H will endure a higher temperature and W is moisture resistant.

9-3 ***Limitation of heat on conductors.*** Except for mechanical abuse, the greatest hazard that conductors must endure is heat. Continued exposure to excessive heat causes insulation to become soft, perhaps to melt, and in extreme cases to burn. This heat comes from two sources: from the ambient air surrounding the conductors or from the current the conductors must carry. Although conducting materials such as copper or aluminum have a low resistivity, there is a point where an increase in current causes excessive heat.

The NEC has developed tables for each specific insulation that determine the current-carrying capacity of each size conductor, that is, the "ampacity." Obviously, it means capacity in amperes.

9-4 ***Materials for conductors.*** Although any metallic substance conducts an electric current, some materials are better conductors than others.

Of all materials the choice has been narrowed to two: copper and aluminum. Copper is by far the most common, especially for small conductors, since it is stronger than aluminum. For larger conductors aluminum might be preferred because of its lower cost and weight. For the same volume, aluminum weighs about one-third as much as copper.

The relative ability of a material to conduct can be determined by its resistivity. This is usually expressed in ohms-circular mil per foot. The resistivity of a material is the resistance in ohms of a sample 1 ft long and 1 mil in diameter, or 1 cmil in cross-sectional area. (See Sec. 2-5.)

Table 9-2 lists the resistivity of different materials at 20°C. From this table it can be noted that silver is the best conductor but that cost makes copper a more economical choice. Aluminum, despite 60% more resistance than copper, is desirable because of its lower weight and cost. We can usually compensate for this increase in resistance by using aluminum conductors one size larger.

Conductors vary in size over a very wide range. Wires of one million cmil (0.7-in diameter) become so cumbersome to install that flat bars are

Table 9-2. Resistivity of some common conducting materials at 20°C used for conductors and resistors

Material	Resistance at 20°C (Ω-cmil/ft)
Silver	9.8
Copper	10.4
Aluminum	17
Tungsten	33
Iron	60
Nickel	50
Manganin	290
Nichrome V	650

often used. A copper bar $\frac{1}{4}$ in. by 4 in. would be able to carry 1000 A. An aluminum bar $\frac{3}{8}$ in. by 4 in. would have approximately the same resistance and also be able to carry 1000 A. A flat material is usually used for this purpose because its increased rectangular surface area provides better cooling than a square or round shape.

9-5 ***Current-carrying capacity—ampacity.*** There are several factors that affect the *ampacity* of a given conductor other than its cross-sectional area. If conductors are in free air and easily capable of dissipating heat, the ampacity would be greater than if concealed in a raceway. If more than three conductors are in the same raceway, their ampacity must be reduced so that the accumulated heat from their surroundings plus the heat from the current in the conductors does not exceed the heat rating of the conductors.

The ampacities given in Tables 9-3 and 9-4 are for more or less normal

Table 9-3. Allowable ampacities of insulated conductors: single conductor in free air

AWG or MCM	Copper		Aluminum	
	60°C, 140°F R, T, TW	75°C, 167°F RH, RHW, TH, THW	60°C, 140°F R, T, TW	75°C, 167°F RH, RHW, TH, THW
14	20	20	Not	Used
12	25	25	20	20
10	40	40	30	30
8	55	65	45	55
6	80	95	60	75
4	105	125	80	100
3	120	145	95	115
2	140	170	110	135
1	165	195	130	155
0	195	230	150	180
00	225	265	175	210
000	260	310	200	240
0000	300	360	230	280
250	340	405	265	315
300	375	445	290	350
350	420	505	330	395
400	455	545	355	425
500	515	620	405	485
600	575	690	455	545
700	630	755	500	595
750	655	785	515	620
800	680	815	535	645
900	730	870	580	700
1000	780	935	625	750

Table 9-4. Allowable ampacities of insulated conductors; not more than three conductors in raceway or cable; based on ambient temperature of 30°C. (86°F)

AWG or MCM	Copper 60°C, 140°F R, T, TW	Copper 75°C, 167°F RH, RHW, TH, THW	Aluminum 75°C, 140°F R, T, TW	Aluminum 75°C, 167°F RH, RHW, TH, THW
14	15	15	Not	Used
12	20	20	15	15
10	30	30	25	25
8	40	45	30	40
6	55	65	40	50
4	70	85	55	65
3	80	100	65	75
2	95	115	75	90
1	110	130	85	100
0	125	150	100	120
00	145	175	115	135
000	165	200	130	155
0000	195	230	155	180
250	215	255	170	205
300	240	285	190	230
350	260	310	210	250
400	280	335	225	270
500	320	380	260	310
600	355	420	285	340
700	385	460	310	375
750	400	475	320	385
800	410	490	330	395
900	435	520	355	425
1000	455	545	375	445

Table 9-5. Correction factors for ambient temperatures over 30°C. 86°F for conductors approved for 60, and 75°C

Ambient Temperature C°	Ambient Temperature F°	Correction Factor for Conductor 60°C, 140°F	Correction Factor for Conductor 75°C, 167°F
40	104	0.82	0.88
45	113	0.71	0.82
50	122	0.58	0.75
55	131	0.41	0.67
60	140	—	0.58
70	158	—	0.35

conditions, when the ambient temperature is not more than 30°C or 86°F. When conductors are consistently subjected to temperatures above this value, Table 9-5 gives a factor by which the ampacity of a given conductor must be multiplied so that its ampacity will be reduced. This will limit the heat generated by the current to a safe level.

Example ***9-1.*** What is the ampacity of No. 6 copper conductors, three in one raceway, with RH insulation? The installation is in a boiler room where the temperature might reach 113°F.

Solution

From Table 9-4 the ampacity is 65 A for three conductors in a raceway. From Table 9-5 the derating factor due to higher than normal temperature is 0.82. Therefore, ampacity is

$$(65)(0.82) = 53.2 \text{ A}$$

Example ***9-2.*** A certain load is rated at 200 A. What size copper conductors are necessary if three RHW conductors are enclosed in one raceway and the ambient temperature is 135°F?

Solution

From Table 9-5 the derating factor for 135° when using RHW conductors is 0.58. Therefore, ampacity must be increased to 200 A/0.58 = 344 A. From Table 9-4 an ampacity of 344 A requires three 500 MCM conductors in a raceway.

9-6 ***Overcurrent protection.*** The proper design of any electrical system must include safeguards to assure that the ampacity of each conductor is not exceeded. This excess of current can range from a small overload to a complete short circuit (load resistance zero). The latter is usually an accidental situation where conductors, carrying the opposite polarities of the supply, come together. This excessive current is called "fault current."

Since it is the heat generated in the conductors that must be kept within limits, it must be remembered that the electric power producing heat is I^2R. This means that if the current in a conductor is permitted to be *double* its ampacity, the heat in the conductor is *four times* its normal rating. If short-circuit conditions were allowed to persist, even for a few seconds, the conductor would become hot enough to melt, cause insulation to burn, and set fire to any surrounding material. The purpose of overcurrent protection is to provide assurance that the circuit will be *interrupted before* any excess current can cause any damage, either to the conductors themselves or the electrical load they supply.

9-7 *Fuses.* There are basically two classifications of devices that accomplish this overcurrent protection function, with several modifications in each type: *fuses* and *circuit breakers*. Fuses are used very extensively, especially in small current ratings. Fuses are defined as overcurrent devices that destroy themselves when they interrupt the circuit. They are made of a low-melting temperature metal and are so calibrated that they melt at a specific current rating. Since they are connected in series with the load, they open the circuit when they melt. All fuses have an *inverse time* characteristic. A fuse rated at 30 A should carry 30 A continuously, but with about 10% overload it would melt in a few minutes and with 20% overload it would melt in less than a minute. A 100% overload would require only a fraction of a second to cause the element to melt and open (or clear) the circuit. Some fuses have a higher time lag so that overloads of short duration, say several seconds, will not cause the fuse to "blow".

Fuses are physically made in several sizes so that a larger capacity fuse cannot be interchanged with a smaller one and vice versa. Plug fuses are made with screw shells up to 30 A for circuits up to 125 V. Cartridge fuses have two voltage classifications, 250 and 600 V, as shown in Table 9-6.

Table 9-6. Classifications of cartridge fuses

Rating (A)	Type of Retainer	Dimensions (in.) 250 V	Dimensions (in.) 600 V
Up to 30	Ferrule	9/16 × 2	13/16 × 5
35–60	Ferrule	13/16 × 3	1 1/16 × 5½
70–100	Knife blade	1 × 5⅞	1½ × 7⅞
110–200	Knife blade	1½ × 7⅝	1¾ × 9⅝
225–400	Knife blade	2 × 8⅝	2½ × 11⅝
450–600	Knife blade	2½ × 10⅜	3 × 13⅜

9-8 *Circuit breakers.* A *circuit breaker* is a device that will interrupt a circuit without injury to itself so that it can be reset and reused over again. Unlike fuses, circuit breakers are not expendable. Circuit breakers may be operated by magnetic means so that they will respond quickly. Other types are thermal in operation and may require a short time interval to operate, depending on the overload current. Circuit breakers may have two or three poles on the same assembly so that an overload on one pole or line will disconnect all conductors (lines) of the power supply (with the exception of the neutral, which is studied further in Chapter 10).

The need for overload circuit breakers of different capacities has caused several manufacturers to offer breakers of different current ratings so that they can be installed in a pressembled cabinet. These, unlike most other wiring devices, are not interchangeable among different manufacturers.

9-9 *Voltage drop calculations and limitations.* In addition to the necessity of preventing conductors from overheating, a problem of voltage drop must be considered when conductors are selected. If conductors are heated by the current they must carry, this would be a power loss. There also must be a drop in voltage along the conductor, between the supply and the load. Under these conditions the system is a series circuit, with the resistance of the conductors in series with the load.

The quality performance of many electrical devices is dependent on the voltage they receive. For example, lighting systems that operate at 95% of their rated voltage (5% drop) produce only 90% of their rated output. This can be shown by the relationship of

$$P = \frac{V^2}{R} = (0.95)^2 = 0.9 \qquad (3\text{-}5)$$

The selection of conductors of cross-sectional area sufficient to keep their resistances within proper limits will limit voltage drop.

The value of the permissible conductor resistance can be easily calculated. The conductor size then can be selected from Table 9-7. This table gives the resistance of copper and aluminum conductors in ohms per thousand feet.

Example 9-3. What size copper conductors are necessary to supply a 150-A load 200 ft from its 115-V power source? Limit the voltage drop to 2% $(0.02 \times 115) = 2.3$ V).

Solution

See Fig. 9-1; R is the resistance of 400 ft of wire. The voltage across R must be limited to 2.3 V. The voltage across the load then will be 112.7 V.

$$R = \frac{V}{I} = \frac{2.3 \text{ V}}{150 \text{ A}} = 0.0153\ \Omega \text{ for 400 ft}$$

This represents the resistance of 400 ft of wire. Since the table uses 1000 ft as a basis for its resistance values, it is necessary to convert this value to the equivalent resistance of 1000 ft

$$R \text{ of 1000 ft } (0.0153) \times \frac{1000 \text{ ft}}{400 \text{ ft}} = 0.083\ \Omega$$

Table 9-7. Properties of conductors

Size AWG or MCM	Area (CM)	Solid or Stranded Number of Wires	Solid or Stranded Diameter of Each Wire (in.)	DC Resistance (Ω/1000 ft) 25°C 77°F Copper	DC Resistance (Ω/1000 ft) 25°C 77°F Aluminum
18	1,620	Solid	0.0403	6.51	10.7
16	2,580	Solid	0.0508	4.10	6.72
14	4,110	Solid	0.0641	2.57	4.22
12	6,530	Solid	0.0808	1.62	2.66
10	10,380	Solid	0.1019	1.018	1.67
8	16,510	Solid	0.1285	0.6404	1.05
6	26,240	7	0.0612	0.410	0.674
4	41,740	7	0.0772	0.259	0.424
3	52,620	7	0.0867	0.205	0.336
2	66,360	7	0.0974	0.162	0.266
1	83,690	19	0.0664	0.129	0.211
0	105,600	19	0.0745	0.102	0.168
00	133,100	19	0.0837	0.0811	0.133
000	167,800	19	0.0940	0.0642	0.105
0000	211,600	19	0.1055	0.0509	0.0836
250	250,000	37	0.0822	0.0431	0.0708
300	300,000	37	0.0900	0.0360	0.0590
350	350,000	37	0.0973	0.0308	0.0505
400	400,000	37	0.1040	0.0270	0.0442
500	500,000	37	0.1162	0.0216	0.0354
600	600,000	61	0.0992	0.0180	0.0295
700	700,000	61	0.1071	0.0154	0.0253
750	750,000	61	0.1109	0.0144	0.0236
800	800,000	61	0.1145	0.0135	0.0221
900	900,000	61	0.1215	0.0120	0.0197
1000	1,000,000	61	0.1280	0.0108	0.0177

The 0.0383 Ω now represents the resistance of 1000 ft of wire, which if used in this problem would keep the voltage drop within the specified limits of 2.3 V.

The nearest size wire to this value listed in Table 9-7 is 300 MCM.

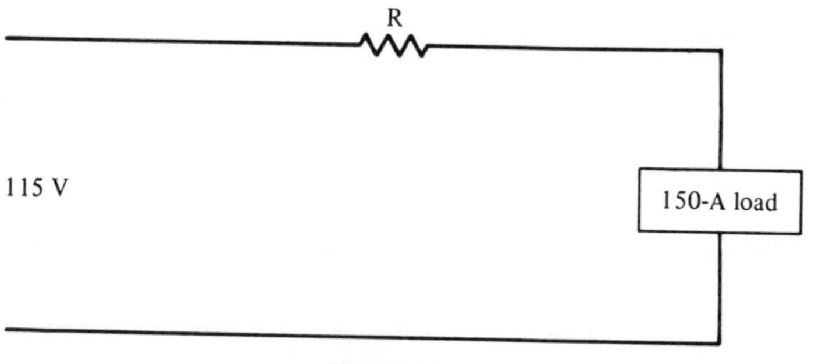

Figure 9-1

This answer must always be checked in the ampacity table to be sure that we have selected a wire large enough for 150 A. *This is most important.* Considering *both* the ampacity tables and the voltage drop, the *largest* conductor is the one to be selected.

There is a more direct method that will give the circular mil area directly from an equation. We simply include all factors that determine the size wire for any load situation.

$$\text{cmil} = \frac{\rho I D^*}{(V_d)} \qquad (9\text{-}1)$$

where cmil is the cross-sectional area in circular mils; p is the resistivity of the material used in ohms per circular-mil-foot; I is the current required by the load in amperes; and D is the distance from the supply to the load, in feet. This distance must be multiplied by two for single-phase, two- or three-wire systems. The neutral conductor of a three-wire, single-phase system is not included in Eq. (9-1) since with a balanced load it would carry no current and, therefore, contribute no voltage drop. If neutral did carry current under unbalanced load conditions, there would be less current in it than in one of the other conductors. When we design a three-wire system it is intended to be balanced when the total load is utilized. If it is unbalanced, some of the load is not in use, resulting in a lower voltage drop. In three-phase systems, D must be multiplied by $\sqrt{3}$ or 1.732 to determine the feeder size in circular mils. V_d in Eq. (9-1) is the allowable voltage drop in volts. Example 9-4 illustrates Eq. (9-1), used to solve Ex. 9-3.

Example **9-4.** What size copper conductors are necessary to supply a 150-A load 200 ft from its 115-V, single-phase power source? Limit the voltage drop to 2%.

Solution

$$\text{cmil} = \frac{\rho I 2D}{V_d} = \frac{(10.4)(150)(2)(200)}{(115)(0.02)} \qquad (9\text{-}1)$$

$$= 272{,}000 \quad \text{or } 272 \text{ MCM}$$

The nearest next largest standard size is 300 MCM. Compare this solution and answer with Ex. 9-3.

---

* 2 D for single-phase; $\sqrt{3}$ D for three-phase.

Example **9-5.** What size copper conductors are necessary to supply a 75-kW, 3-phase, 230-V load that has a power factor of 0.85? Allow a 2% voltage drop. The distance to the load is 200 ft.

Solution

$$I = \frac{P}{E\sqrt{3}\ \text{pf}} = \frac{75{,}000}{(230)(\sqrt{3})(0.85)} \qquad (5\text{-}4)$$

$$= 221 \text{ A}$$

This current is calculated for a voltage of 230 V that introduces a slight error since the load voltage will be slightly below this value. However, this is a usual procedure and the error is slight.

$$\text{cmil} = \frac{(\rho ID)(\sqrt{3})}{V_d} = \frac{(10.4)(221)(200)\sqrt{3}}{(230)(0.02)} \qquad (9\text{-}1)$$

$$= 172{,}000 \quad \text{nearest size is 0000 AWG}$$

This answer always must be checked as to its ampacity according to Table 9-3 or 9-4 with proper consideration given to the temperature rating of the insulation.

It is difficult to make a blanket statement as to the proper voltage drop to permit. This is usually a part of the design specification. In the previous examples, if 1% were specified it obviously would make the required conductors twice as large. It is debatable if this added expense could be justified in permitting such a small voltage drop and less wasted power.

The NEC permits a voltage drop of 5%, although this includes the entire length of the system to the last outlet. This is discussed in more detail in Chapter 11.

9-10 ***Skin effect on conductors.*** The table of conductor sizes and resistances makes reference to dc resistance. This may seem odd since alternating current is practically universal. However, when alternating current flows in a conductor it has a tendency to flow toward the outside of the conductor rather than uniformly throughout its cross section. This is called "skin effect," and it increases as the frequency of the ac current flow is increased. Its effect, however, is slight in small conductors but must be considered in very large conductors. To compensate for this, we must multiply the dc resistance by the factor in Table 9-8.

It is noted from this table that for sizes smaller than No. 00 AWG, skin effect requires little consideration, but for large sizes it is important,

Table 9-8. Multiplying factors for converting dc resistance to 60-Hz ac resistance (skin effect)

Size AWG or MCM	For Nonmetallic Sheathed Cables or in Air		For Metallic Sheathed Cables or in Metallic Raceways	
	Copper	Aluminum	Copper	Aluminum
Up to 3	1	1	1	1
2	1	1	1.01	1
1	1	1	1.01	1
00	1.001	1	1.02	1
00	1.001	1.001	1.03	1
000	1.002	1.001	1.04	1.01
0000	1.004	1.002	1.05	1.01
250	1.005	1.002	1.06	1.02
300	1.006	1.003	1.07	1.02
350	1.009	1.004	1.08	1.03
400	1 011	1.005	1.10	1.04
500	1.018	1.007	1.13	1.06
600	1.025	1.010	1.16	1.08
700	1.034	1.013	1.19	1.11
750	1.039	1.017	1.21	1.12
800	1.044	1.017	1.22	1.14
1000	1.067	1.026	1.30	1.19

especially when conductors are enclosed in metal raceways. The skin effect on copper conductors is about twice as much as on aluminum conductors.

Example 9-6. What is the resistance of 800 ft of 500 MCM copper wire if it is used to supply a 60-Hz ac load? The conductors are to be enclosed in a metal raceway.

Solution

From Table 9-7, the dc resistance of 500 MCM copper wire is 0.0216 Ω per 1000 ft.

$$\text{Resistance of 800 ft} = (0.0216)\frac{800}{1000} = 0.0173\ \Omega$$

From Table 9-8, the multiplying factor to compensate for the skin effect is 1.13 or an increase in resistance of 13%.

$$\text{Resistance} = (0.0173)(1.13) = 0.0195\ \Omega$$

9-11 ***Function of raceways.*** The role of conductors in an electrical system has been thoroughly explored and defined. It should be quite obvious that these conductors cannot be installed without giving consideration to mechanical protection. The NEC states that conductors must be enclosed or protected and provides several methods by which this can be assured.

A *raceway* is defined as a *channel* for holding wires, cables, or bus bars. The channel may be (1) in the form of a pipe called *conduit* in electrical language, (2) a thinner wall conduit called *electrical metallic tubing*, or (3) a square sheet metal *duct* of which one side has a removable cover. We shall investigate in Chapter 10 how the electrical current characteristics of these raceways serve another useful and necessary function. All raceways are mechanically installed as a complete system with all necessary outlet boxes and fittings. Afterward the conductors are installed (pulled through) the raceway.

Cables are defined as complete assemblies consisting of conductors and raceways *as a unit*. The raceway is actually a covering that may be either metallic or nonmetallic. The NEC has approved several wiring methods. The type of raceway employed classifies each method. A brief description of the major methods follows (Secs. 9-12 and 9-13).

9-12 ***Metal-clad cable.*** The code permits the manufacture and use of two types of metal-clad cable. MC is used for large loads since its smallest size is No. 4 AWG copper or No. 2 AWG aluminum. AC is used for smaller loads and is available in sizes as small as No. 14 AWG. The metal covering for these cables is a steel spiral wrapping that forms a flexible raceway. It is manufactured as a complete assembly with the conductors installed.

AC is commonly known as *BX cable*. It cannot be buried in concrete or used in a damp or wet location. It is available as either a two- or three-conductor assembly. Two-conductor cables have black and white conductors. The third conductor in a three-conductor assembly is red.

Where AC cable terminates, the conductors must be protected from the sharp edges of the armor by proper bushings and fittings.

A cable with lead-covered conductors is available for wet locations. Its type letters are ACL.

9-13 ***Nonmetallic sheathed cable.*** This type of cable has a nonmetallic covering of fabric or plastic. When required, it also may have a smaller wire for grounding purposes. This wire may be uninsulated, but if it is insulated it must be green in color. This function is investigated in detail in Chapter 10. It is available in size Nos. 14 to 1 AWG in copper and Nos. 12 to 2 AWG in aluminum conductors.

Both metallic and nonmetallic cable are used quite extensively for wiring buildings of frame construction, whereby the cable is concealed in the hollow spaces in walls and ceilings. A complete line of boxes and connectors are made to accommodate each wiring method.

Because of NEC specifications, all cables, boxes, and various fittings are interchangeable among all manufacturers.

Nonmetallic sheathed cable is called "Romex" and also is available as a two- or three conductor assembly.

9-14 ***Electrical metallic tubing (EMT).*** This type of raceway, more commonly called EMT, has a thin wall that does not permit threading. Connectors and couplings are secured either by compression rings or set screws. For many years the largest size permitted has been 2 in. Recently this maximum size has been increased to 4 in. The size is in reference to the same inside diameter as that of comparable standard conduit. EMT makes an excellent raceway for conductors. It may be buried in concrete if proper fittings are used. It cannot be used, however, where it is subject to continuous moisture.

EMT up to 1 in. size can be easily bent without reducing the inside diameter with the assistance of specially formed bending tools for each particular size.

9-15 ***Rigid conduit.*** Electrical conduit has all the outward appearances of plumber's pipe used for steam, gas, or water. Electrical conduit, however, must have a smooth, enameled interior to facilitate the pulling and installation of wires. It is made in 10-ft lengths (as in EMT). Connections to boxes are made with locknuts and bushings after conduit is threaded. Conduit may be used in the most severe cases where the possibility of mechanical injury or the presence of moisture presents a problem.

9-16 ***Wireways.*** Wireways are sheet-metal troughs with removable covers. They cannot be concealed, but are very useful in maintaining a complete raceway system when many devices must be interconnected in a limited area. Wireways have numerous knockouts so that other wiring methods can be extended at any time. Wireways are also available for various types of raceways designed for specific applications. One system of underfloor raceways used in concrete floors provides a flexible system of wiring outlets at any time. Figure 9-2 illustrates this concealed, versatile raceway system.

Another system has a fabricated assembly of duct and bus bars with openings so that a switch may be plugged into the bus bars at any point throughout its length. This gives versatility to factory installations. Figure 9-3 illustrates this exposed system.

Some manufacturers make available a complete line of fittings and raceways designed for surface work and somewhat decorative in appearance.

Fig. 9-2. Underfloor raceway. *Courtesy Square D Company.*

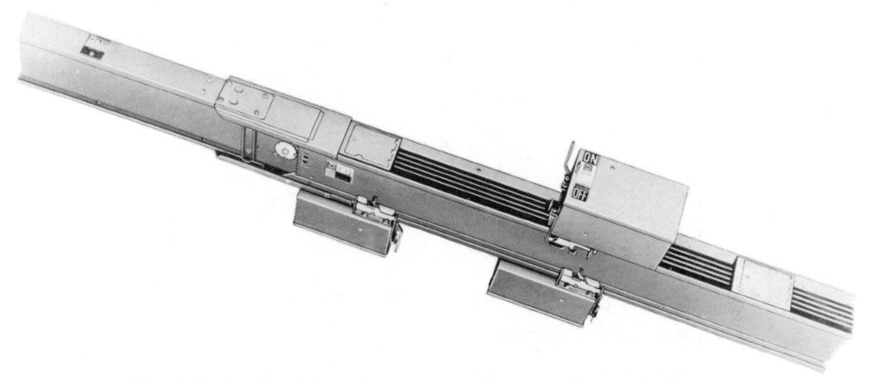

Fig. 9-3. Overhead bus duct system. *Courtesy Square D Company.*

These are called "surface extensions." They are not designed specifically to wire a complete building but for adding extensions to existing circuits where outlets are needed.

9-17 ***Selection of raceways.*** A conduit or tubing system must be installed completely before conductors are inserted. This is quite logical because one of the major advantages of these raceways is the ability to replace the conductors. Therefore, if the system is installed in such a way that the original conductors can be installed, their replacement is also possible. The NEC limits the total bends between outlets or fittings (any place where conductors are accessible) to 360° or fourquarter bends.

In anticipation of larger future loads, conduits larger than necessary may be installed. Table 9-9 lists the maximum number of conductors permitted in the standard sizes of conduit or tubing for new installations.

Recently the NEC has given approval to a conductor with very thin

Table 9-9. **Maximum number of conductors in trade sizes of conduit tubing, new work**

	½ in.		¾ in.		1 in.		1 ¼ in.		1½ in.		2 in.		2½ in.		3 in.		3½ in.		4 in.		4½ in.		5 in.		6 in.	
	A	B	A	B	A	B	A	B	A	B	A	B	A	B	A	B	A	B	A	B	A	B	A	B	A	B
14	4	8	6	15	10	24	18	43	25	58	41	96	58	137	90		121		155		197					
12	3	6	5	11	8	18	15	32	21	43	34	71	50	102	76	158	103		132		168					
10	1	4	4	7	7	11	13	20	17	27	29	45	41	65	64	100	86	134	110	172	140		173			
8	1	2	3	4	4	6	7	11	10	16	17	26	25	37	38	58	52	78	67	100	85	127	105	157	152	
6	1	1	1	2	3	4	4	7	6	9	10	16	15	23	23	35	32	47	41	61	52	78	64	96	93	139
4	1	1	1	1	1	2	3	4	5	6	8	9	12	14	18	21	24	29	31	37	40	48	49	59	72	85
3			1	1	1	2	3	3	4	5	7	8	10	12	16	18	21	24	28	31	35	40	44	50	63	72
2			1	1	1	1	3	3	3	4	6	7	9	10	14	15	19	20	24	26	31	34	38	42	55	61
1			1	1	1	1	1	2	3	3	4	5	7	7	10	11	14	15	18	20	23	25	29	31	42	45
0					1	1	1	2	2	2	4	4	6	6	9	9	12	13	16	16	20	21	25	26	37	38
00					1	1	1	1	1	2	3	3	5	5	8	8	11	11	14	14	18	18	22	22	32	32
000					1	1	1	1	1	1	3	3	4	4	7	7	9	9	12	12	15	15	19	19	27	27
0000							1	1	1	1	2	2	3	3	6	6	8	8	10	10	13	13	16	16	23	23
250							1	1	1	1	1	2	3	3	5	5	6	6	8	8	11	11	13	13	19	19
300							1	1	1	1	1	1	3	3	4	4	5	5	7	7	9	9	11	11	16	16
350							1	1	1	1	1	1	1	2	3	3	5	5	6	6	8	8	10	10	15	15
400									1	1	1	1	1	2	3	3	4	4	6	6	7	7	9	9	13	13
500									1	1	1	1	1	1	3	3	4	4	5	5	6	6	8	8	11	11
600											1	1	1	1	1	2	3	3	4	4	5	5	6	6	9	9
700											1	1	1	1	1	2	3	3	3	3	4	4	6	6	8	8
750											1	1	1	1	1	1	3	3	3	3	4	4	5	5	8	8
800											1	1	1	1	1	1	2	2	3	3	4	4	5	5	7	7
900											1	1	1	1	1	1	1	2	3	3	4	4	4	4	7	7
1000											1	1	1	1	1	1	1	1	3	3	3	3	4	4	6	6

Table 9-10. Reduction of ampacities when more than three conductors are in the same raceway or cable

Number of Conductors	Per Cent of Values in Tables 9-3 and 9-4
4–6	80
7–24	70
25–42	60
more than 43	50

nylon insulation, type THWN. This permits more conductors in a particular tubing. Column A in Table 9-9 is for types R, RH, RW, T, TW, and THW. Column B gives the number of THWN conductors permitted.

When considering the maximum current for a conductor, Tables 9-3 and 9-4 are applicable only when no more than three conductors are in the same raceway or cable. Most of the design problems developed in this text seldom require us to exceed this limit of three conductors. However, frequently it would be economical to group a larger number of conductors in one raceway; therefore, the entire table as given in the NEC is reproduced in Table 9-9.

As more conductors are concentrated in one raceway it will be necessary to derate the ampacities of Tables 9-3 and 9-4 to limit the heat. Table 9-10 lists this derating factor.

Although four conductors are necessary for a feeder from a three-phase, four-wire supply, their ampacities need not be derated according to Table 9-10. At full balanced load the neutral current is zero, and therefore it contributes no heat. If it is unbalanced, the line currents decrease as neutral current increases and the heating effect does not change.

If a portion of the feeder load is fluorescent or discharges lighting this situation will change. Tests have shown that a balanced load of fluorescent ballasts connected in wye to a three-phase feeder causes a current approximately equal to the line currents to flow in the neutral. The neutral current will then not be zero; therefore, a derating is necessary.

The NEC permits the neutral of a single-phase or three-phase feeder to be reduced in size when the calculated current is more than 200 A. This reduction may be 70% of the current over 200 A if no discharge (fluorescent) lighting is supplied. No reduction is permitted for any portion of the load that consists of discharge lighting.

PROBLEMS

9-1 What size RH copper conductor is necessary to carry each of the following currents if no more than three conductors are in the same raceway: (a) 235 A, (b) 92 A, and (c) 160 A?

9-2 What size TW aluminum conductor is necessary to carry each of the following currents if no more than three conductors are in the same raceway: (a) 350 A, (b) 125 A, and (c) 70 A?

9-3 What size RH copper conductor is necessary to carry the following currents, if no more than three conductors are in the same raceway but are subject to an ambient temperature of 120° F: (a) 220A; (b) 75 A; (c) load of 100 kW at 230 V, single phase, and unity pf; (d) load of 100 kW at 230 V, three phase, and unity pf; and (e) load of 100 kW at 230 V, three phase, and with pf of 0.7?

9-4 If a particular load requires 300 MCM copper conductors, what size aluminum conductors could be used so as not to increase the voltage drop?

9-5 What is the resistance of 400 ft of No. 14 AWG copper wire?

9-6 Calculate the resistance of 1200 ft of 300 MCM aluminum wire.

9-7 Calculate the resistance of $\frac{1}{2}$ mi of 750 MCM aluminum wire.

9-8 A light-duty copper extension cord is made of No. 18 AWG conductors and has a resistance of 6.5 Ω/1000 ft. If a portable saw requires 10 A and 115 V, what is the longest cord (two wires) that can be used if the supply voltage is 120 V? The voltage at the saw must not go below 115 V.

9-9 Calculate the minimum size aluminum-type RH conductors that could be used for each of the following loads (not over three conductors in one raceway): (a) 350 A; (b) 240 A; (c) 150 A; (d) 50 kW at unity pf, 230 V, single phase, and two wire, (e) 50 kW at unity pf, 230 V, single phase, and three wire, and (f) 50 kW at unity pf, 230 V, and three phase.

9-10 Calculate the minimum size copper type T conductors that can be used for the following loads (no more than three conductors in a raceway): (a) 75 kVA, 230 V, and three phase; (b) 75 kW at a pf of 0.8, 230 V, and three phase; (c) 75 kVA, 460 V, and three phase; and (d) 75 kW at a pf of 0.8, 460 V, and three phase.

9-11 (a) An overhead line supplying a building is to be of copper conductors with an insulation rated at 60°C. If the distance does not contribute a voltage drop problem, what size conductors are necessary if the maximum load is 200 A at 230 V and single phase?
(b) If the distance is 300 ft, what size conductors are necessary if the voltage drop is not to exceed 3 V?

9-12 What size aluminum conductors will replace a copper conductor that is carrying 150 A and not cause a larger voltage drop? Use RH conductors in each case.

9-13 (a) What is the longest distance a 20-A branch circuit using No. 12 AWG copper conductors can extend and not lose more than 2% of the 115-V supply voltage if the current that must be carried is 16 A? Assume the load is concentrated at the end of the circuit.
(b) How far can a circuit extend if the conductors are aluminum and not exceed a 2% voltage drop?

9-14 (a) What is the longest distance a 20-A branch circuit using No. 12 AWG copper conductors can extend if the supply voltage is 277 V? Allow a 2% voltage drop.
(b) How far can a circuit extend if the conductors are aluminum, and not exceed a 2% voltage drop?

9-15 What size conduits are necessary for each of the following groups of conductors:
(a) three 0000 AWG type RH, (b) two No. 10 AWG type TW,
(c) three 350 MCM type RH, (d) ten No. 4 AWG type TW,
(e) six 250 MCM type RHW, (f) thirty No. 14 type THWN,
(g) fifty No. 12 type THWN, and
(h) one hundred No. 14 type THWN?

9-16 (a) If six 250 MCM copper conductors type RH were to be installed in the same conduit, what current would they be permitted to carry?
(b) What size conduit is necessary?
(c) If two raceways are used with three conductors in each, what size is necessary?
(d) If the supply is 230 V and three phase, what is the maximum power the system can deliver to a load in (a) and (c)? Assume unity pf.

9-17 What is the resistance of 1000 ft of 750 MCM copper conductors if used to carry 60 Hz ac when installed in a metal raceway?

9-18 What is the resistance of 400 ft of 300 MCM copper conductor if installed in a metal raceway when used to carry a 60-Hz current?

9-19 If a group of No. 4 AWG THW copper conductors are to be exposed to an ambient temperature of 130°F, what current would they be permitted to carry? Not over three conductors are in a raceway.

9-20 The wiring on a large boiler is to be exposed to a constant ambient temperature of 110°F. To what current should each of the following conductors be limited (not more than three conductors in a raceway): (a) No. 12 AWG copper type T, (b) No. 8 AWG copper type THW, and (c) No. 10 AWG aluminum type THW?

9-21 A group of twenty No. 12 type THW copper conductors are to be installed in one conduit.

(a) What current may each conductor carry?
(b) What size conduit shall be used?
(c) What size conduit is necessary if nylon-insulated type THWN conductors are used?

9-22 For control wiring it is necessary to install fifty No. 14 AWG type THW conductors in one raceway.
(a) What is the maximum current each conductor may carry?
(b) What size conduit is necessary?
(c) If type THWN conductors are used, what size conduit is necessary?

Chapter Ten

Grounding of Systems and Equipment

10-1 ***Grounding defined.*** The *grounding* of electrical systems is one of the most important details that must be understood by those responsible for electrical installations. By the term "grounding" we mean being connected to the earth. (In fact the British use the expression "earth" instead of ground.) The implication of connecting something to the earth is somewhat nebulous, however, because how do we secure a wire to mother earth? A better term such as a "common connection" might better illustrate the principle in question. Then if we can somehow and somewhere be sure the common connection is affixed to the ground, the mission is accomplished.

The wiring system of an automobile is a good example of how loosely the term "grounding" is sometimes used. All electrical devices in this case, such as lights, motors, or radio, have one terminal connected directly to the battery and the other to the metal frame or body. Of course one terminal of the battery also must be connected to the frame so that current from all the devices can get back to the power source. This common connection, the body of the vehicle, is sometimes called a *ground.* A car is really not grounded at all since 4 or 5 in. of rubber keep the vehicle from touching the ground.

We now have the option to connect one wire from the car body or frame to something that is grounded. This would effectively ground one terminal of every electrical device in the vehicle. Obviously, the actual grounding of the automobile body is hardly practical. An exception to this might be the braided strap some motorists use to reduce static electrical shocks when stepping from the vehicle.

We now can ascertain that the grounding of the vehicle does not have any effect on the operation of the car's electrical equipment. This is one important fact about the grounding of electrical supplies: Under normal conditions the electrical load will function properly whether or not one terminal is grounded.

Figure 10-1a shows the current flow through the various devices in an automobile. Note the common connection of all devices to the negative terminal of the battery. This would be called a *negative ground installation*, but actually nothing is grounded. The term "ground" alludes to the common terminal.

Figure 10-1b shows a typical electrical load with current flowing from a supply with one terminal grounded. The only difference between the two

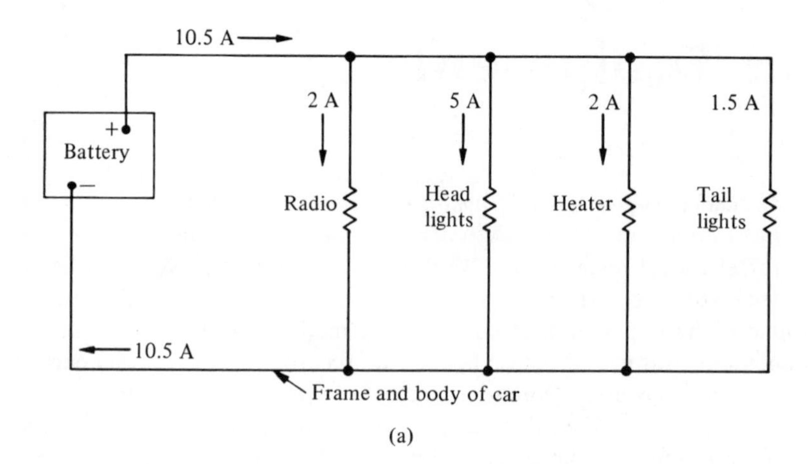

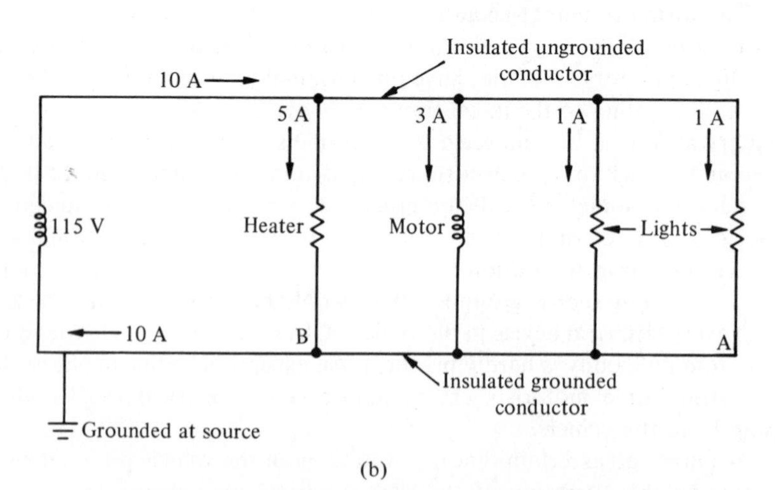

Fig. 10-1. (a) Current flow in an automobile with car body carrying current back to source. (b) Typical grounded system with an insulated conductor carrying current back to source.

systems is that the automobile wiring uses conductors for only the positive terminal of the system. The circuit of Fig. 10-1b has insulated conductors for both terminals of the ac supply. This is a very important NEC rule: Insulated conductors must be used to carry current to a device and also to the return route back to the supply.

To clarify the path for current flow in this chapter, we shall show the power supply as it actually exists in most installations. In this case a 115-V supply will be one coil of a transformer and a 230-V supply will be two of these coils in series. The midpoint is the neutral of the three-wire system. This arrangement enables us to follow the path of the current from the supply to the load and back to the supply. In Sec. 3-3 this basic fact of current flow was emphasized. If current cannot get back to the source it will not leave.

10-2 ***Criteria for system grounding.*** Affixing a connection from one terminal of a power supply to a good low-resistance path to ground is called a *system ground.* Obviously, this means that the system is grounded. The next logical question is "Are all systems grounded?" This is where the National Electric Code leaves the issue somewhat in doubt.

First we are told that a system *must* be grounded if, after grounding, the voltage to ground is less than 150 V. "The voltage to ground" means the potential from any point in the system to any other object that itself is grounded. In buildings these grounded objects would be any piping for water, gas, or sewage, since they eventually are connected to other pipes buried in moist earth. Such buried pipes become the best way to obtain a grounded object.

The steel frames of buildings and any metal objects that are secured to, or even touch, the metal frames, also are grounded. Concrete floors are not necessarily conductive. When moist and in contact with earth, a concrete floor could be considered to be a conductive path to ground. Figure 10-2 illustrates the voltage-to-ground principle.

The 115–230-V, three-wire power supply is generally used when a single-phase source is required. The choice of which of the three conductors is to be grounded must be the neutral. Should we ground one of the other conductors, the voltage to ground would be 230 V. This would be an NEC violation. Figure 10-3 illustrates this situation.

When three-phase systems are installed in buildings the question of grounding must be considered. Section 5-5 shows how a four-wire, three-phase, 120–208-V supply is obtained. The neutral conductor must be grounded. The voltage to ground from any of the other three conductors is 120 V.

When single-phase loads are connected from any line wire to neutral (there will be three such possibilities), each single-phase load is supplied with a grounded conductor. This is an NEC requirement. All single-phase systems

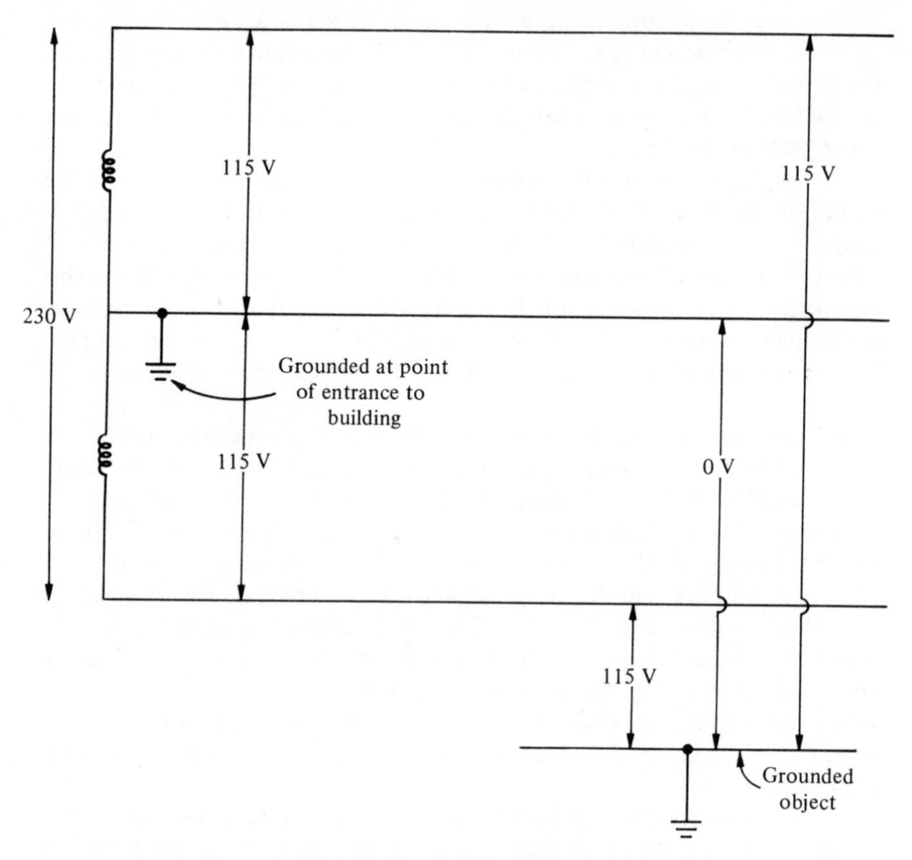

Fig. 10-2. Voltages to ground in a grounded single-phase, three-wire system.

supplying lighting circuits must be grounded. Figure 10-4 shows how this is accomplished.

Next let us consider the 230-V, three-wire, three-phase system. This would not be used for lighting but for power. The NEC does not require this power supply to be grounded. It does *recommend* that such a system be grounded if the voltage to ground can be less than 300 V. The grounding of any one conductor of this system would fulfill this recommendation. The NEC also states that higher voltage systems may be grounded. This may be interpreted to imply that 460-V, three-phase systems may be grounded. Since these are advisory rather than mandatory rules, the question of grounding 230-or 460-V, three-phase supplies depends on the discretion of the local utility company.

Let us summarize the question of when to ground or not to ground one conductor of a system. If the voltage to any other grounded object can

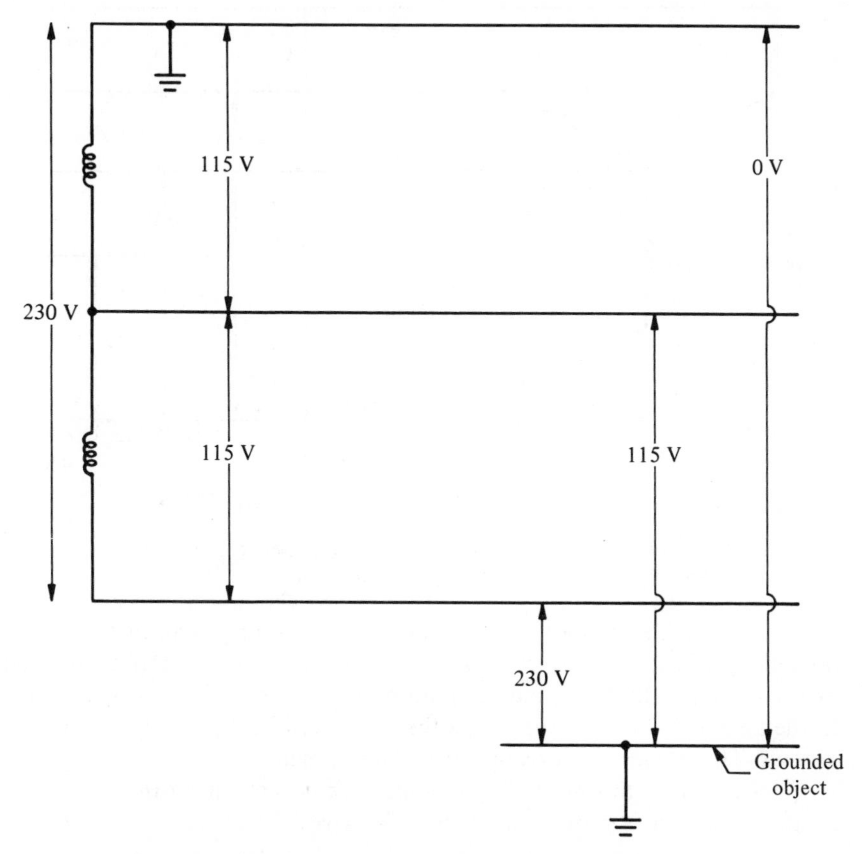

Fig. 10-3. Voltages to ground in a three-wire, single-phase system when neutral is not grounded—a code violation.

be kept under 150 V, then we definitely must ground the system. This includes all domestic electrical systems and also any system that uses incandescent lamps.

Any other system of a higher voltage would be power circuits for operating motors or industrial equipment. The grounding of these supplies is up to the discretion of the local authorities.

The principal reason for system grounding is to limit the voltage to ground inside the building from rising above a safe value because of a fault outside the building. An example of this might be charges if outside wires were struck by lightning. The ground acts like a lightning rod. For this reason the NEC requires the grounding to be accomplished immediately at the point of entrance of the service conductors. Also, a faulty transformer could cause voltages to rise inside the building if one conductor of the system were not grounded.

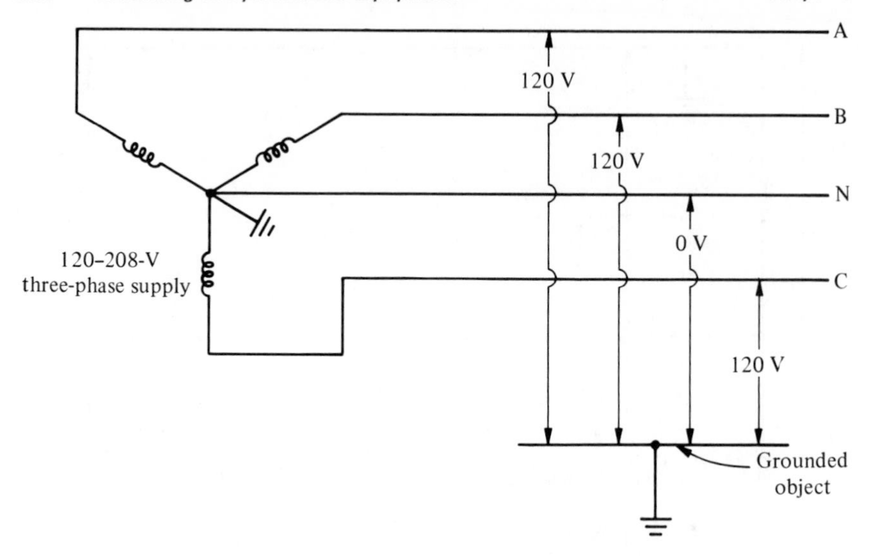

Fig. 10-4. A three-phase system supplying power to three single-phase loads. Each load has a grounded conductor.

The method employed to ground a power supply is to install a conductor of sufficient size from the main service switch to a suitable grounded object. This usually is an underground piping system such as a water pipe. In the absence of such a water pipe the NEC permits the use of an artificial ground. This could be a driven rod or buried plates.

The size of this grounding conductor is dependent on the size of the conductor that it is grounding. Table 10-1 gives these sizes.

Table 10-1. Service and grounding conductors for grounded systems

Size of Service Conductor		Size of Grounding Conductor	
Copper	Aluminum	Copper	Aluminum
2 or smaller	0 or smaller	8	6
1 or 0	00 or 000	6	4
00 or 000	0000 or 250 MCM	4	2
Over 000 to 350 MCM	Over 250 MCM to 500 MCM	2	0
Over 350 MCM to 600 MCM	Over 500 MCM to 900 MCM	0	000
Over 600 MCM to 1100 MCM	Over 900 MCM to 1750 MCM	00	0000
Over 1100 MCM	Over 1750 MCM	000	250 MCM

10-3 ***Definition of voltage to ground.*** Throughout our studies of interior wiring systems the term "voltage to ground" will be referred to frequently. The NEC states that if a system is to be grounded, the voltage to ground must be the lowest possible potential. This is one of the principal reasons for grounding the neutral of a three- or four-wire system.

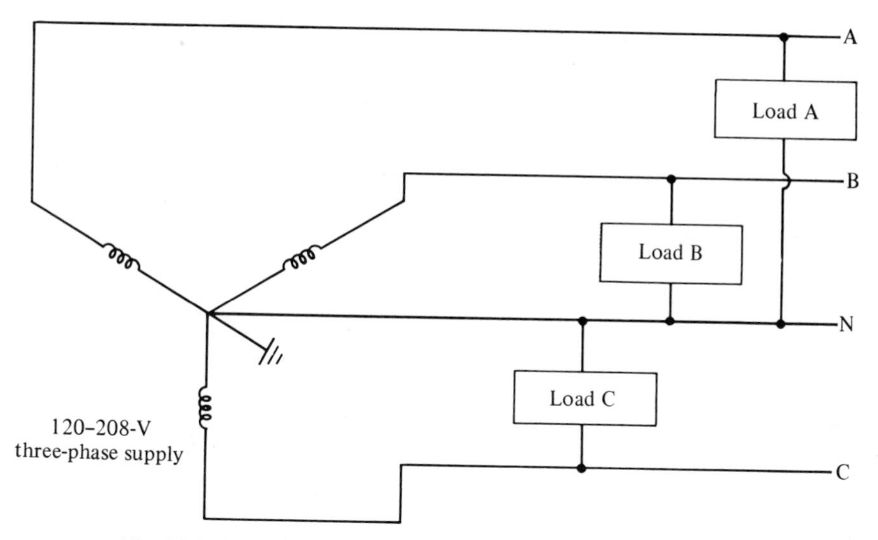

Fig. 10-5. Voltages to ground in a four-wire, three-phase system.

After a system is grounded, the voltage to ground from any ungrounded conductor must be the same as the voltage to the grounded conductor. Figure 10-2 illustrates this for a 115–230-V, single-phase system and Fig. 10-5 for a 120–208-V, three-phase system.

Let us now consider a 230-V, three-phase, three-wire system. This power supply has no neutral; therefore, the question of grounding is optional. Assuming the system is not grounded, what is the voltage to ground?

According to Fig. 10-6, the voltage to ground should be zero from any conductor. There should be no measurable voltage to any grounded object from any conductor of the system since there is no electrical connection from any system conductor to ground. It appears then that in an ungrounded system the voltage to ground would be zero. However, this will not be true. As an electrical system becomes extensive in terms of conductors and raceways, the alternating current in the conductors produces a capacitive effect. This results in a voltage generation between the conductors and the raceways.

Since we shall note later that the raceways themselves are grounded, this induced voltage actually is a "voltage to ground." It should be noted, however, that this induced voltage is a static voltage, and if a load such as a group of lamps were connected between any conductor and ground very little current could flow. This is because of the very high impedance between the conductors and the raceways. This is a situation whereby a voltage supply is not a good current supply.

As an electrical system becomes more involved with many loads, such as lighting units, heaters, and windings of motors, it becomes easier for an

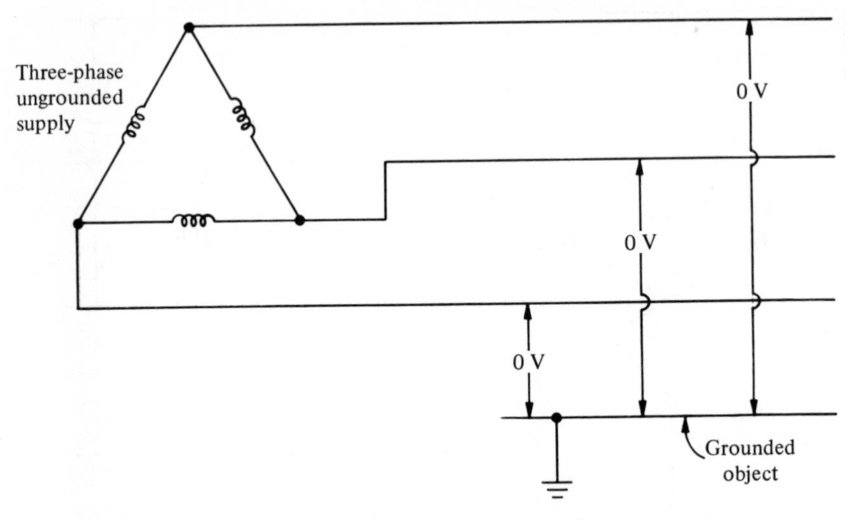

Fig. 10-6. Ungrounded system showing possibility of no voltage between its conductors and ground.

unintentional ground to occur. Sometimes excessive moisture in a motor winding can cause an inadvertent ground.

Electrically, this would appear to be the same as if the system were grounded. Now the voltage to ground might be the same value as the maximum voltage between conductors. For this reason the NEC states that the voltage to ground in an ungrounded system must be considered to be the maximum voltage of the system. It is our responsibility then to make certain that an ungrounded system is kept clear of unintentional grounds. How this is done is discussed later in this chapter.

In the event of an accidental ground the system would still function normally. If two of these accidental grounds should occur on two different

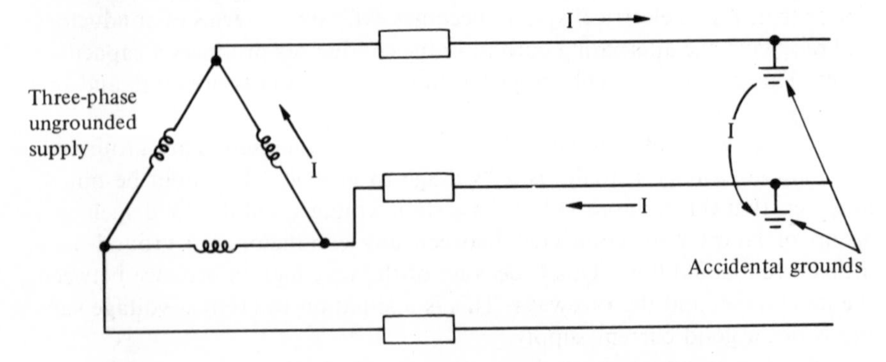

Fig. 10-7. Two accidental grounds causing a short circuit in an ungrounded system.

conductors of the system, current then would flow between these two points. If the resistance of these grounds is low enough, current would flow to cause overcurrent devices to open the circuit. Figure 10-7 illustrates this possibility.

10-4 ***Identification of grounded conductors.*** Lighting systems operating from a single-phase, 115–230-V, three-wire supply always must have one conductor of the circuit grounded. The principal reason for this NEC requirement soon becomes obvious: One terminal of an incandescent lamp is the screw shell. Should this screw shell be connected to an ungrounded conductor, a voltage of 115 V would then exist between the shell and any grounded object. If a person touching or standing on a grounded object should insert or remove an incandescent lamp and simultaneously touch this screw shell he would be exposed to a shock hazard. Under some conditions this could be fatal. Figure 10-8 illustrates this voltage between ground and the screw shell should it be incorrectly connected to the ungrounded conductor of the system.

It is of extreme importance that all screw shells of lighting systems be grounded by being connected to the grounded conductor of the power supply. This is now the "common connection" mentioned in Sec. 10-1. It must be noted, however, that an insulated conductor must be used. This grounded conductor can be grounded at only one point: where the power supply conductors enter the building.

Any further contact with any grounded object after this initial grounding would be multiple grounding. This is forbidden by the NEC, and the resultant hazards are discussed later in this chapter.

To properly identify this grounded conductor throughout the system,

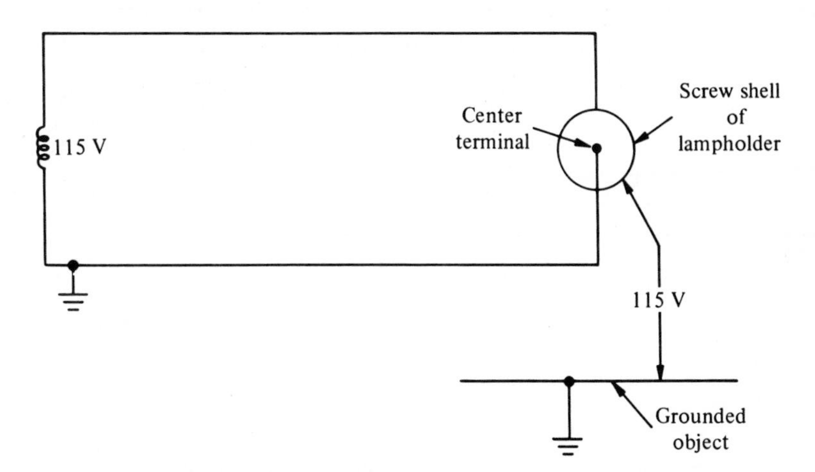

Fig. 10-8. Shock hazard when screw shell of lampholder is connected to ungrounded conductor.

the NEC requires that it have a white or natural gray outer covering. For conductors larger than No. 6 AWG this identification need be only at the terminal connection. This may be done by painting or using white tape. Once the grounded conductor is identified by a white covering, any ungrounded conductors are usually black. For circuit identification other colors except green may be used. Green is reserved for still another important function discussed later.

To complete this identification of grounded terminals, lampholders should have their screw shells identified by a nickel- or zinc-colored screw.

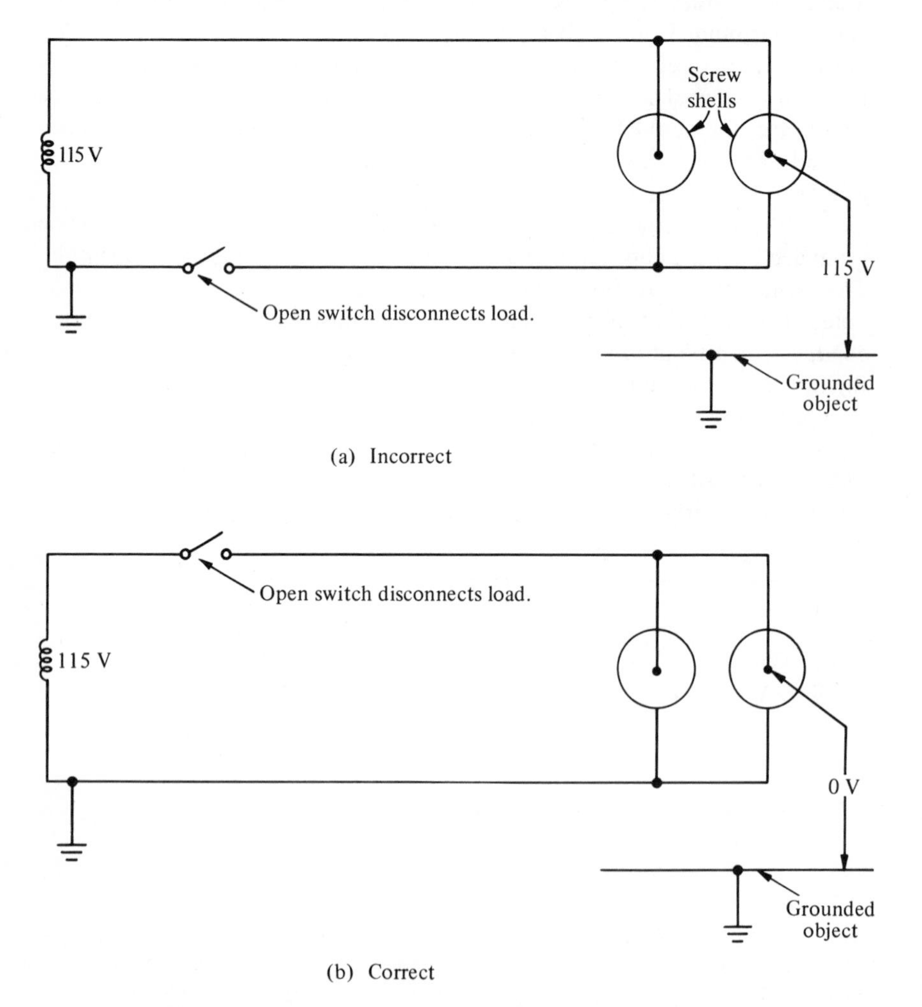

Fig. 10-9. Single-pole switch controlling a lighting load (a) incorrectly (voltage to ground still exists at load) and (b) Correctly (no voltage to ground at load).

The center terminal usually has a brass colored screw. For fixtures or for any type of lampholder that has wire leads instead of terminals, the screw shell lead should be of a white or natural gray color.

The grounded conductor of a system usually should not be opened under any conditions. The only exception to this is if all conductors can be opened at the same time. A single-pole switch placed in the grounded conductor will operate the load satisfactorily, but the ungrounded conductor is still intact and a voltage to ground is still available at the load. If switches should be connected in this fashion, a shock hazard is possible with the load supposedly disconnected from the supply. Figure 10-9 shows the incorrect and the correct methods of installing single-pole switches in grounded circuits. The NEC requirement in this case is quite simple: Provide switches in all ungrounded conductors.

The same precaution must be followed for overcurrent devices. A circuit breaker may have sufficient poles to disconnect all conductors of the system including the one that is grounded. This is satisfactory since any overload on one conductor will cause the breaker to open all poles. It is unnecessary to provide a pole for the grounded conductor.

When fuses are used for overcurrent protection they may never be installed in a grounded conductor. This is because a fuse only affects the conductor in which it is inserted. Should the fuse in a grounded conductor interrupt the circuit but the fuse in the ungrounded conductor be intact, the circuit would be inoperative but a voltage to ground would still exist. This is also an NEC violation. Figure 10-10 illustrates this important point.

Since only one fuse is necessary to provide overcurrent protection in a two-wire grounded circuit, panel boards and distribution equipment can be much smaller.

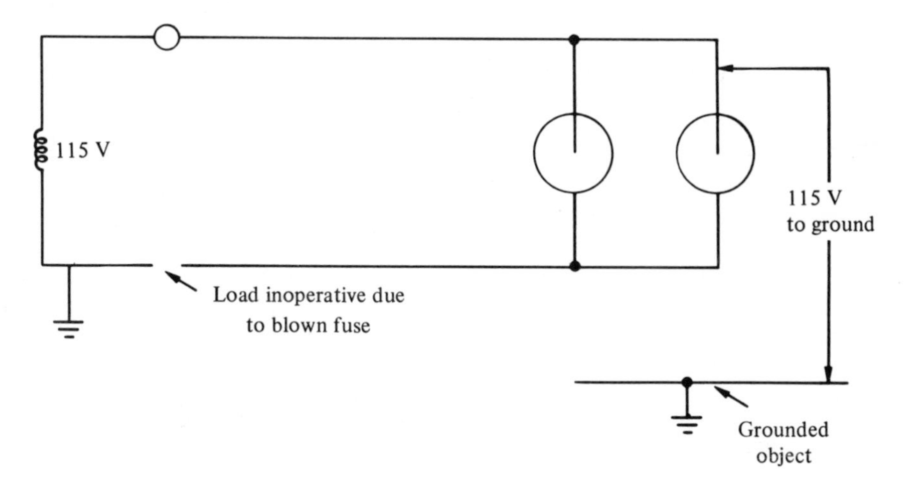

Fig. 10-10. Hazard of double fusing a grounded circuit.

The omission of any switches or fuses in a grounded conductor leads to the term *solid neutral* since the grounded conductor is also the neutral conductor of a three-wire, single-phase supply. This also applies to a three-phase, four-wire supply.

10-5 ***Principles of equipment grounding.*** This chapter thus far has discussed only the grounding of one of the current-carrying conductors of a power supply. If at this point the student does not understand this portion, he should go back to the beginning of the chapter and restudy the material before continuing.

Chapter 9 discussed the function of raceways in an electrical system. These raceways, and all boxes, fittings, switch gear, and enclosures for fuses or circuit breakers must be grounded. The NEC requires that all metallic equipment associated with an electrical system be grounded. (The only exception to this rule is an installation that is so isolated that no person could touch it and a grounded object simultaneously.)

This type of grounding is called *equipment grounding*. It must be carried out regardless of the voltage of the system. The higher the voltage the greater the hazard from the faulty equipment grounding. The principal reason for equipment grounding is to prevent any metallic object that a

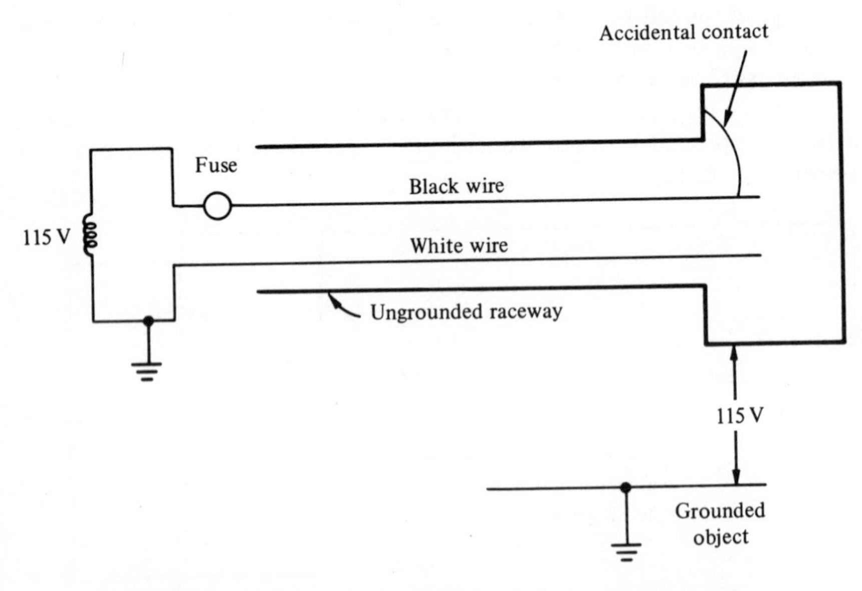

Fig. 10-11. Breakdown between ungrounded conductor and raceway or outlet box. Note voltage between entire raceway system and ground.

person might touch from becoming "alive." More specifically, if an ungrounded conductor of the system by fault should touch any metallic equipment, junction box, motor frame, or fitting, then that equipment would have a voltage between it and ground. If the equipment is not grounded a very serious shock hazard exists. Figure 10-11 shows this voltage between a raceway and ground. The raceway could be connected to a box containing a switch. Then a switch plate secured to the switch is also a shock hazard. A person now touching the switch plate is exposed to the same danger as touching the ungrounded conductor itself.

By grounding the metal equipment any contact with the ungrounded wire now is equivalent to a short circuit between the two conductors of the system. This causes the fuse to blow, thereby preventing the metal equipment from maintaining a "live" voltage to ground. The grounding of all metal that could in any possible way contact an ungrounded conductor prevents that equipment from becoming "alive". Our expression "alive" alludes to a voltage above ground or zero potential.

Figure 10-12a shows the path of current flow due to an accidental ground on the ungrounded conductor. Note that the current does not actually flow "to ground". It flows from the supply through the accidental contact to the common connection where the system and equipment ground are connected together, as illustrated in Fig. 10-12b. Thus the circuit is completed back to the supply. This current is called a "fault current" since it only flows if there is a fault in the circuit.

It is very important to make certain that the equipment ground path, back to its connection with the system ground, has a low resistance. This permits overcurrent devices to open the circuit.

In this respect the equipment-grounding resistance path decreases as the conductor size increases: The larger conductors require larger raceways. Larger raceways have lower resistance. As conductor size increases, the overcurrent devices increase in capacity.

The NEC also requires that all equipment be grounded to the same electrode (usually a water pipe) that grounds the system. This is accomplished automatically by the grounding conductors of Table 10-1 as they are connected together in the main service switch. However, this system-grounding conductor is installed in a conduit or tubing. This raceway and grounding conductor are secured to the water pipe with a special fitting. This raceway may be considered to be the grounding conductor that grounds the equipment. This dual grounding function is shown in Fig. 10-13.

Table 10-2 lists the proper equipment-grounding conductor to be installed when a service entrance conductor (system) is not grounded. Note that this may be either a conductor or a raceway.

When a conductor must be used to ground any raceway or equipment

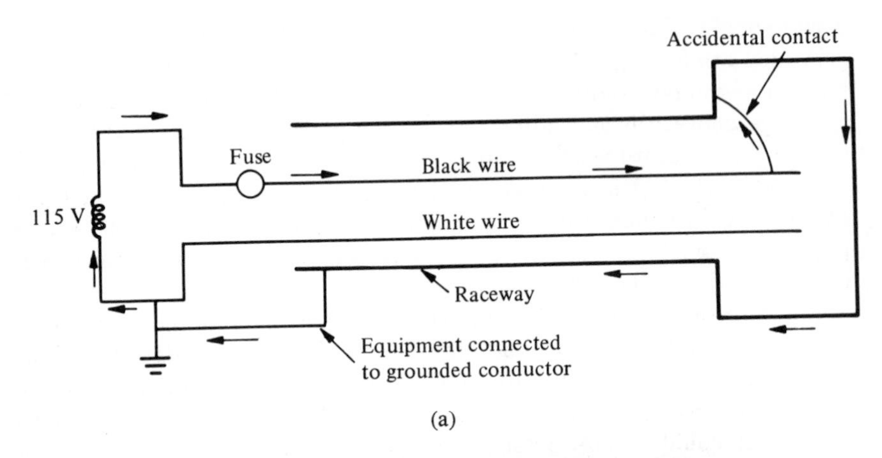

(a)

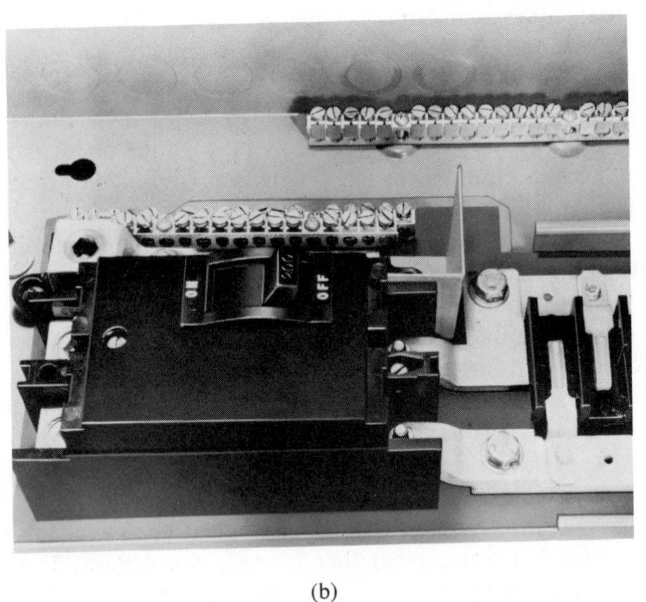

(b)

Fig. 10-12. (a) Path of current flow due to an accidental ground on the ungrounded conductor. Fuse should blow and interrupt fault current. (b) Common connection of system and equipment ground. *Courtesy of Square D Company.*

beyond the service entrance equipment, Table 10-3 lists the proper size. This depends on the rating of the overcurrent device ahead of the equipment to be grounded.

An important exception to this table should be noted: A No. 18 AWG copper conductor may ground equipment rated at 15 or 20 A if it is included

Table 10-2. Service equipment grounding conductor for ungrounded systems

Size of Service Conductor		Size of Grounding Conductor			
Copper	Aluminum	Copper	Aluminum	Conduit	Electrical Metallic Tubing
2 or smaller	0 or smaller	8	6	½	½
1 or 0	00 or 000	6	4	½	1
00 or 000	0000 or 250 MCM	4	2	¾	1¼
Over 000 to 350	Over 250 MCM to 500 MCM	2	0	¾	1¼
Over 350 MCM to 600 MCM	Over 500 MCM to 900 MCM	0	000	1	2
Over 600 MCM 1100 MCM	Over 900 MCM to 1750 MCM	00	0000	1	2

Table 10-3. Size of equipment grounding conductors for grounding interior raceway and equipment

Rating of Overcurrent Device Ahead of Equipment (A)	Size of Grounding Conductor Copper, AWG	Aluminum, AWG
15	14	12
20	12	10
30	10	8
40	10	8
60	10	8
100	8	6
200	6	4
400	3	1
600	1	00
800	0	000
1000	00	0000

as a cable assembly. This permits Nos. 12 and 14 romex (NM cable) to use a No. 18 grounding conductor.

Should the neutral be grounded at a distribution point beyond the main switch, some of the neutral current can flow back to the source via the raceway. Figure 10-14a illustrates this situation. To show this current flow, we assume an unbalanced neutral current of 40 A. The resistance of the conduit is assumed to be three times that of the neutral. Therefore, one-fourth of the current will flow through the conduit. Figure 10-14b shows a parallel circuit of conductor and raceway. The currents divide inversely to the resistance of the two paths. This is an NEC violation. No current *except fault current* must flow through metallic equipment.

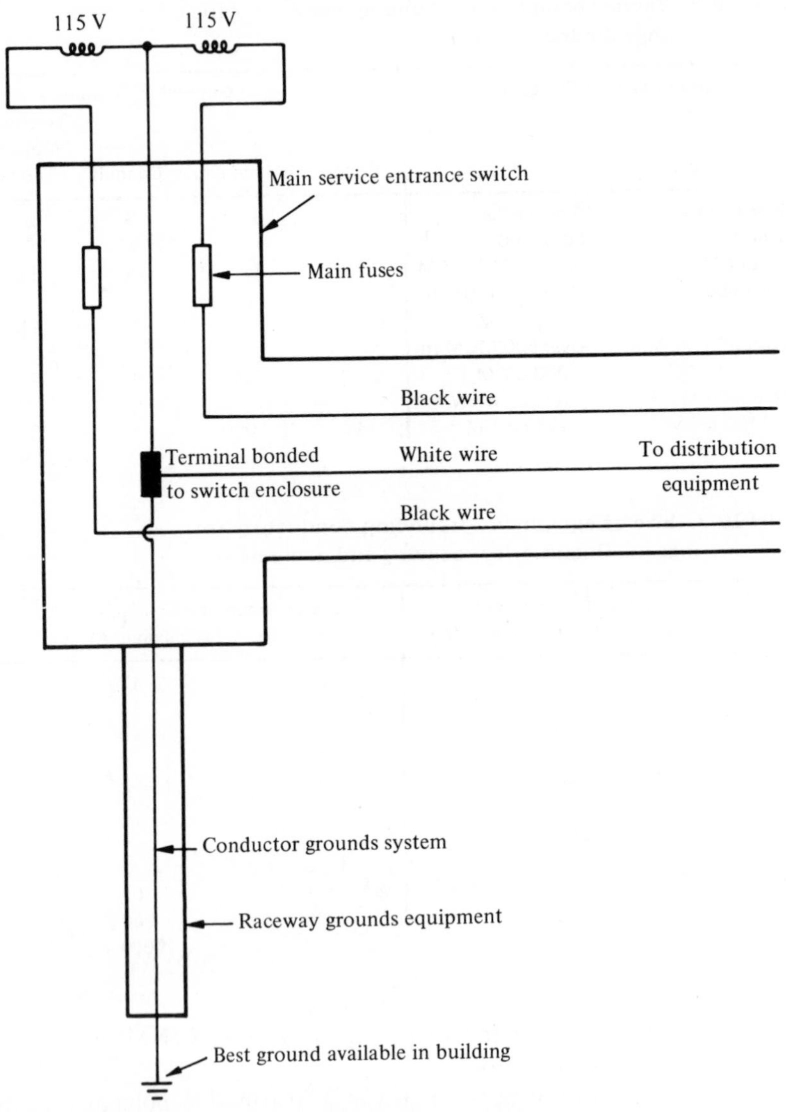

Fig. 10-13. System and equipment grounding terminals in service entrance switch.

Of course it is difficult to evaluate the resistance of the conduit system and to compare it with the resistance of the neutral conductor. However, regardless of how high it might be in comparison, it will carry some current in this particular situation. Again, the important issue here is not to permit the grounded conductor to be grounded beyond the initial grounding.

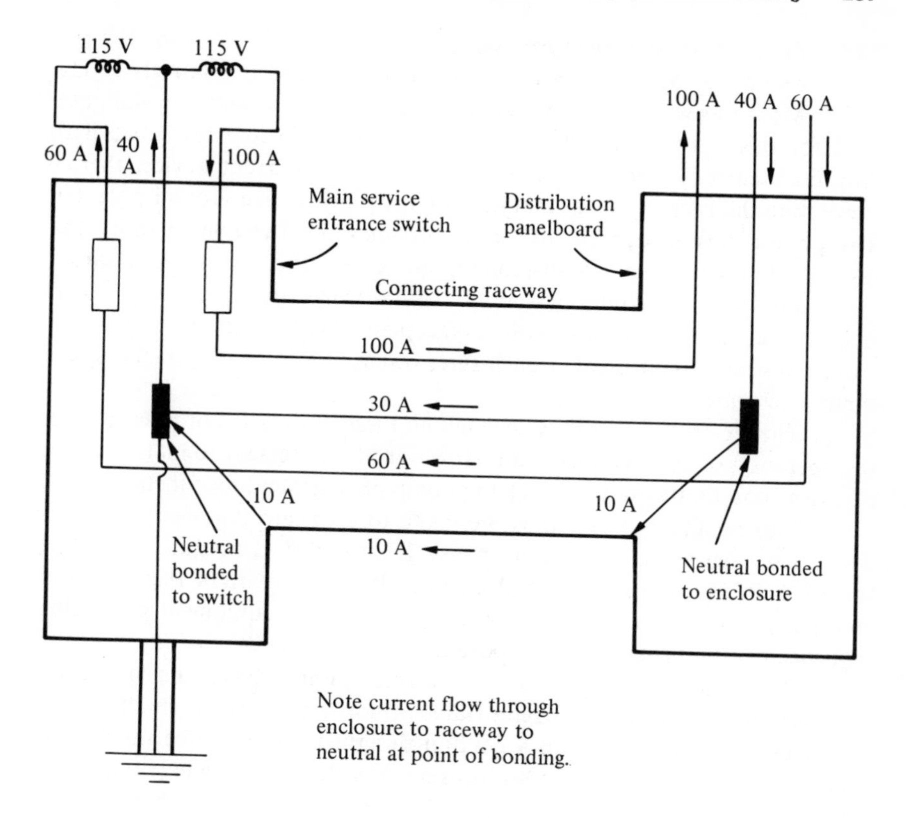

(a)

3R = conduit
10A
40 A
40 A
30 A
R = neutral conductor

(b)

Fig. 10-14. (a) Neutral current finding a path through raceway due to multiple grounding of neutral. (b) Equivalent circuit of raceway and neutral showing current division.

10-6 ***Methods of equipment grounding.*** For many wiring systems the procedure for maintaining the equipment grounded is relatively simple.

Most raceways are either conduit or tubing. The steel or aluminum used for this purpose if *installed correctly* provides a low resistance path. Any fault current then is able to operate the overcurrent device, thereby preventing the raceway from maintaining a voltage above ground potential. The phrase "installed correctly" now becomes the variable quantity that distinguishes between a safe installation and one that is a possible hazard.

Equipment includes not only the raceways that enclose the conductors but also every box, switch, faceplate, fitting, motor, motor controller, fuse or circuit breaker enclosure, frame of electrical appliances, and meter enclosure.

The electrical continuity through all these devices is not necessary for the operation of any electrical load. However, it is necessary to make positive electrical continuity to carry out the equipment grounding function.

This means that care must be taken to ascertain that every device is connected securely by whatever means is provided. Locknuts, bushings, threaded fittings, connectors, and clamps all have a part in this electrical continuity. Since they are all connected in series, the equipment grounding sequence can be broken by one loose fitting or locknut.

When using armored cable as a wiring method the cable armor itself provides the path for fault current. Cable armor, especially the smaller sizes, has much less resistance than a conduit designed for the same number of conductors. To help reduce this resistance a flat, uninsulated aluminum ribbon is included in the cable assembly. This uninsulated ribbon lies between the paper-covered conductors and the armor. Because of this arrangement the cable armor is in constant contact with the aluminum throughout its length. Since they are in parallel, the cable armor resistance is reduced. Armored cable is usually secured to outlets with connectors or clamps. Care must be exercised to assure they are securely fastened at the outlets and are not subject to any strain so they may later be allowed to drop out of contact.

Nonmetallic cable has no armor to provide for equipment grounding. An extra conductor is provided for this purpose. This wire may be uninsulated, but if it is insulated it must have a *green covering*. *Green* has been reserved for equipment *grounding* conductors in cables and cords or if installed separately. This ground wire must be secured in an approved manner to each outlet box and cabinet.

The grounding of the metal parts of portable appliances presents the problem of making them free from shock hazards. A portable appliance is a device that uses a cord with an attachment cap rather than being permanently wired into a system. Table and floor lamps as used in the home may be plugged into their receptacles either way. As a result, the screw shells have a 50–50 chance of being grounded.

It is the opinion of many people in the electrical industry that the NEC was in error many years ago, when portable appliances and lamps were first introduced, not to insist on polarized attachment caps. Many a person has received a shock from a defective floor lamp in which the screw shell insulation has broken down and is in contact with the metal frame. The polarized cap would not ground the metal frame, but it would assure the grounding of the screw shell. However, even this protection would be lost if some inexperienced person connected the attachment cap in reverse.

The 1959 NEC made the first overture toward equipment grounding for portable tools and washing machines. The grounding was done by a separate green conductor, requiring a three-wire cord. The receptacle and attachment cap were designed to include a separate prong for grounding purposes. This is called a U-shaped grounding terminal. The grounding of the appliance is completed through the receptacle to the box and then to the raceway back to the source. The other end of the green wire is secured to the metal frame of the device. Unfortunately these attachment caps do not fit the old-style two-wire receptacles. Figure 10-15 shows this arrangement.

Large appliances such as electric ranges and dryers have their frames connected to the neutral conductor of the supply. These are the only instances where this is permitted.

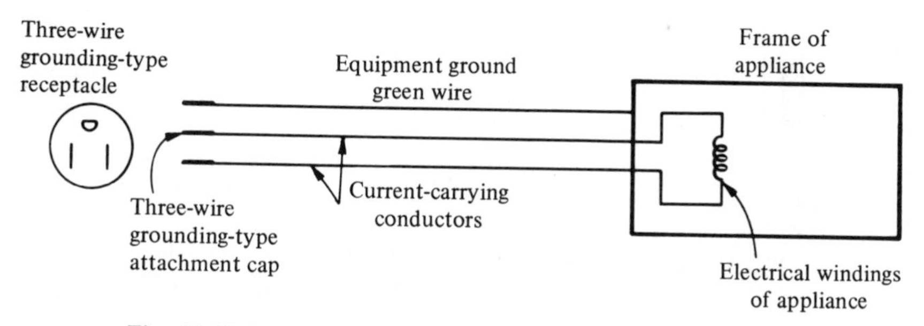

Fig. 10-15. Method of assuring equipment ground for portable appliances with grounding-type receptacle and plug.

10-7 ***Fault currents—magnitudes and paths.*** The electrical resistance of raceways and their associated connectors and fittings is always low. This provides a path for a fault current to flow back to the point where the grounded conductor is bonded to the equipment.

However, let us consider some examples where because of poor workmanship this is not possible, and then evaluate the consequences. Figure 10-16 represents a three-wire, 115–230-V supply fused at 100 A. At some point in the system there is a loose connection between the raceway and a cabinet. It might be difficult to assign a particular ohmic value to a loose connection, but let us assume it is equivalent to about 2 Ω. The resistance of

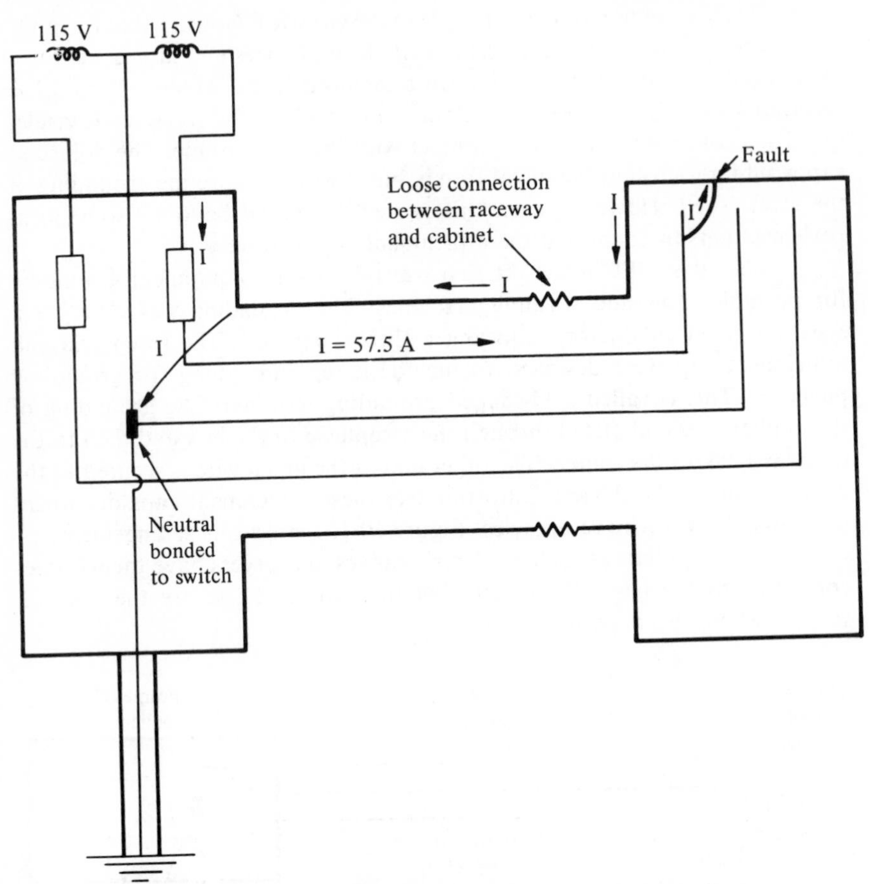

Fig. 10-16. A loose equipment ground connection preventing a fault current from performing its function of operating overcurrent device.

the raceway, if a short length, is practically zero. If an ungrounded conductor should contact the cabinet, the fault current would be

$$I = \frac{E}{R} = \frac{115}{2} = 57.5 \text{ A}$$

Instead of 2-Ω resistance, the loose connection might be considered to be such a poor contact that not more than 57.5 A could pass through that point. This would not be enough current to melt the 100-A fuse. The fault current would continue to flow, creating two distinct hazards: First, the raceway between the fault and the loose connection would have a voltage between itself and ground; this is a shock hazard. Second, the heating effect

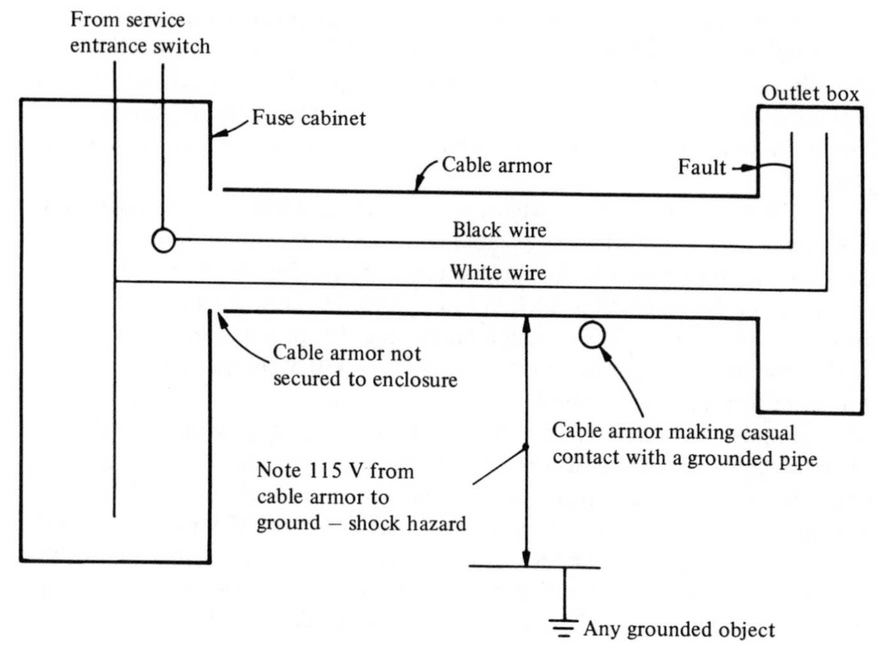

Fig. 10-17. Fire and shock hazards resulting from ungrounded cable armor.

or the arcing at either the fault or the loose connection could start a fire if combustible matialer were adjacent.

Next let us consider a section of armored cable supplying a lighting circuit and fused at 15 A. Assume at some point in the circuit the cable armor does not make contact with the box to which it should be secured. Figure 10-17 illustrates this situation. Now there is no positive equipment ground for the metallic equipment beyond this point. Any accidental breakdown of insulation between the ungrounded conductor and any box, fitting, or raceway causes that section of equipment to have a voltage to ground. There is then a shock hazard, since there is no path for fault current to permit the 15-A fuse to blow. However, sections of armored cable in a building are undoubtedly going to come in contact with some grounded object such as a water or heating pipe. If this contact is intermittent or casual, the equipment ground path is now a poor one. Arcing may result where the cable armor contacts the pipe, and a fire hazard would then exist.

Like the spare tire of the motorist, which is always carried but seldom used, so the equipment grounding path of an electrical system must always be intact and standing by. Then it can provide the path for fault current should it ever exist.

10-8 *Ground detectors.* When three-phase systems are not grounded, the possibility of an unintentional ground on one phase always exists. This will not affect the operation of any of the electrical loads. There are two objections to allowing this ground to remain: the shock hazard to ground from the other two ungrounded conductors or the resultant short circuit should another ground occur elsewhere in the system on another phase.

For these reasons an extensive three-phase wiring system consisting of many machines and their associated control devices should have a ground detection system. This can be done quite inexpensively by using three sets of lamps so arranged that each set is connected between one phase and ground. Each set must have enough lamps in series to agree with the voltage of the system. If voltage is 440 V, at least four 110-V lamps should be connected in series in each group.

Figure 10-18 illustrates a ground detector. It may be connected anywhere in the system that all three phases are available. It should be located where it can be readily seen by maintenance personnel. A direct contact to ground on any phase will cause that particular set of lamps to go out completely. A higher resistance path to ground will cause that group to become dim. A normal situation free of grounds will cause all lamps to be of the same brilliancy. This will not be of full brilliancy. The sets of lamps

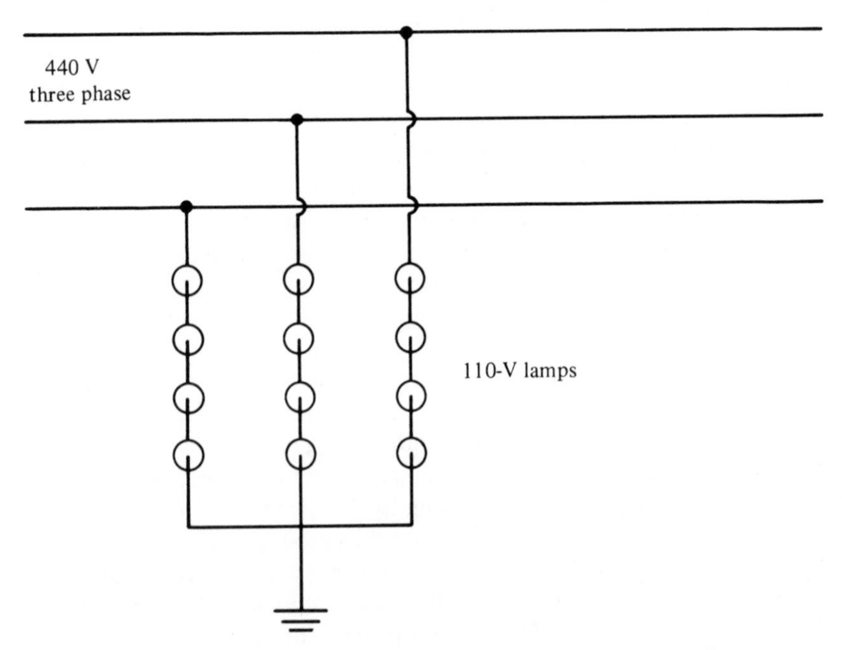

Fig. 10-18. A ground detector system on a three-phase supply.

are connected in wye. Therefore, the voltage across each set is $440/\sqrt{3}$ or 254 V. Voltage across each lamp is 63.5 V. The number of lamps could be reduced by using lamps of higher voltage rating.

PROBLEMS

10-1 Determine the voltage to ground from each conductor of a three-wire, 115–230-V supply.

10-2 How can the neutral of a 115–230-V, three-wire supply be determined?

10-3 Draw a complete sketch, including the transformer secondary, showing the result of grounding a conductor other than the neutral of a three-wire, 115–230-V service entrance (a) if this is the only service on the distribution transformer and (b) if there are other services on the same transformer with the neutral grounded.

10-4 (a) Refer to Fig. 10-1b. Draw a sketch and explain how the circuit would function if the circuit were grounded again at point *A*.
(b) Repeat part (a) if the circuit is broken at point *B*.

10-5 Determine the size of the grounding conductor for each of the following conditions: (a) a No. 6 AWG service conductor, (b) a 300 MCM aluminum service conductor, (c) 100-A service entrance equipment, and (d) 200-A service entrance equipment.

10-6 Explain the principle of grounding the screw shells of lampholders.

10-7 Explain by sketch any hazard when using a single-pole switch in a grounded conductor. Does it operate the load?

10-8 Explain the principal reason for grounding electrical equipment.

10-9 A single-pole switch is controlling a 115-V lighting load. Determine the voltage to ground from each switch terminal if (a) the lights are on and (b) the lights are off.

10-10 A No. 16 AWG three-wire cord 150 ft long has a green wire as the equipment ground for a 115-V portable tool.
(a) Calculate the resistance of the equipment ground.
(b) Is this low enough to protect the portable tool from a voltage to ground in the event of an accidental contact between the frame and the ungrounded conductor of the cord?

10-11 What are the results of an accidental ground on one phase of a three-phase feeder if (a) one phase is grounded at the service entrance and (b) the supply is ungrounded at the service entrance?

10-12 A three-wire, single-phase, 115–230-V feeder is supplying a 30-kW unity pf load. The neutral is incorrectly bonded to the panel board frame. Assuming the raceway has twice the resistance of the neutral, draw a sketch showing current flow if (a) the load is balanced and (b) 15 kW is operating on one side but only 5 kW on the other side of the three-wire system.

10-13 The secondary of a wye-connected transformer bank is the power supply for a 120-V lighting load.

(a) If a line conductor instead of the neutral is grounded, show by a sketch the voltages to ground from each of the four conductors.

(b) If grounding is omitted, show by sketch the possible voltages to ground from each of the four conductors.

(c) If the neutral is grounded, show by sketch the voltages to ground from each of the four conductors.

(d) Justify the validity of (c).

Chapter Eleven

Branch Circuit and Feeders

11-1 ***Branch circuit definition and principles.*** Any building, from a modest cottage to a skyscraper, has one electrical fact in common: a multitude of electrical appliances or devices. These are of various sizes and shapes, from the few watts required for an electric clock to giant motors of hundreds of horsepower. Although lighting a large area requires a considerable amount of electric power, it consists of small individual lamps.

Providing the proper conductors and overcurrent devices for each individual device requires careful attention to the proper design of the system. This situation could be simplified if each lamp, each convenience receptacle, and each appliance were supplied by two conductors and protected by its own fuse. This, however, would be impractical—perhaps as impractical as an airline supplying a separate airplane for each passenger. As we can observe, the correct number of passengers in the appropriately sized airplane for the correct distance to be traveled leads to an economical operation.

Similarly, if we can group a number of devices to be supplied by a circuit with one fuse, we can make the distribution of electrical power both practical and economical. An extreme case could be to install two wires through the building and tap everything off them and rely on only one fuse. This arrangement would not provide suitable overcurrent protection for small devices. Also, any fault resulting in a blown fuse would cause the entire system to be out of operation.

The NEC has established a system called *branch circuits* to alleviate this problem. A branch circuit is defined as that part of the wiring system beyond the *last fuse*. The rating of that fuse or circuit breaker now determines the capacity of that branch circuit. We then select the appropriate size conductors with ampacity to agree with the overcurrent rating. By following the NEC requirements of providing the proper branch circuit for the appropriate loads, we will not have a situation in which any electrical device is not properly protected. Also, by providing many circuits any fault will shut down only a small portion of the total load.

The NEC classifications of branch circuits are 15, 20, 30, 40, and 50 A. The following three section describe what electrical loads may be connected to these circuits.

11-2 ***Permissible loads on* 15- *and* 20-A *branch circuits.*** Each and every branch circuit must be so designed that its conductors and the devices it supplies with current will not be damaged by a short circuit or a ground. For example, the small wire in a fluorescent fixture may have an ampacity much smaller than the branch circuit conductors. If this fixture were connected to a high-ampacity circuit, a fault may cause severe heating or fire before the overcurrent device could operate. For this purpose, No. 18 or 16 may be tapped from No. 14 or 12 conductors but not larger, as the smaller fixture wires would not be suitably protected.

Individual lighting units have a wide variation in power rating. The code permits the manufacture of several sizes of lamp sockets. Aside from decorative or other special-purpose lamps, the medium-base lampholder may be used for lamps up to 300 W. The mogul or heavy-duty socket may be used for lamps from 300 to 1500 W.

For general lighting in any type of occupancy, 15- and 20-A branch circuits are extensively used. The code permits any type of lampholder to be used on these circuits. The overcurrent limit is low enough to properly protect any individual lighting unit. Also, it is high enough to permit the grouping of enough lighting units to make an economical installation. The maximum current permitted for a continuous load is 12 A for the 15-A branch circuit, 16 A for the 20-A branch circuit, or 80% of their ratings.

This is a good design. As the current approaches the rating of the circuit, it is approaching the point of overcurrent operation. A well-designed system should never cause a blown fuse.

The current rating of a permanently installed appliance cannot exceed 50% of its branch circuit rating if lighting units or portable appliances are connected to the same branch circuit.

The rating of a portable appliance cannot exceed 80% of its circuit rating. Portable appliances make electrical systems difficult to design. This is especially true in residences where appliances constitute a large portion of the electrical load. The NEC requires three 20-A branch circuits in the kitchen, dining room, and pantry areas so that appliances will have sufficient branch circuit capacity. In addition, single houses or multifamily dwellings that do not provide common laundry facilities also must have a 20-A circuit specifically for laundry equipment in each unit.

For the use of portable appliances, receptacles may be rated at 15 or 20 A. By providing different-sized receptacles and attachment caps or plugs for each branch circuit rating, portable appliances designed for one particular branch circuit cannot be used on another. However, to simplify appliance use in residences, the 15- and 20-A receptacles and plugs may both be used on 20-A branch circuits.

The NEC permits a maximum current of 15 or 20 A on these branch circuits. When the load is continuous, however, this current is derated to 80% of capacity. When designing branch circuits, the maximum current that can be allocated to these circuits is 12 or 16 A.

11-3 ***Permissible loads on* 30-A *branch circuits.*** As individual lighting units become larger—750, 1000, and 1500 W—their current requirements are too high for the 15- or 20-A circuits. One 15-A branch circuit would be necessary for each 1000-W lamp. The NEC permits only the permanently connected mogul or heavy duty lampholders to be installed on 30-A circuits in other than dwelling units. The medium-sized lampholder is not permitted on any branch circuit rated over 20 A.

When portable appliances are to be used on a 30-A circuit the appliance cannot be rated over 24 A. Receptacles for portable appliances must be rated for 30 A. This makes it impossible to use such an appliance on a smaller or higher current rating branch circuit.

11-4 ***Permissible loads on* 40- *and* 50-A *branch circuits.*** The 40- and 50-A branch circuits are used to provide for fixed cooking units or infrared heating units. Lighting units employing the heavy-duty lampholders that use 300 W and larger lamps also may be connected to 40- and 50-A

Table 11-1. Branch circuit requirements

Branch Circuit Rating	15 A	20 A	30 A	30 A	50 A
Conductors, copper, minimum size AWG					
Circuit wires	14	12	10	8	6
Taps	14	14	14	12	12
Fixture wires	18	18	14	12	12
Overcurrent Protection	15 A	20 A	30 A	40 A	50 A
Outlet devices					
Lampholders	Any type	Any type	Heavy duty	Heavy duty	Heavy duty
Receptacle rating	15 A	15 or 20 A	30 A	40 or 50 A	50 A
Maximum load	15 A	20 A	30 A	40 A	50 A
Maximum continuous load	12 A	16 A	24 A	32 A	40 A

circuits. However, this is not permitted in dwellings and is more logically used in industrial applications.

The branch circuit specifications and requirements are tabulated in Table 11-1, on page 243.

11-5 ***Voltage drop in branch circuits.*** In ordinary practice the electrical load on a branch circuit is distributed throughout its length. This makes voltage drop calculations somewhat complex. We can alleviate this problem by designing a lighting system so that the branch circuits extend a limited distance from the distribution center.

The NEC permits a maximum of 3% voltage drop to the farthest outlet in a branch circuit. A further stipulation is made that a maximum voltage drop of 5% is permitted when the feeders are included. The complicated procedure to assure this 3% voltage drop is illustrated in Example 11-1.

---

*Example **11-1.*** A 20-A branch circuit is supplying its maximum constant load of 16 A. The load consists of eight lighting units requiring 2 A each. They are located 20 ft apart. What is the voltage at the last outlet? (The supply voltage is 115 V.) As each lighting unit receives a slightly lower voltage as the circuit is extended, a slight error is introduced in assuming that every unit receives 2 A. Refer to Fig. 11-1.

Solution

The resistance of No. 12 AWG conductors between units is

$$R \text{ per } 1000 \text{ ft} = 1.62\ \Omega \qquad \text{from Table 9-7}$$

$$R \text{ per } 40 \text{ ft} = \frac{1.62}{25} = 0.065\ \Omega$$

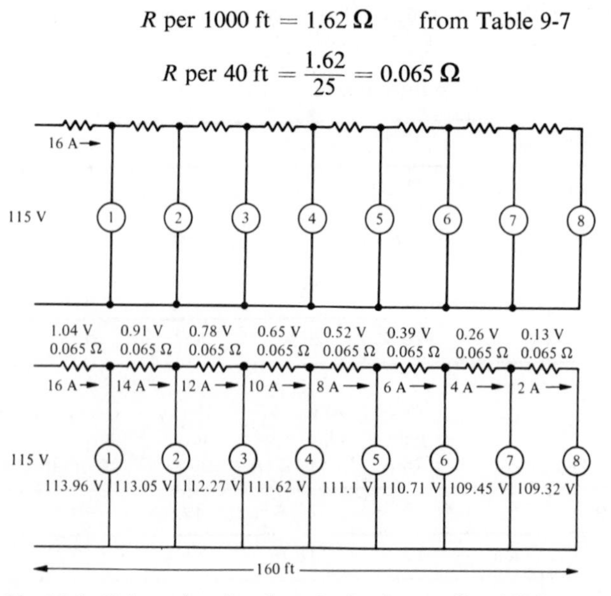

Fig. 11-1. Voltage drop in a branch circuit extending 160 ft.

Note in Fig. 11-1 that each 0.065-Ω resistor represents both conductors between units.

Voltage drop in first section $= IR = (16)(0.065) = 1.04$ V

Voltage at first unit $= 115 - 1.04 = 113.96$ V

Voltage drop in second section $= (14)(0.065) = 0.91$ V

Voltage at second unit $= 113.96 - 0.91 = 113.05$ V

Voltage drop in third section $= (12)(0.065) = 0.78$ V

Voltage at third unit $= 113.05 - 1.13 = 112.92$ V

Voltage drop in fourth section $= (10)(0.065) = 0.65$ V

Voltage at fourth unit $= 112.92 - 0.65 = 112.27$ V

Voltage drop in fifth section $= (8)(0.065) = 0.52$ V

Voltage at fifth unit $= 112.27 - 0.52 = 111.75$ V

Voltage drop in sixth section $= (6)(0.065) = 0.39$ V

Voltage at sixth unit $= 111.75 - 0.39 = 111.36$ V

Voltage drop in seventh section $= (4)(0.065) = 0.26$ V

Voltage at seventh unit $= 111.36 - 0.26 = 111.10$ V

Voltage drop in eighth section $= (2)(0.065) = 0.13$ V

Voltage at last unit $= 111.10 - 0.13 = 110.97$ V

This branch circuit, which extends 160 ft, is obviously poorly designed since the voltage drop is 5% at the last outlet.

If the circuit of Ex. 11-1 were permitted to extend only half the distance, or about 80 ft, all the individual voltage drops would be about half the values in Ex. 11-1. The total voltage drop would then be about 2.5%.

When lighting systems are designed the distribution centers should be so located that the branch circuits are less than 100 ft in length. This is a general rule that must be considered for circuits operating at 115 V. As the operating voltage is increased, the length could be increased proportionally. For example, a lighting circuit at 277 V could extend $2\frac{1}{2}$ times as far and maintain the same voltage drop. There will be more voltage drop, but we have more voltage to start with.

11-6 ***Division of load into branch circuits.*** The branch circuit has been defined as that part of the electrical system beyond the final overcurrent device. However, when designing the overall system the number and size of various branch circuits must be considered first. The total load must be evaluated to assure proper branch circuit design.

In residences this problem is complicated because of the combination of fixed lighting units and portable appliances on many household circuits. This is usually simplified by allocating six or eight outlets to a circuit unless the circuit is for a specific appliance. The NEC requires at least one circuit for lighting only for each 500 ft^2 of area. As domestic appliances also must be supplied, the area allocated for each circuit must be reduced.

Large lighting loads in industrial and commercial buildings are usually fixed units, and branch circuits can be properly designed since the power or current for each circuit can be determined.

To assure proper branch circuit design the following procedures may be followed:

1. Determine the load current I required by each lighting unit. If incandescent (resistive),

$$I = \frac{P}{E} \tag{3-4}$$

   If fluorescent (inductive),

$$I = \frac{P}{(E)(\text{pf})} \tag{4-5}$$

   The NEC requires that inductive lighting units have their current rating indicated on a nameplate or some obvious place.

2. Determine how many of these units can be accommodated on the appropriate branch circuit not exceeding 80% of the branch circuit rating.
3. Calculate the total number of branch circuits required.
4. We have already determined that 115-V circuits should extend a maximum of 100 ft—with 75 ft preferred. A large area to be illuminated then will require more than one load center or overcurrent enclosure. Distances will determine the number and location of load centers. Branch circuits operating at 277 V may extend 200 ft and not have excessive voltage drop.

---

*Example **11-2.*** A supermarket 100 ft × 150 ft is to be illuminated by 215 fluorescent fixtures operating at 115 V. Each lighting unit requires 2.2 A.

(a) Calculate the number of 20-A branch circuits necessary.
(b) What would be an appropriate number of branch circuit panel boards?
(c) If lighting units have 0.8 pf, calculate the total power of the lighting load.

(d) Calculate the electric bill for a 12-hr day if the energy cost 2 cents per kWh.

Solution

(a) Number of units per 20-A branch circuit (at 80%):

$$\frac{16\text{ A}}{2.2\text{ A}} = 7.27 \quad \text{or a maximum of 7}$$

Number of branch circuits:

$$\frac{215}{7} = 30.7 \quad \text{or a minimum of 32}$$

(Panel boards have an even number of circuits.)
(b) Refer to Fig. 11-2. If one panel board were located in the center of the store at point *A*, the longest branch circuit would extend more than 100 ft. Electrical raceways are always installed at right angles.

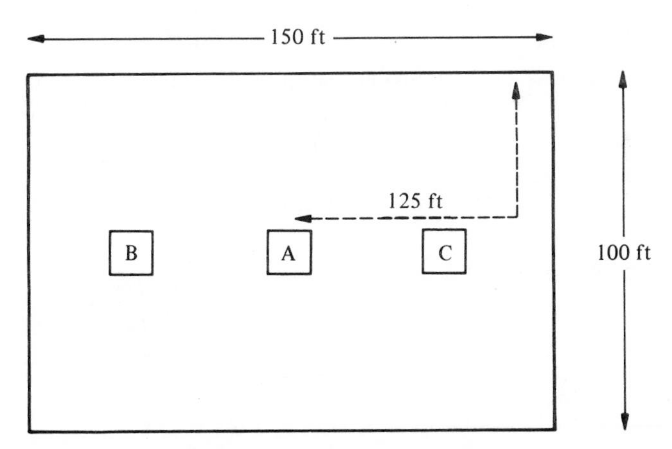

Fig. 11-2. Solution to Ex. 11-2(b).

Two locations at points *B* and *C* would make this a better installation. Each panel board must provide at least 16 circuits.
(c) Total current: (215)(2.2 A) = 473 A.
Total power:

$$P = (EI)(\text{pf}) \tag{4-5}$$

$$= (115)(473)(0.8) = 43{,}500\text{ W} \quad \text{or } 43.5\text{ kW}$$

(d) Cost of lighting per 12-hr day:

$$\text{Total cost} = \frac{(P)(C)(t)}{1000} \qquad (3\text{-}8)$$

$$= \frac{(43{,}500)(0.02)(12)}{1000}$$

$$= \$10.40$$

---

*Example **11-3**.* A factory assembly area requires 60 kW of general illumination using 150 W incandescent lamps. The area measures 200 × 100 ft. Design the branch circuit equipment. Use 20-A branch circuits at 115 V.

Solution

$$\text{Current for each lamp} = I = \frac{P}{(E)(\text{pf})} \qquad (4\text{-}5)$$

$$= \frac{150}{(115)(1)} = 1.3 \text{ A}$$

$$\text{Lamps per 20-A branch circuit} = 80\% \text{ of } 20 \text{ A} = 16 \text{ A}$$

$$\frac{16 \text{ A}}{1.3} = 12.3 \quad \text{or 12 lamps maximum}$$

$$\text{Total lamps required} = \frac{60{,}000 \text{ W}}{150 \text{ W}} = 400 \text{ lamps}$$

$$\text{Number of 20-A branch circuits} = \frac{400 \text{ lamps total}}{12 \text{ lamps/ckt}}$$

$$= 33.4 \quad \text{or 34 minimum}$$

Figure 11-3 shows one possible arrangement of panel boards that would limit the length of the branch circuits to a maximum of 75–100 ft. Each panel board must have provisions for at least 12 branch circuits.

It should be noted that incandescent lighting units of this type use lampholders that can accommodate lamps up to 300 W. Should more illumination become necessary, larger lamps could be installed, but this would overload the branch circuits. The 34-circuit total is the absolute minimum and makes no provision for any additional load. Well-designed branch circuits should have provisions for additional load. This can hardly be arbitrary, but is part of design specifications.

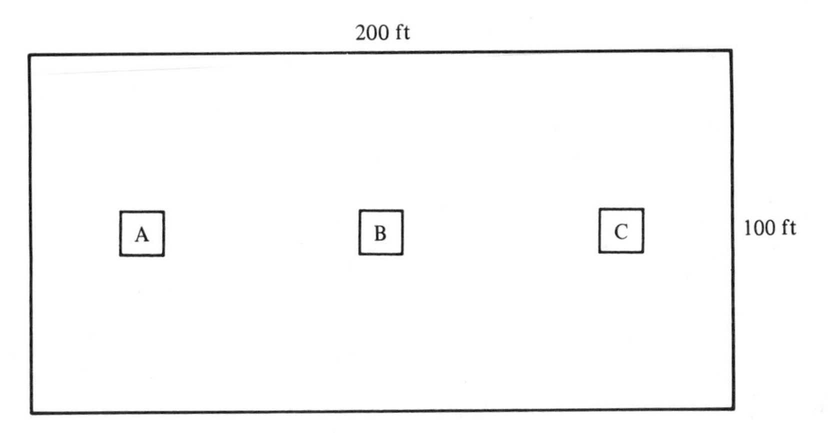

Fig. 11-3. Panel locations for Ex. 11-3.

---

*Example **11-4.*** Repeat Ex. 11-3 but make provisions for a possible 30% increase in load on each branch circuit.

Solution

We must revert to the 1.3 A for each lamp and increase this by 30% (a factor of 1.3):

$$(1.3 \text{ A})(1.3) = 1.69 \text{ A}$$

$$\text{Lamps per 20-A branch circuit} = \frac{16 \text{ A}}{1.69} = 9.5 \quad \text{or 9 lamps maximum}$$

Total lamps remain at 400 but can be larger in power rating.

$$\text{Number of 20-A branch circuits} = \frac{400 \text{ lamps}}{9 \text{ lamps/ckt}}$$

$$= 44.5 \quad \text{or 45 minimum}$$

If we follow the layout of Fig. 11-3 each location should then provide a 22-circuit panelboard.

11-7 ***Selection and types of distribution equipment.*** Electrical systems in residential buildings, single-family houses, or multiunit buildings have the same problem with regard to providing suitable branch circuits. Circuits of various sizes must be provided: 15 A for general lighting,

20 A for laundry and kitchen appliances, 30 A for water heaters and large air conditioners, and 50 A for ranges.

To provide these different circuits in a single enclosure, several manufacturers provide circuit breakers of different capacities that can be inserted into a preassembled enclosure. These breakers are available in single pole for 115-V circuits or double pole for 230-V circuits. Also, they are available in a system of fuse assemblies that perform the same function. A main switch also may be incorporated into the same enclosure. A fused panel board is illustrated in Fig. 11-4.

The principal function of panel boards is to provide each branch circuit with its own fuse or circuit breaker. They may also be equipped with a switch for each circuit so they can be the control point for an electrical load.

The type of load determines the branch circuit ratings to be selected, according to Table 11-1. It is a good design practice to have more circuits than are necessary to provide either for a reduction in load on each branch

Fig. 11-4. Panel board for fuses. *Courtesy of Square D Company.*

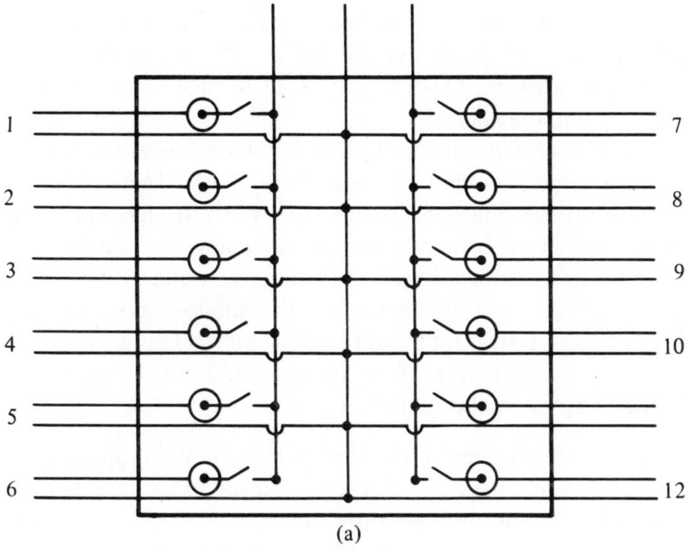

(a)

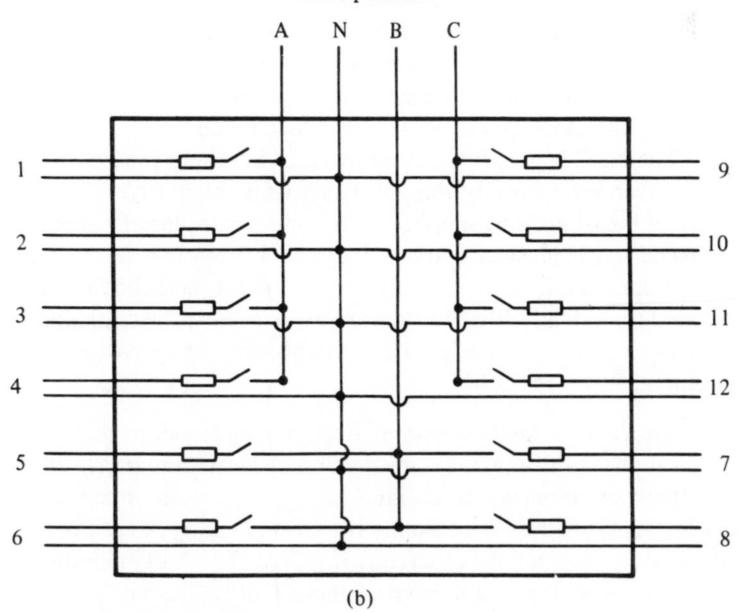

(b)

Fig. 11-5. (a) Three-wire, 115–230-V, single-phase panel. (b) Four-wire, 277–480-V, three-phase panel.

circuit or for an added load in the future. Many electrical installations become inadequate in a few years.

Branch circuits designed for 115 V are usually connected to panel boards that are supplied with current by means of three-wire, single-phase feeders. This is a 115–230-V supply. Only half the current is necessary as compared to a two-wire, 115-V supply. Half of the circuits are connected to each line wire and the neutral.

Commercial and industrial areas in which longer branch circuits are necessary use circuits designed to operate at 277 V. These panel boards are connected to a three-phase, four-wire, 277–480-V feeder. One third of the circuits are connected between each line wire and the neutral.

Figure 11-5 shows a diagram of a three-wire and a four-wire, 12-circuit panel board. Plug fuses are shown on the single-phase panel board, but cartridge fuses are shown on the three-phase one. Plug fuses cannot be used on circuits operating above 150 V to ground. Switches are shown on all circuits. SN refers to solid neutral.

11-8 ***Feeder definition and principles.*** A large building must have an electrical system capable of delivering power to its various loads over a considerable distance. It is the function of the feeders to deliver this power to the panel boards from which the branch circuits originate. Actually, there is no practical limit to the length a feeder may be extended. As this distance increases the voltage drop increases. We then can increase the circular-mil area by using a larger feeder and thereby decreasing the resistance of the feeder. This lowers the voltage drop to tolerable limits. There is a point, however, when increased feeder size becomes expensive, and we should investigate the advantages of using a higher voltage.

The NEC (with a few exceptions) requires every conductor of a system to have overcurrent protection according to its ampacity. Each branch circuit panel board must be supplied with power by means of a separate feeder. This feeder itself must be protected against overcurrent at its source.

A large building with many branch circuit panel boards must have a switchboard or load center to provide overcurrent protection and a disconnecting means for each feeder. This switchboard can be part of the entrance equipment.

11-9 ***Feeder size—load evaluation.*** Each branch circuit panel board must be supplied with power by means of a separate feeder. This feeder must have an ampacity to provide for the maximum anticipated load. In most installations this is not simply the current rating of each branch circuit multiplied by the number of circuits involved. The NEC permits the feeder load to be calculated on the basis of the actual total branch circuit load, in amperes.

To assure adequate feeder and branch circuit capacities for lighting in buildings, the NEC provides a minimum requirement of watts per square foot for various occupancies. Table 11-2 lists these requirements. A branch circuit provides this power for a small area. A feeder must do the same for the larger area it serves.

The NEC recognizes that as some types of occupancies become larger the possibility of using the entire load according to Table 11-2 becomes

Table 11-2. General lighting load by occupancies

Type of Occupancy	Load per ft^2 (W)
Armories and auditoriums	1
Banks	2
Barber shops and beauty parlors	3
Churches	1
Clubs	2
Court rooms	2
Dwellings (other than hotels)	3
Garages: commercial (storage)	½
Hospitals	2
Hotels and motels or apartment houses without cooking facilities for tenants	2
Industrial commercial (loft) buildings	2
Lodge rooms	1½
Office buildings	5
Restaurants	2
Schools	3
Stores	3
Warehouse storage	¼

Table 11-3. Demand factors applied to lighting loads

Type of Occupancy	Portion of Lighting Load to which Demand Factor Applies (W)	Feeder Demand Factor (%)
Dwellings other than hotels	Up to 3000	100
	3001–120,000	35
	Remainder over 120,000	25
Hospitals	Up to 50,000	40
	Remainder over 50,000	20
Hotels and motels and apartments without cooking facilities for tenants	Up to 20,000	50
	20,001–100,000	40
	Remainder over 100,000	30
Storage warehouses	Up to 12,500	100
	Remainder over 12,500	50
All others	Total watts	100

unlikely. Table 11-3 lists the demand factors that may be applied to the load that is calculated on the basis of watts per square foot. It obvious that the feeder to one apartment of a large apartment building must be large enough for the entire load even if this should occur for a few minutes a day. However, a feeder to supply a number of apartments can be smaller than necessary for the total maximum load of all apartments.

A store or a school would most likely operate all its lighting simultaneously; therefore, a 100% demand factor must be used.

When large appliances are used in multifamily residential buildings, the NEC recognizes that these appliances are not used simultaneously nor at full rating. The feeders for a group of apartments or the service entrance conductors for the entire building are calculated accordingly. Table 11-4

Table 11-4. Demand loads for household electric ranges. Not more than 12-kW

Number of Ranges	Maximum Demand (kW)
1	8
2	11
3	14
4	17
5	20
6	21
7	22
8	23
9	24
10	25
11	26
12	27
13	28
14	29
15	30
16	31
17	32
18	33
19	34
20	35
21	36
22	37
23	38
24	39
25	40
26–40	15-kW plus 1-kW for each range
Over 41	25-kW plus ¾-kW for each range

For ranges over 12-kW rating up to 27-kW the demand shall be increased by 5% for each kilowatt above 12-kW.

lists the maximum demand expressed in kilowatts for electric ranges if not more than 12-kW rating each. Table 11-5 lists the demand factor for household electric clothes dryers expressed in percent of total kilowatt rating of all dryers.

For all residential buildings 3000 W must be included for small appliances for each unit. If no laundry facilities are provided, an additional 1500 W must be added. The demand factors in Table 11-3 apply to the sum of the lighting load and the small appliance load.

Table 11-5. Demand factors for household electric clothes dryers

Number of Dryers	Demand Factor (%)
1	100
2	100
3	100
4	100
5	80
6	70
7	65
8	60
9	55
10	50
11–13	45
14–19	40
20–24	35
25–29	32.5
30–34	30
35–39	27.5
40 and over	25

Example ***11-5.*** The electrical load of Ex. 11-2 is 473 A at 115 V, divided between two branch circuit panel boards. Calculate the size of the feeder necessary for each panel. The supply is 115–230 V, three wire, and single phase.

Solution

$$\text{Current to each panel at 115 V} = \frac{473}{2} = 236.5 \text{ A}$$

$$\text{Current if supply is 115–230 V (see Ex. 4-9)} = I = \frac{236.5}{2} = 118.25 \text{ A}$$

From Table 9-4 the proper feeder would be three copper conductors size No. 0 type TW or No. 1 type TH. If aluminum conductors, the size would be No. 000 type TW or No. 0 type TH.

Example ***11-6.*** An apartment building consists of 20 apartments, each 900 ft^2 in area and equipped with an 11-kW range. The power supply is 115–230 V.

(a) Calculate the load for each apartment and the proper size feeder.
(b) Calculate the entire building load and the proper size feeder or service entrance conductors.

Solution

(a) The lighting load for each apartment is

$(900\ ft^2)(3\ W/ft^2)$	=	2700 W
Small appliance load	=	3000 W
Total lighting and appliance load	=	5700 W

The application of demand factor is

3000 W at 100%	=	3000 W
2700 W at 35%	=	945 W
		3945 W
Electric range load	=	8000 W
Total load for each apartment	=	11,945 W

The current required for each apartment is

$$I = \frac{P}{E} = \frac{11{,}945}{230} = 52\ \text{A} \qquad (3\text{-}4)$$

The feeder size to panel for each apartment from Table 9-4 is No. 6 type TW copper or No. 4 type TW aluminum.

(b) The total load of 20 apartments is as follows: The lighting and small appliance load is

$$(20\ \text{apartments})(5700\ \text{W}) = 114{,}000\ \text{W}$$

The application of demand factor is

3000 W at 100%	=	3,000 W
111,000 W at 35%	=	39,000 W
Net computed load without ranges	=	42,000 W
Load for 20 ranges from Table 11-4	=	25,000 W
Total computed load	=	67,000 W

$$I = \frac{P}{E} = \frac{67{,}000}{230} = 291\ \text{A} \qquad (3\text{-}4)$$

The size of the feeder from Table 9-4 is 500 MCM type TW copper or 700 MCM type TW aluminum.

It should be noted that the feeders selected are the minimum size for the calculated load. Using a larger size for an anticipated increase in load is usually written into electrical specifications. Also, this example makes no provisions for "house" lights or equipment or elevators that are metered separately.

Observing the results of (a) and (b) we must provide each apartment with a feeder sufficient for 52 A. However, demand factors as allowed by the NEC recognize that the maximum anticipated current would never exceed 291 A. The feeder is selected accordingly.

---

*Example **11-7**.* A single-family residence has the following electrical load: The area is 2000 ft^2 with 14-kW of space heating, an 11-kW range, a 5-kW clothes dryer, a 1.2-kW dishwasher, and a 2.5-kW water heater. Calculate the load in amperes and the size of service equipment.

Solution

Lighting load (2000)(3)	= 6,000 W
Small appliance load	= 3,000 W
Laundry circuit	= 1,500 W
Load to which demand factor can be applied	= 10,500 W

The demand factor application is

3000 W at 100%	= 3,000 W
7500 W at 35%	= 2,620 W

The load that must use 100% demand is

Space heating	= 14,000 W
Range	= 8,000 W
Clothes dryer	= 5,000 W
Dishwasher	= 1,200 W
Water heater	= 2,500 W
Total calculated load	= 36,320 W

The current required is

$$I = \frac{36{,}320}{230} = 158 \text{ A} \tag{3-4}$$

200-A service equipment is necessary, using No. 000 RH copper or No. 0000 aluminum conductors. (This is a special provision by NEC.)

Example ***11-8.*** A 60-kW incandescent lighting load is connected to a 120–208-V, three-phase supply. The load is balanced. Calculate the feeder size.

Solution

The current in each line wire is

$$I_L = \frac{P}{\sqrt{3}\, E_L \cos\theta} \qquad (5\text{-}4)$$

$$= \frac{60{,}000}{(1.732)(208)(1)} = 167 \text{ A}$$

From Table 9-4, one of the following conductors might be selected: No. 000 type TW or No. 00 type THW copper or 250 MCM type TW or No. 0000 type TWH aluminum. Four conductors are necessary.

Example ***11-9.*** A large factory assembly area is illuminated with 277-V fluorescent fixtures. The source is 277–480 V and three phase. The load consists of 200 units requiring 1.2 A each. They each have a power factor of 0.8.

(a) What size feeder (four wires) is necessary?
(b) Draw a sketch of the branch circuit panel and feeder.
(c) What is the cost of operating this lighting load per month of 20 days, 10 hr/day? (The energy is 0.02/kWh.)
(d) If the feeder is 200 ft long, what is the cost per month of power lost in feeders?

Solution

(a) The current to load if the supply is 277 V and single phase is

$$I = (200)(1.2) = 240 \text{ A}$$

The use of a 277–480-V supply permits one-third of the load to be connected to each phase, between a line wire and neutral (wye connected). The current per line is

$$I = \frac{240}{3} = 80 \text{ A}$$

From Table 9-4 select No. 4 AWG copper type THW conductor. (Four are necessary.)

(b) Refer to Fig. 11-6. Using 20-A circuits, each circuit can accommodate

$$\frac{16 \text{ A}}{1.2 \text{ A}} = 13.3 \quad \text{or 13 fixtures (maximum)}$$

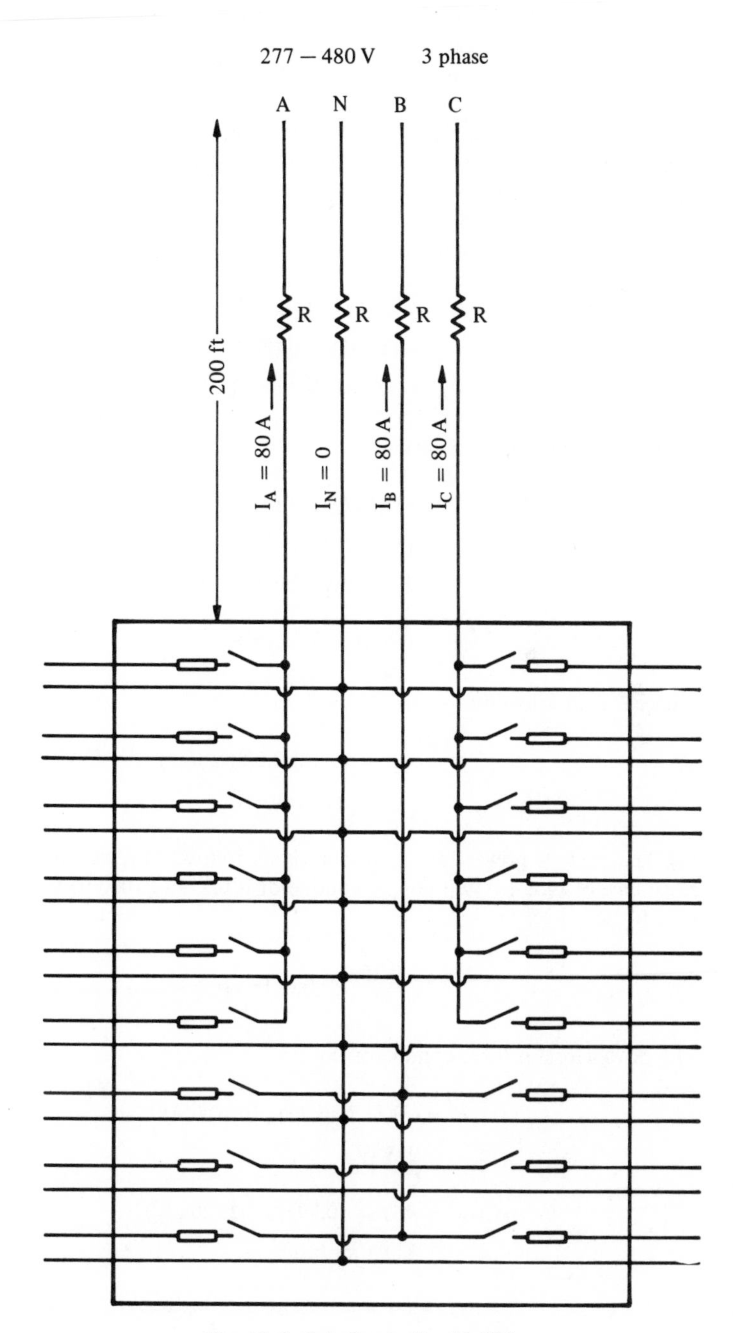

Fig. 11-6. Solution to Ex. 11-9(b).

The number of branch circuits is

$$\frac{200 \text{ fixtures}}{13 \text{ fixtures/ckt}} = 15.4 \quad \text{or 16 circuits}$$

A minimum of an 18-circuit panel would be necessary, six on each phase (from each line wire to neutral). To assure balanced line currents, 16 circuits could have 11 fixtures each and two circuits would have 12, making a total of 200 lighting units.

(c) To calculate the total cost per month using line voltage and current,

$$\begin{aligned} P &= \sqrt{3}\,(E_L I_L \cos\theta) \qquad (5\text{-}4)\\ &= (1.732)(480)(80)(0.8)\\ &= 53{,}200 \text{ W} \quad \text{or } 53.2 \text{ kW} \end{aligned}$$

Using phase voltage and current,

$$\begin{aligned} P &= (3)(E_p I_p)(\cos\theta) \qquad (5\text{-}3)\\ &= (3)(277)(80)(0.8)\\ &= 53{,}200 \text{ W} \quad \text{or} \quad 53.2 \text{ kW} \end{aligned}$$

$$\text{Total cost} = PCt \qquad (3\text{-}8)$$

where P is in kilowatts.

$$\begin{aligned} \text{Total cost} &= (53.2)(0.02)(20)(10)\\ &= \$212/\text{month} \end{aligned}$$

(d) The cost of power loss in feeders is as follows: From Table 9-7 the resistance of No. 4 AWG copper conductor is 0.259 Ω/1000 ft. Each feeder (200 ft) has

$$\frac{0.259}{5} = 0.0518\ \Omega$$

The power loss in three conductors is

$$\begin{aligned} P &= (3)(I^2R) = (3)(80)^2(0.0518) \qquad (3\text{-}6)\\ &= 995 \text{ W} \end{aligned}$$

$$\begin{aligned} \text{Cost} &= PCt = (0.995)(0.02)(20)(10) \qquad (3\text{-}8)\\ &= \$3.98/\text{month} \end{aligned}$$

11-10 ***Voltage drop in feeders.*** The calculations of feeders in Sec. 11-9 make no allowances for voltage drop. We must design any system with sufficient feeder circular-mil area to assure that the voltage drop does not exceed specifications under the maximum possible computed load. Usually systems operate below this level, making the voltage drop problem less serious.

The NEC requirements permit a maximum of 3% voltage drop in a branch circuit, with the further stipulation that it cannot be more than 5% when including the feeder.

Actually this is a rather generous provision. If all circuits operated with a 5% loss in voltage we are wasting 5% of the power that is being paid for. By good design the branch circuit voltage loss can be minimized by limiting their extended distance to less than 100 ft. We can design feeders to operate under maximum computed load with no more than 2% voltage drop. We then can have an even more efficient system than the NEC permits.

We must remember that NEC requirements are minimum and that individual specifications for a particular building may be more demanding than those of the NEC.

Consideration of voltage drop gives us two provisions that must be satisfied: First and foremost, the ampacity of the conductor must be sufficient for the anticipated load. If the distance the feeder must be routed is excessive, an increase in circular-mil area is necessary. To illustrate an excessive distance, consider Ex. 11-10 carefully.

Example ***11-10.*** A group of electric drying ovens used in a manufacturing process require 60 kW of power at 230 V, single phase. The routing of the feeder requires it to be 400 ft long. Refer to Fig. 11-7.

(a) Calculate the feeder size according to ampacity Table 9-4, using copper THW conductors.
(b) Calculate the voltage drop in the feeder of (a).
(c) Calculate the feeder size of sufficient circular-mill area to limit the voltage drop to 2%.

Solution

(a)

$$I = \frac{P}{(E)(\text{pf})} = \frac{60{,}000}{(230)(1)} \qquad (4\text{-}5)$$

$$= 261 \text{ A}$$

(Note: Heating elements have unity pf.) From Table 9-4 select a 300 MCM conductor. (Only two conductors are necessary.)

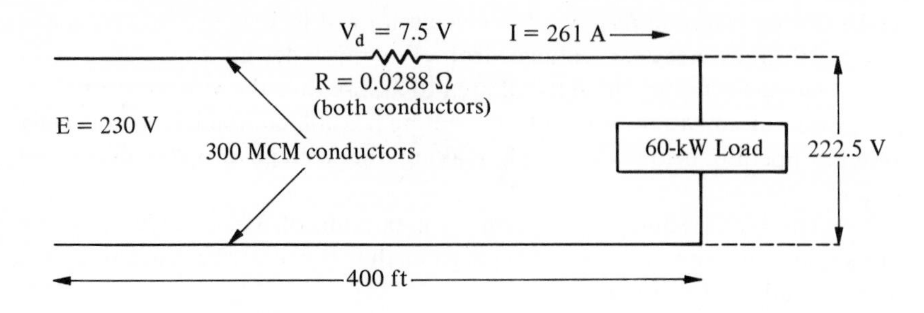

(a) and (b)

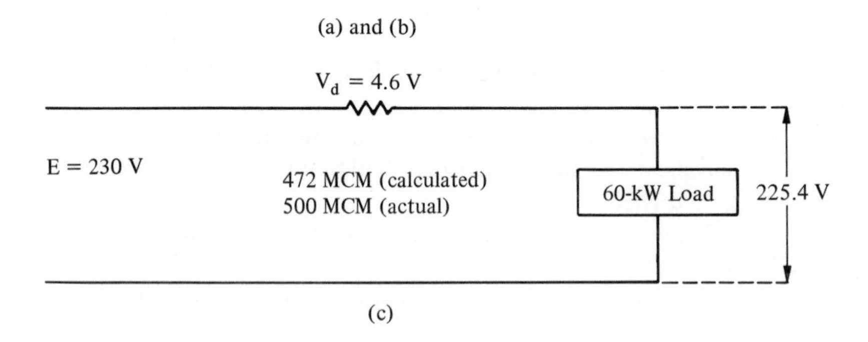

(c)

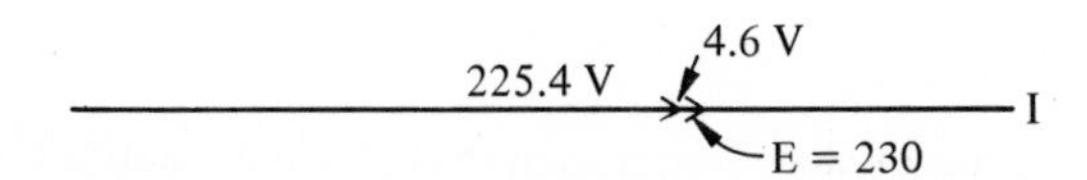

(d) Vector Addition of Voltage Drop and Load Voltage when **pf** = 1.0

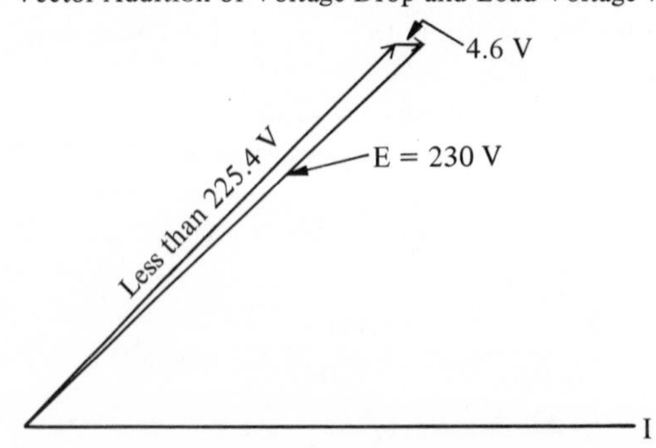

(e) Vector Addition of Voltage Drop and Load Voltage when **pf** = 0.7.

Fig. 11-7. Diagram and solution to Ex. 11-10.

(b) From Table 9-7, the resistance of 1000 ft of 300 MCM conductor is 0.0360 Ω. The resistance of 800 ft is

$$R = (0.036)(0.8) = 0.0288\ \Omega$$

$$= \frac{\rho L}{\text{cmil}} \qquad (2\text{-}5)$$

$$= \frac{(10.4)(800)}{300{,}000} = 0.0277\ \Omega$$

(There is a slight difference of 3.8%, probably because of consideration given to the stranding of cable in Table 9-7.)

$$\text{Voltage drop} = IR = (261\ \text{V})(0.0288\ \Omega) = 7.5\ \text{V}$$

$$\text{Per cent voltage drop} = \frac{7.5\ \text{V}}{230\ \text{V}}(100) = 3.26\%$$

(c) A direct approach to calculating the feeder size would use Eq. (9-1). This includes all factors that contribute to conductor circular-mil area.

$$\text{cmil} = \frac{(\rho_{ID})(2)}{V_{\text{drop}}} = \frac{(10.4)(261)(400)(2)}{(0.02)(230)} = 472{,}000\ \text{cmil} \qquad (9\text{-}1)$$

The next largest standard size listed in Table 9-7 is 500 MCM. The additional 28 MCM permits the voltage drop to be slightly less than 2%.

Figure 11-7 shows a series circuit of conductor resistance and the load. The voltage drop and load voltage are added arithmetically to equal the supply voltage. Actually this is only possible when load pf is unity because conductor voltage drop is always in phase with the current.

As the pf of the load becomes smaller, the conductor voltage drop will have less effect in reducing voltage at the load. The effect is slight when the pf is above 0.8. The vector addition of Fig. 11-7d illustrates Ex. 11-10c. The vector of Fig. 11-7e shows the addition if the pf were 0.7. Obviously at unity pf the voltage drop will have its greatest effect on the load.

Consideration of the results in Ex. 11-10 leads to some important observations. Part (a) shows that the load requires 300 MCM feeders when considering only the ampacity table. This, however, causes almost a 3.27% voltage drop—nearly twice the tolerable limit. If the feeder were half as long, 200 ft, then the conductor resistance and resultant voltage drop would also be half, or less than the required 2%.

This is not to imply that 200 ft is the point that feeders must be made larger in circular-mil area to limit voltage drop. This situation will change as smaller feeders are used for smaller loads and as higher voltages are used.

Each situation must be examined in detail. The use of Eq. (9-1) is of great assistance in determining the proper feeder for any situation.

In Ex. 11-11 we investigate the effect of a three-phase system on feeder size. In Ex. 11-12 we use a higher voltage.

Example ***11-11.*** Design the feeders for the load of Ex. 11-10 using a three-wire, 230-V, three-phase supply. Assume that if 20 kW can be connected to each phase in delta the load will be balanced. Repeat (a), (b), and (c). Refer to Fig. 11-8.

Solution

(a) Calculate the feeder size according to Table 9-4. The current to the load is

$$I = \frac{P}{(1.732)(E)(\text{pf})} \tag{5-4}$$

$$\frac{60{,}000}{(1.732)(230)(1)} = 150 \text{ A}$$

From Table 9-4 select a No. 0 AWG conductor (Three conductors are necessary.)

(b) The voltage drop in the feeder for (a) is as follows: From Table 9-7, the resistance of 1000 ft of No. 0 AWG conductor is 0.102 Ω. The resistance of each conductor (400 ft) is

$$(0.102)(0.4) = 0.0408\ \Omega$$

The voltage drop in each conductor is

$$V_{\text{drop}} = IR = (150 \text{ V})(0.0408\ \Omega) = 6.12 \text{ V}$$

These three-line conductor voltage drops are in a wye configuration; therefore their combined voltage drop, between the supply and the load, is $(6.12)(\sqrt{3}) = 10.5$ V. The voltage available at each of the three loads in delta is $230 - 10.5 = 219.5$ V. This voltage drop may be calculated directly, using Eq. (9-1):

$$V_{\text{drop}} = \frac{(\rho_{ID})(\sqrt{3})}{\text{cmil}} \tag{9-1}$$

$$= \frac{(10.4)(150)(40)(1.732)}{105{,}600} = 10.2 \text{ V}$$

(From Table 9-7, a No. 0 AWG conductor is 105,600 cmil in area.)

$$\text{Per cent voltage drop} = \left(\frac{10.2}{230}\right)(100) = 4.45\%$$

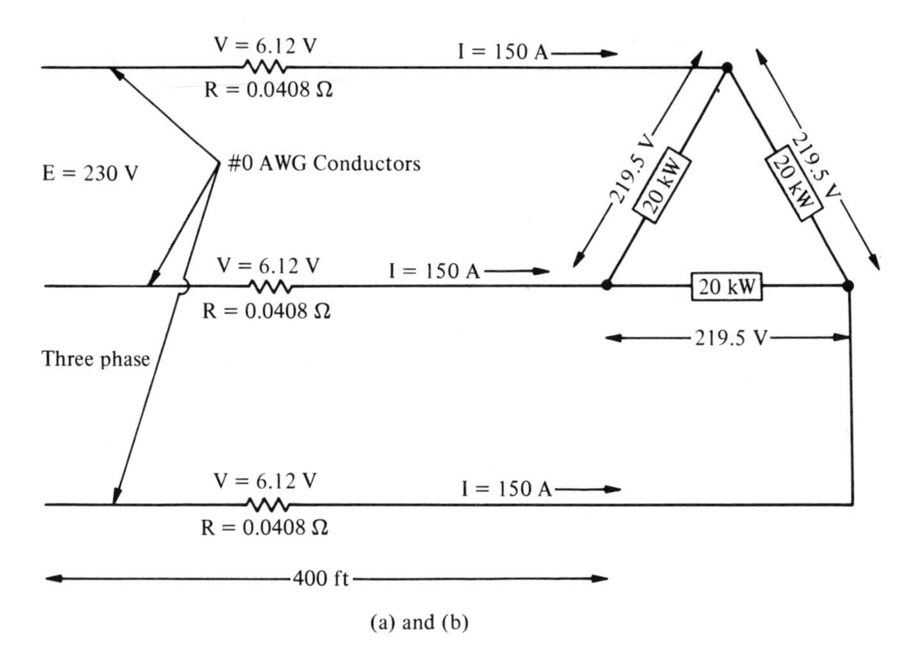

(a) and (b)

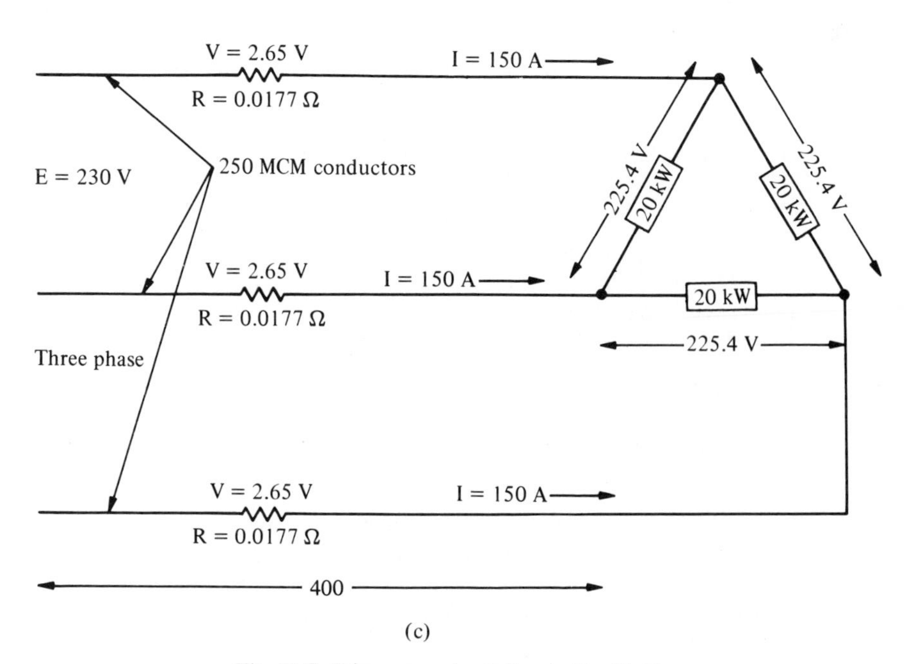

(c)

Fig. 11-8. Diagram and solution to Ex. 11-11.

(c) Calculate the correct feeder to limit voltage drop to 2%. Permitting a 2% voltage drop at the load will cause the load voltage to be

$$230 - 4.6 = 225.4 \text{ V}$$

The voltage drop in each conductor is

$$V = \frac{4.6}{\sqrt{3}} = 2.65 \text{ V}$$

The resistance of each conductor (400 ft long) is

$$R = \frac{V_{\text{drop}}}{I} = \frac{2.65 \text{ V}}{150 \text{ A}} = 0.0177 \ \Omega$$

The resistance per 1000 ft is

$$\frac{400 \text{ ft}}{1000 \text{ ft}} = \frac{0.0177 \ \Omega}{R}$$

$$R = \frac{1000}{400}(0.0177) = 0.044 \ \Omega$$

From Table 9-7, the conductor with the next lowest resistance is 250 MCM. Using Eq. (9-1),

$$\text{cmil} = \frac{(\rho ID)(\sqrt{3})}{V_{\text{drop}}} \qquad \text{(9-1)}$$

$$= \frac{(10.4)(150)(400)(1.732)}{(0.02)(230)} = 235{,}000 \text{ cmil}$$

The next largest standard size is 250 MCM.

---

*Example **11-12**.* Design the feeders for the load of Ex. 11-10 using a single-phase, 460-V supply. Heaters are now rated at 460 V but the same 60 kW. Repeat (a), (b), and (c). Refer to Fig. 11-9.

Solution

(a) Calculate the feeder size according to Table 9-4. The current to the load is

$$I = \frac{P}{(E)(\text{pf})} = \frac{60{,}000}{460(1)} = 130 \text{ A} \qquad \text{(4-5)}$$

From Table 9-4 select a No. 1 AWG conductor. (Only two conductors are necessary.)
(b) Calculate the voltage drop in (a). From Table 9-7 the resistance of 1000

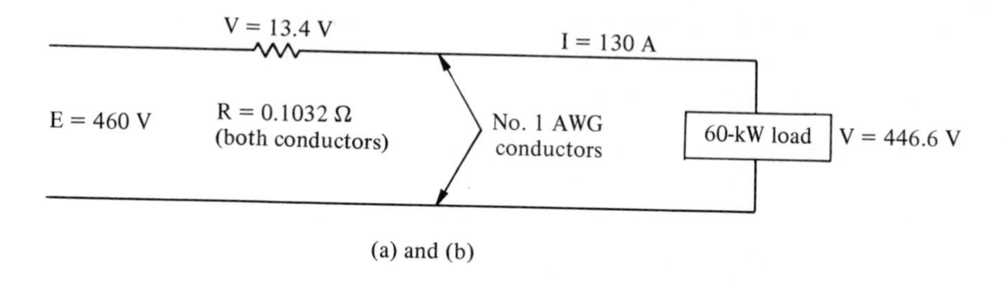

(a) and (b)

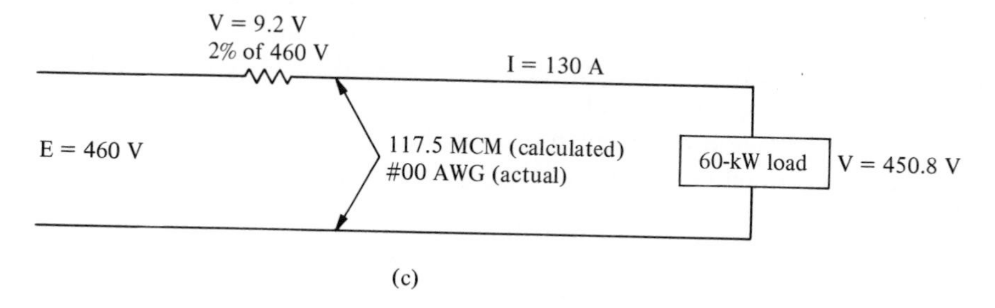

(c)

Fig. 11-9. Diagram and solution to Ex. 11-12.

ft of No. 1 AWG conductor is 0.129 Ω. The resistance of 800 ft is

$$R = (0.129)(0.8) = 0.1032\ \Omega$$

$$\text{Voltage drop} = IR = (130\ \text{A})(0.1032\ \Omega) = 13.4\ \text{V}$$

$$\text{Per cent voltage drop} = \frac{13.4}{460}(100) = 2.92\%$$

(c) Calculate the feeder size that will limit the voltage drop to 2%.

$$\text{cmil} = \frac{(\rho ID)(2)}{V_{\text{drop}}} \tag{9-1}$$

$$= \frac{(10.4)(130)(400)(2)}{(0.02)(460)} == 117{,}600\ \text{cmil}$$

The next largest size in Table 9-7 is No. 00 AWG.

To summarize the three previous examples, the least expensiveinstallation with regard to wiring materials would be as in Ex. 11-12, using a 460-V, three-phase supply.

It also should be noted that when a three-phase system is used instead of a single-phase system of the same voltage, conductors slightly over half the cross-sectional area (circular mils) are necessary; of course three conductors are necessary rather than two when using a single-phase system.

11-11 ***Feeders and subfeeders.*** Although each panel board is served by a separate feeder, it is not necessary that each feeder originate from the service entrance location. This would create a problem in large buildings by requiring many feeders to extend long distances.

One large feeder may be installed part of the distance to another switchboard. From this point subfeeders may be routed to the individual panel board. The switchboard must provide overcurrent protection according to the ampacity of each feeder. In many instances this will be in excess of the calculated load because of larger feeders due to voltage drop.

The current that must be carried by the feeder will be the sum of the calculated currents of the subfeeders.

The voltage drop calculations now become more complex since the 2% drop permitted in feeders must apply to all conductors from the service entrance to the panel boards.

---

*Example **11-13**.* Three 115-230-V panel boards are supplied with three subfeeders. A is 40 ft, B is 150 ft, and C is 200 ft in length. The load on A is 15 kW with a pf of 1.0; load B is 25 kW with a pf of 0.8; and load C is 25 kW with a pf of 0.7.

The three subfeeders are supplied with a feeder 150 ft in length. Calculate the size of the feeder and three subfeeders. Refer to Fig. 11-10. The problem may be simplified by allowing a 1% voltage drop in each subfeeder and 1% in the feeder. Use aluminum RH conductors.

Solution

$$I = \frac{P}{(E)(\text{pf})} \tag{4-5}$$

$$I_A = \frac{15{,}000}{(230)(1)} = 65 \text{ A}$$

$$I_B = \frac{25{,}000}{(230)(0.8)} = 136 \text{ A}$$

$$I_C = \frac{25{,}000}{(230)(0.7)} = 155 \text{ A}$$

$$\begin{aligned} I_A &= 65 + \text{j}\,0 \\ I_B &= 109 - \text{j}\,81.7 \\ I_C &= 108.5 - \text{j}\,110 \\ \hline I_T &= 282 - \text{j}\,191.7 \\ &= 342 \angle -34.2° \text{A} \end{aligned}$$

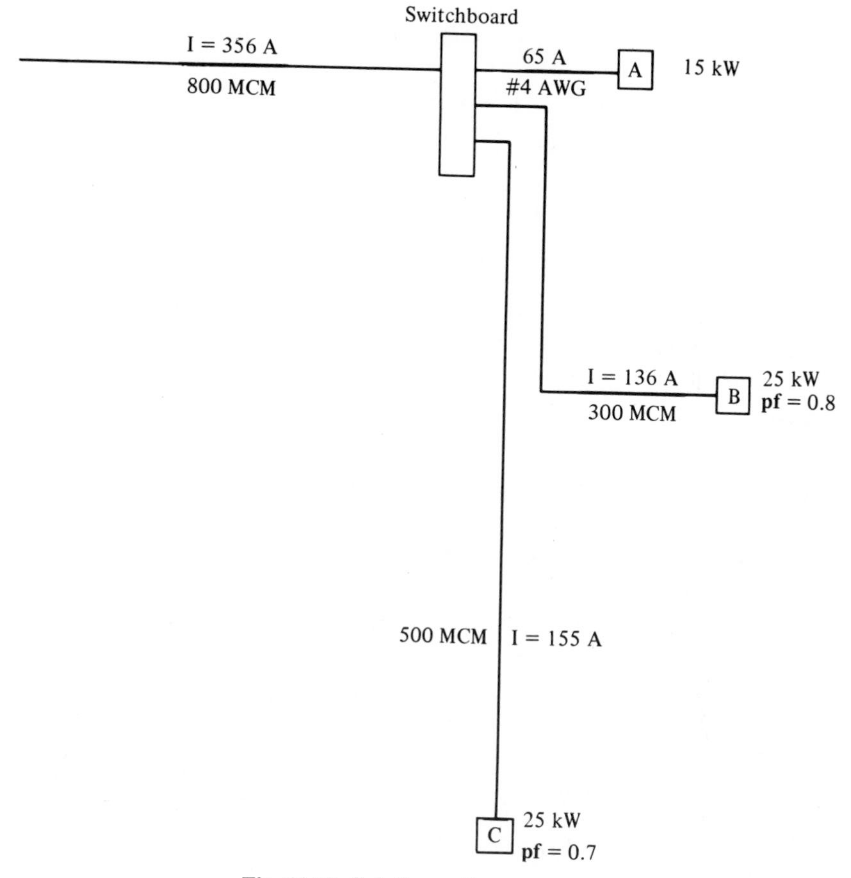

Fig. 11-10. Solution to Ex. 11-13.

Subfeeder for panel board A

$$\text{cmil} = \frac{\rho ID(2)}{V_{\text{drop}}} \tag{9-1}$$

$$= \frac{(17)(65)(40)(2)}{(0.01)(230)} = 38{,}400 \text{ cmil} \quad \text{or} \quad \text{No. 4 AWG}$$

According to Table 9-4, the ampacity of No. 4 AWG aluminum is only 55 A. Number 3 AWG must be used (it will cause a lower voltage drop); No. 4 AWG TH would have sufficient ampacity but the specifications stated TW conductors. The subfeeder for panel board B is

$$\text{cmil} = \frac{(17)(136)(150)(2)}{(0.01)(230)} = 300 \text{ MCM} \tag{9-1}$$

The subfeeder for panel board C is

$$\text{cmil} = \frac{(17)(155)(200)(2)}{(0.01)(230)} = 457 \text{ MCM} \tag{9-1}$$

$$\text{Nearest largest standard size} = 500 \text{ MCM}$$

To calculate the feeder,

$$\text{cmil} = \frac{(17)(356)(150)(2)}{(0.01)(230)} = 790 \text{ MCM} \tag{9-1}$$

$$\text{Nearest largest size} = 800 \text{ MCM}$$

The switchboard must have overcurrent protection for A set for 65 A. For B it may be up to 230 A. Panel board C may have protection as high as 310 A.

11-12 ***Selection of service equipment.*** Voltage drop calculations in feeders and branch circuits are made from the point of entrance of the service entrance conductors. The NEC requires the main disconnect switch to be located immediately adjacent to that point; therefore, the service entrance conductors themselves are relatively short in length. Any voltage drop problem before this switch or circuit breaker is the responsibility of the utility company.

The sizes or classifications of service equipment were outlined in Chapter 8.

The current rating of service entrance equipment (conductors and main switch) is determined by adding all the feeder loads. Consideration also must be given to future loads to prevent the installation from becoming obsolete

Fig. 11-11. Current transformers in cabinet. *Courtesy of Square D Company.*

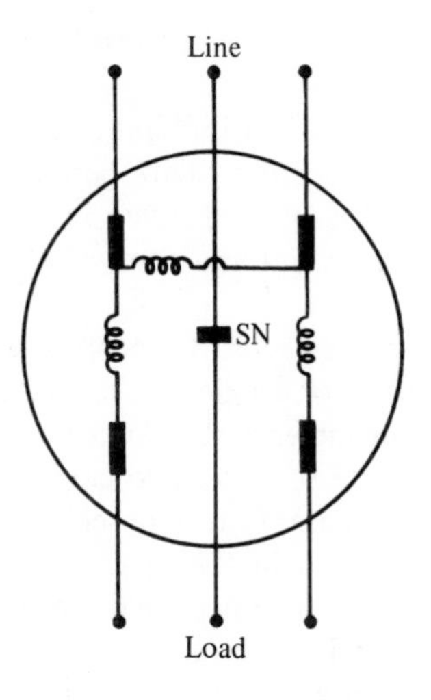

(a) Single Phase, Three Wire

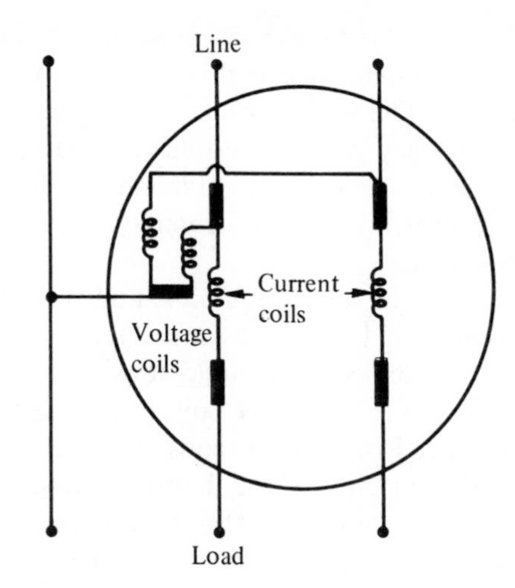

(b) Three Phase, Three Wire

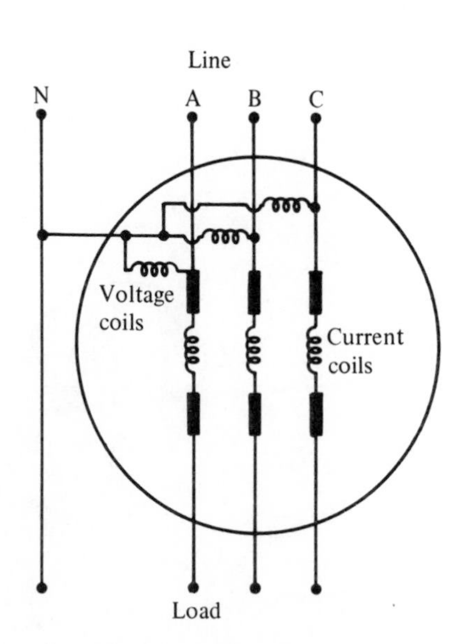

(c) Three Phase, Four Wire

Fig. 11-12. Kilowatt-hour meter connections.

in a short time. For example, the NEC requires a 100-A service in a residence if the initial calculated load is 10 kW or more.

The NEC permits as many as six service switches in a building; however, they must be grouped together in the same location. This permits separate metering for different types of loads such as water heaters in residences. Many small commercial buildings require single- and three-phase services.

The service equipment must include the metering devices to record the energy of the load as measured in kilowatt-hours. For single-phase, 115–230-V services this would consist of a kilowatt-hour meter with two current coils and one voltage coil. For three-phase, three-wire services, two current coils and two voltage coils are necessary. For three-phase, four-wire services three current coils and three voltage coils are included in the meter. For services rated over 200 A, current transformers are used to reduce the physical size of the kilowatt-hour meter. These are illustrated in Fig. 11-11.

The various coil configurations determine how the meter will be connected to the service equipment. Figure 11-12 shows these coils on the various types of meters. Most kilowatt-hour meters are plugged into meter

Fig. 11-13. A 100-A meter socket. *Courtesy of Square D Company.*

Fig. 11-14. A three-phase meter with circuit breaker. *Courtesy Square D Company.*

sockets or troughs. The dots (·) indicate the connections that must be brought into the meter trough.

Figure 11-13 illustrates a 100-A meter socket for a single-phase, three-wire service. Figure 11-14 shows a three-phase, four-wire socket equipped with seven jaws and including a circuit breaker for the main disconnect in the same enclosure.

For large power requirements (above 600 A) a switchboard is usually provided. This contains a circuit breaker as a main disconnecting means and a circuit breaker of proper capacity for the various feeders that carry power to panel boards or other switchboards throughout the building.

Figure 11-15 illustrates a typical switchboard. Suitable current transformers are included to facilitate metering of the load.

Fig. 11-15. Switchboard. *Courtesy of Square D Company.*

PROBLEMS

11-1 What branch circuit rating and conductor size would be used for each of the following devices: (a) a 1200-W, 115-V appliance, (b) a 2000-W, 115-V appliance, (c) a 2000-W, 230-V appliance, and (d) a 115-V electric clock using 4 W?

11-2 Describe the branch circuit necessary for each of the following electrical loads: (a) a laundry outlet, (b) an oil burner with a $\frac{1}{4}$-hp, 115-V motor, (c) a 3-kW, 115–230-V clothes dryer, (d) a 200-W lamp, and (e) eight 200-W lamps.

11-3 Describe the branch circuits necessary for each of the following electrical loads: (a) lighting for an area of 1200 ft^2, (b) a lighting load consisting of two hundred 150-W lamps, (c) a lighting load consisting of two hundred 200-W lamps, (d) a lighting load consisting of one hundred 500 W lamps, and (e) a lighting load of seventy-five 100 W lamps.

11-4 Determine the number of 20-A branch circuits necessary for each of the following: (a) a 40 kW fluorescent lighting load consisting of

200-W units with 0.8 pf and operating at 120 V from a 120–240-V, single-phase supply, (b) the same load but using a 120–208-V, three-phase supply, and (c) the same kilowatt load but with lighting units operating at 277 V from a 277–480-V supply.

11-5 Calculate the maximum distance a 230-V, 30-A branch circuit should extend when supplying a 5-kW clothes dryer. Limit the voltage drop to 2%.

11-6 Calculate the voltage drop in a 115-V, 15-A branch circuit supplying the largest single appliance permitted if the circuit is 75 ft long.

11-7 A portable power saw requires at least 110 V to operate satisfactorily for a specific load. What size portable cord 150 ft in length must be used from a 115 V outlet?

11-8 A particular power saw under a specific load draws 8 A at 115 V, 9 A at 110 V, and 11 A at 100 V. The saw is to be used 200 ft from a 118-V supply. Calculate the proper size portable cord necessary if the current is not to exceed 11 A.

11-9 A 115-V, 20-A branch circuit wired with No. 12 AWG conductors extends 125 ft. A 3-A lighting unit is connected at each 25-ft interval. Calculate (a) the voltage at the last outlet and (b) the voltage at the last outlet if the entire 15-A load were connected at the end of the circuit.

11-10 Calculate the longest distance a 115-V, 15-A branch circuit may extend if it is supplying the largest appliance permitted by the NEC. The voltage drop must be limited to 3%.

11-11 An assembly area uses 115-V, 200-W soldering irons. If five are connected to each branch circuit extending 100 ft, calculate the necessary branch circuit conductors. The voltage at the last outlet must be not less than 113 V.

11-12 Using 20-A circuits, design a system using one panel board for fifty 115-V, 200-W soldering irons. The longest circuit is to extend 100 ft in length, with a voltage of at least 113 V.

11-13 Determine the appropriately sized panel boards and the branch circuit rating for each of the following loads (the supply is 115–230 V): (a) 100 fluorescent lighting units, each with a nameplate rating of 1.7 A, (b) one hundred 500-W incandescent lighting units, and (c) two hundred 100-W incandescent lighting units and (d) make extra provisions if the lamps in (c) are increased to 150 W.

11-14 A store 200 × 150 ft is to be provided with 250 fluorescent units requiring 1.9 A each at 115 V. Use 20-A branch circuits not over 80 ft in length. Design an appropriate panel board arrangement.

Because of the store's function, the panel boards must be located on outside walls.

11-15 A production area measuring 150 × 75 ft requires one hundred 100-W soldering irons and 200 W for illumination at each location. Design a panel board layout using 20-A branch circuits. Individual circuits may contain both appliances and lighting but should not extend over 90 ft from the panel boards.

11-16 An aircraft assembly plant 1000 × 500 ft in area is illuminated with 2500 fluorescent units operating at 277 V and requiring 1.4 A each. Design a panel board layout using 30-A branch circuits and limit their length to 250 ft. The lighting units are 50 ft from the floor, and this distance must be considered in the length of branch circuits.

11-17 Determine the minimum branch circuit panel board necessary for a two-story residence 36 × 28 ft with the following appliances: at 230 V, a 9-kW range, a 4-kW clothes dryer, and a 3-kW water heater; at 115 V, a 1.2-kW dishwasher and two 10-A air conditioners. In addition to lighting circuits required by the NEC, provide for three extra 15-A circuits for receptacles for other appliances.

11-18 An apartment building has 50 electric ranges, each rated at 10 kW. Determine the maximum anticipated load for the ranges.

11-19 A 30-apartment building has individual laundry facilities in each apartment. The clothes dryers are rated at 3.5 kW. Determine the calculated load for the dryers.

11-20 Determine the total load in kilowatts and the current for a 25-apartment building. Each apartment is 900 ft^2 in area and has a 9-kW range. The public laundry area has four washers and four 4-kW clothes dryers.

11-21 A hospital has a total floor area of 35,000 ft^2. Determine the lighting load in kilowatts according to the minimum NEC requirements.

11-22 A 15-story office building measures 50 × 75 ft. Determine the lighting load in kilowatts.

11-23 A 20-story hotel measures 40 × 100 ft. Determine the lighting load in kilowatts.

11-24 An all-electric single-family residence has two stories and measures 40 × 30 ft. The permanently installed equipment is as follows: an 11-kW range, a 4-kW clothes dryer, a 3-kW water heater, a 1.2-kW dishwasher, and a 20-kW space heater.

(a) Calculate the total load in kilowatts.

(b) Determine the size of the service entrance equipment.

11-25 From the ampacities of Table 9-4 determine the distance a No. 00 AWG feeder may extend. Limit its voltage drop to 2% of the 120–240-V supply, using the following conductors: (a) type T copper, (b) type TH copper, (c) type T aluminum, and (d) type TH aluminum.

11-26 From the ampacities of Table 9-4 determine the distance a No. 00 AWG feeder may extend. Limit its voltage drop to 2% of the 120–208-V, three-phase supply, using the following conductors: (a) type T copper, (b) type TH copper, (c) type T aluminum, and (d) type TH aluminum.

11-27 From the ampacities of Table 9-4 determine the distance a No. 00 AWG feeder may extend. Limit its voltage drop to 2% of the 277–480-V, three-phase supply, using the following conductors: (a) type T copper, (b) type TH copper, (c) type T aluminum, and (d) type TH aluminum.

11-28 Three panel boards evenly distribute a 75-kW, 115–230-V, unity pf load. They are located 100, 150, and 200 ft from the main switchboard and use aluminum conductors.

(a) Calculate the size feeders necessary for each panel board. Limit the voltage drop in each feeder to 2%.

(b) Calculate the size feeder and three subfeeders needed if the feeder extends 100 ft to the location of the first panel board. Limit the total voltage drop in the feeder and each subfeeder to 2%.

Chapter Twelve

Illumination

12-1 ***Unit of illumination—the lumen.*** Previous references to lighting loads have been in terms of current or power. A lighting system cannot be designed only on the basis of the power expended. The basic NEC illumination requirement that feeders in a building must be of sufficient size to provide 3 W/ft^2 for illumination does not dictate any *level* of illumination. It is the purpose of this chapter to start with a desired level of illumination and then to calculate the necessary power and current.

Light *sources* are usually either incandescent or fluorescent *lamps. Luminaires* are the devices designed to *hold* these lamps. Many are decorative in nature, and as such are referred to as lighting fixtures. A luminaire may be somewhat decorative, but its principal function is to *distribute* and *diffuse* the light from the lamps *efficiently*. Glare must be kept to a minimum.

There are too many variables that affect lighting to guarantee that a specific amount of power expended will result in a desired level of illumination. Some of these variables are the type of light source, the construction of the luminaire used, the reflection from the surrounding areas, the cleanness of the light sources over a period of time, and the distances from the light source to the work plane. The work plane is the point where the level of light intensity is to be measured.

To properly design a lighting system we shall use two units of illumination that are interrelated: the *lumen* and the *foot-candle*. Like the unit of mechanical work, the horsepower (originating from the premachine age), the footcandle seems to emerge from the candle era. One footcandle (fc) is the amount of light at a point 1 ft from a candle (lighted of course).

By placing the candle in the center of a sphere with a radius of 1 ft, it should be obvious that every point on the sphere is illuminated to the level of 1 ft, since each point is 1 ft from the light source. It is emphasized that foot-candles are a measure of illumination at a point; no area is involved. The *intensity* of illumination is measured with a photoelectric device called a *foot-candle meter*.

To properly define areas to be illuminated by light sources, we must define the *lumen* (lm): the amount of light falling on each square foot of our sphere. Now we have the involved area. The foot-candle meter indicates 1 fc at any point within the square foot (or any point on the sphere). Actually, the meter reads "lumens per square foot," but the surface of the meter is not 1 ft^2 in area. If we wish to have a level of illumination of 50 fc, then we must provide 50 lm for every square foot of area involved. Thus, 1 fc = 1 lm/ft^2.

Our efforts in this chapter are directed to the design problems associated with allocating a specific number of lumens to each square foot of work plane.

Distance is an important consideration. Light varies *inversely* as the *square* of distance from the light source; twice the distance will reduce the illumination to one-fourth.

Figure 12-1 shows our sphere with a radius of 1 ft with the light source of one candle in the center. A hole with an area of 1 ft^2 is cut so that exactly 1 lm can emerge from the sphere. If we place a rounded screen 2 ft from the light source the direct rays will fall upon 4 ft^2. We have only 1 lm from the source; therefore each square foot located 2 ft from the source receives only $\frac{1}{4}$ lm.

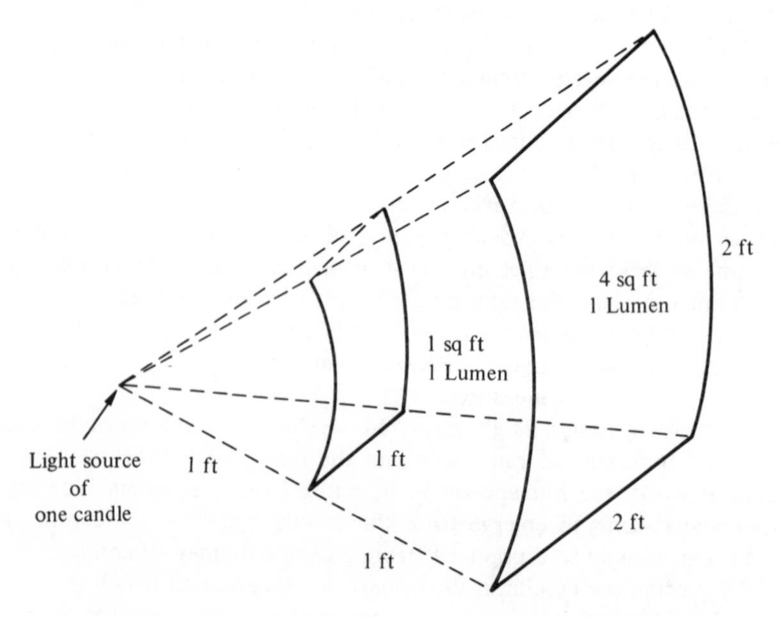

Fig. 12-1. Light intensity varies with the square of the distance from the light source.

12-2 ***Light sources—incandescent.*** The tungsten filament is the principal light source in *incandescent* lamps. Tungsten has a very high melting point, 6120°F, and therefore may operate at a high temperature from 2000 to 3000°F.

According to the temperature vs resistance graph of Fig. 2-1 the resistance of the cold tungsten filament is very small compared to its resistance at rated voltage. This low resistance permits a very large current to flow when the light is turned on. This high current causes the lamp to glow very quickly, resistance rises rapidly, and rated current is maintained. These changes take place in less than 0.2 sec, thus causing the lamp to glow practically instantly.

The high inrush current, although short in duration, requires that switches in incandescent lighting circuits have heavy contacts. This type of switch is called a *T-rated switch.* Its listed current rating is for the lamp's transient inrush current. For example, a 100-W tungsten filament requires 0.87 A at 115 V, but its starting current is about 9 A. Such a lamp requires a switch with a 10-A T rating.

Incandescent lamps are purely resistive and therefore operate at unity power factor. They can be used in practically any environment if suitably protected from the weather.

Table 12-1 lists some of the common sizes of tungsten filament lamps, giving their lumen output and expected life. Lamps deteriorate with use: The blackened inner surface of the bulb results from deposits of tungsten, causing the lumen output to decrease. The mean (average) lumens listed in Table 12-1 are reasonable values to use in light calculations.

To accurately design lighting systems, the voltage received by lamps is extremely important. A 1% decrease below rated voltage results in a 3% decrease in lumen output, although the life of the lamp is increased by about 20%.

Table 12-1. Characteristics of incandescent lamps

Watts Standard Base	Initial Lumens	Mean Lumens	Life (hr)
25	265	230	1000
40	470	415	1000
60	840	795	1000
100	1,640	1,540	750
150	2,700	2,420	750
200	3,800	3,400	750
300	5,750	5,150	1000
Mogul base			
300	5,750	5,150	1000
500	9,900	8,750	1000
750	16,700	14,800	1000
1000	23,300	20,080	1000

Figure 12-2 shows, graphically, the relationships among rated voltage and current, power, and lumen output. All are expressed in per cent of rated values.

Figure 12-3 shows the change in power and lumen output as the lamp ages. The 100% point on this graph would be the life in hours of Table 12-1. From this graph we note that at the end of its life expectancy an incandescent lamp is producing only about 83% of its initial lumen output, but its power consumption has decreased to only 95%. This means that relamping regularly is necessary if lighting levels are to be maintained.

12-3 ***Light sources—fluorescent.*** Of all light sources, fluorescent lamps are the most extensively used. They provide more lumens per watt than incandescent lamps. Light is distributed over a larger lamp surface area. When installed in properly designed luminaires, they can distribute their lumen output in a manner that may approach the ideal situation in which

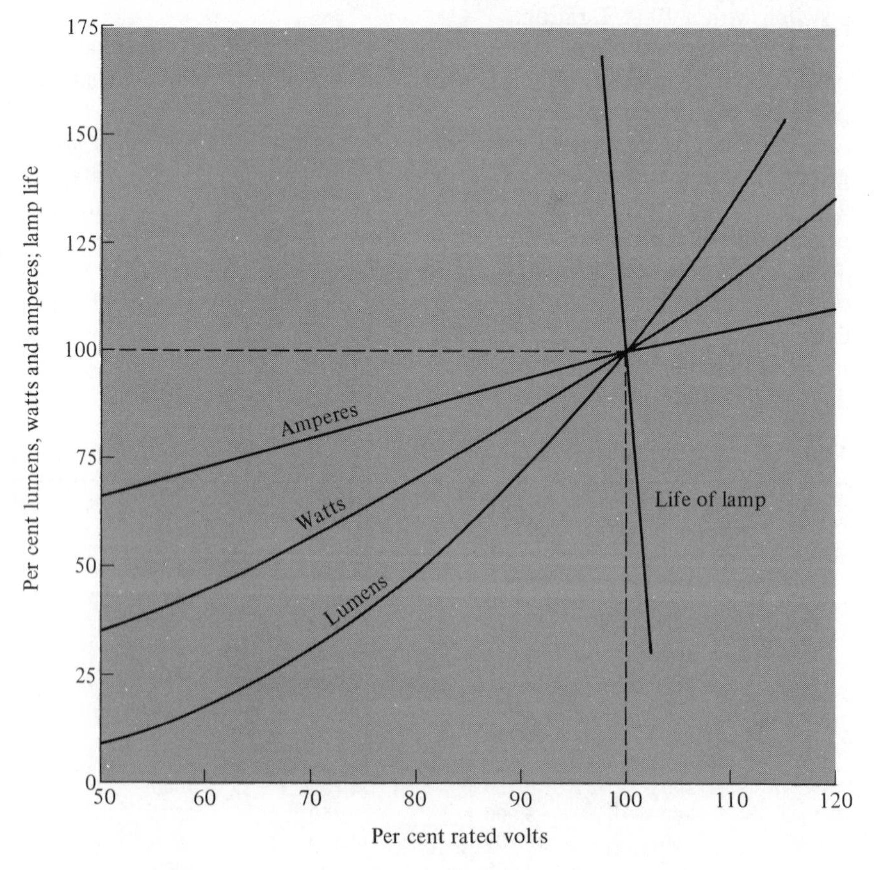

Fig. 12-2. Voltage characteristics of incandescent lamps.

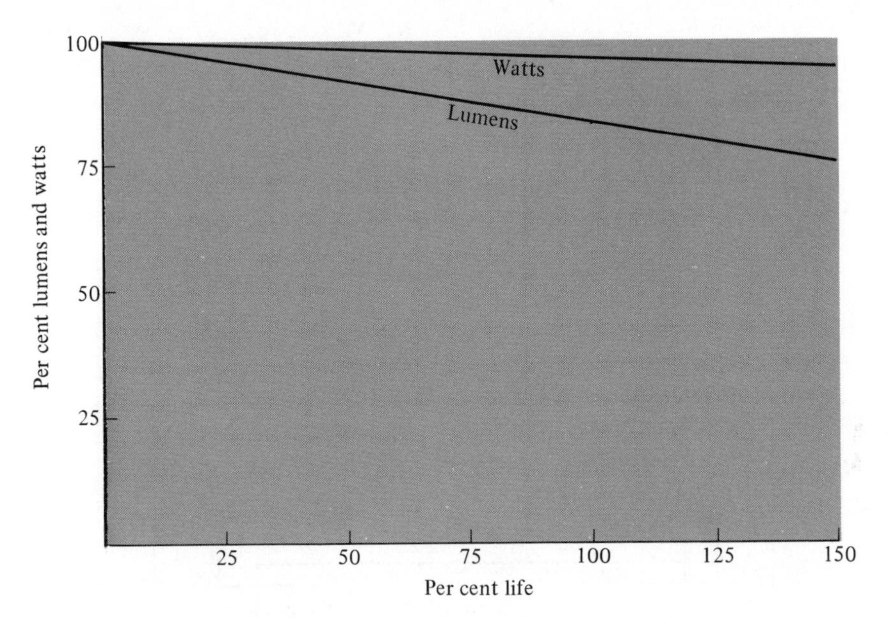

Fig. 12-3. Loss of light due to use of incandescent lamps.

the entire ceiling area emits light. Unlike incandescents, they cannot operate satisfactorily under adverse conditions of humidity or temperature unless suitably protected in a separate enclosure.

All fluorescent lamps operate on the same principle. Each end has an electrode called a *cathode*. The inside of the tube is covered with phosphor. This coating will fluoresce (glow) when current flows from the electrode at one end through mercury vapor to the electrode at the other end. Lamps must be used with proper auxiliary equipment, transformers, and ballasts to limit current after conduction in the lamps has started.

Two basic types of lamps are used with a few modifications of auxiliary starting devices. One type has two contact pins in each end; the other, only one. Many luminaires use ballasts that must use starters. Figure 12-4 shows a diagram of a luminaire using two 40-W lamps.

An autotransformer raises the voltage to a proper starting voltage. The starters permit current flow through the tungsten filaments at each end. After a few seconds the heated filament emits electrons, and as the starter automatically opens, a high voltage causes the current to flow from the filament through the mercury vapor to the lamp coating. Once conducting, the lamp current will be limited by the reactance of the ballast.

The capacitor in series with one lamp enables the current in one lamp to be out of phase with that of the other lamp. This causes the lamps to go on and off at different times with respect to the ac input voltage. If the lamps operated together and no phase shifting was used, a *stroboscopic* effect would

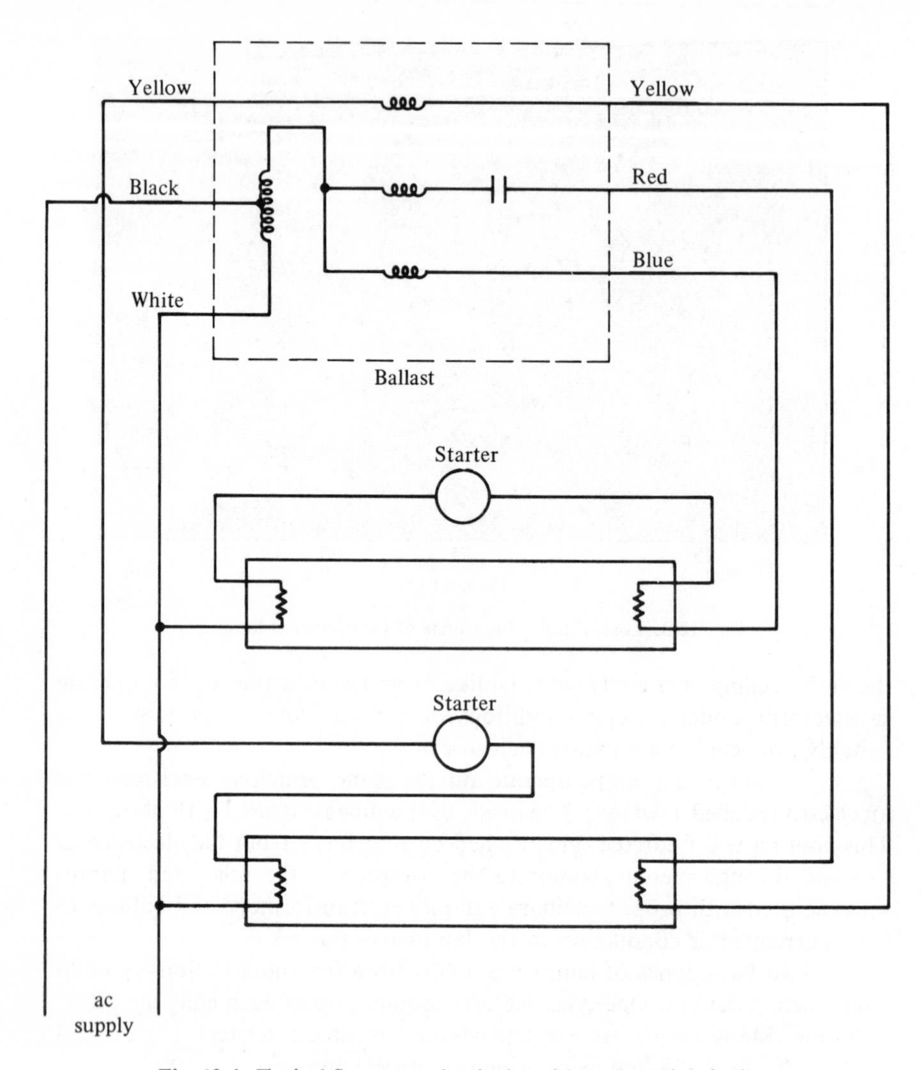

Fig. 12-4. Typical fluorescent luminaire with preheated cathodes.

be noted on moving objects. This effect is of no concern when incandescent lamps are used because the filament continues to glow as the ac goes through zero when reversing. The capacitor also assists the fluorescent unit to operate at a higher power factor, from about 0.9 to 0.95. Without the capacitor, the highly inductive ballasts produce a power factor for the luminaire of 0.6–0.75. For large lighting loads both of these effects would be most undesirable, and therefore capacitors are commonly used.

Table 12-2 lists the most popular sizes of fluorescent lamps and pertinent data. Their diameters are given in eighths of an inch. For example, T-12

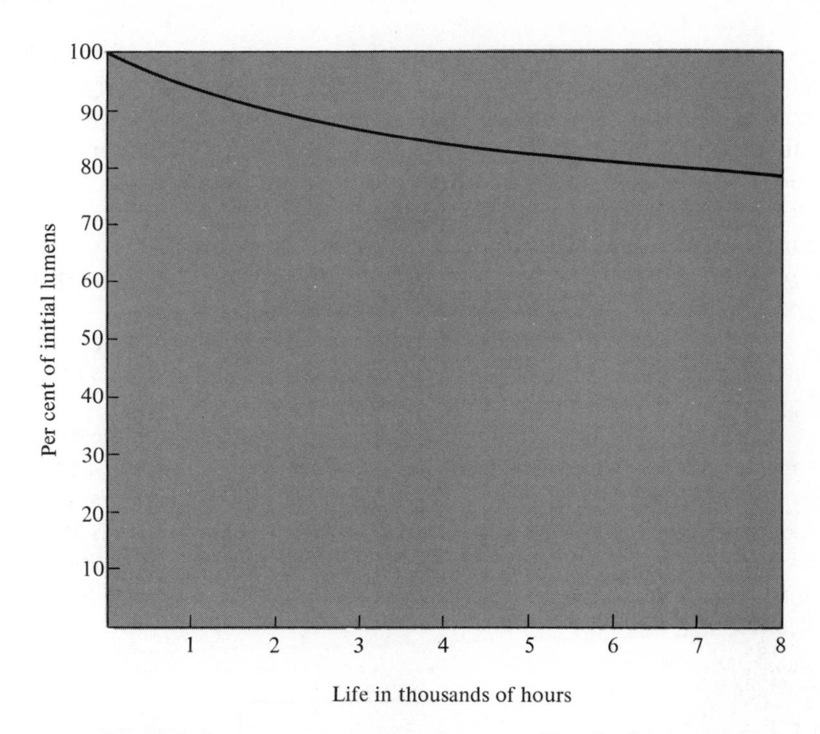

Fig. 12-5. Lumen output vs. age of tungsten filament fluorescent lamps.

Table 12-2. Fluorescent lamp data

Lamp	Length Including Sockets (in.)	Initial Lumens after 100 hr
20-W T-12	24	
White and warm white		1110
Cool white		1080
Daylight		910
40-W T-12	48	
White and warm white		3000
Soft white		1900
Cool white		2900
Daylight		2350
85-W T-12	72	
Cool white		5550
105-W T-12	96	
Cool white		7900
Warm white		7800

is a diameter of $\frac{12}{8}$ in or $1\frac{1}{2}$ in. The life of fluorescent lamps is approximately 7500 hr for average use, about 3 hr per start. Frequent on and off cycles shorten life. Figure 12-5 shows the change in the lumen output as the tungsten cathode lamp continues in operation. *Rapid start* lamps also are made to be used with auxiliary equipments that do not require starters.

A very popular and efficient lamp called the *slimline* has only one contact at each end. The auxiliary equipment causes the lamp to start in about 2 sec. Figure 12-6 shows a diagram of a two-lamp luminaire using this type of lamp. Table 12-3 lists the most frequently used types.

Fluorescent lamps, unlike incandescent ones are *not* rated by voltage. The transformers and ballasts are designed for specific voltages. The most frequently used voltages are 115, 208, 230, and 265 V.

The performance of fluorescent lamps also will vary with the deviation from the rated voltage of ballasts and transformers. Figure 12-7 shows the per cent of rated lumen output for fluorescent lamps as voltage changes. The change in power input to ballasts, transformers, and lamps also is shown.

Figure 12-7 may be considered to be the operating range of fluorescent units. Moderate voltage deviations from the rated voltage will have little effect on lamp life.

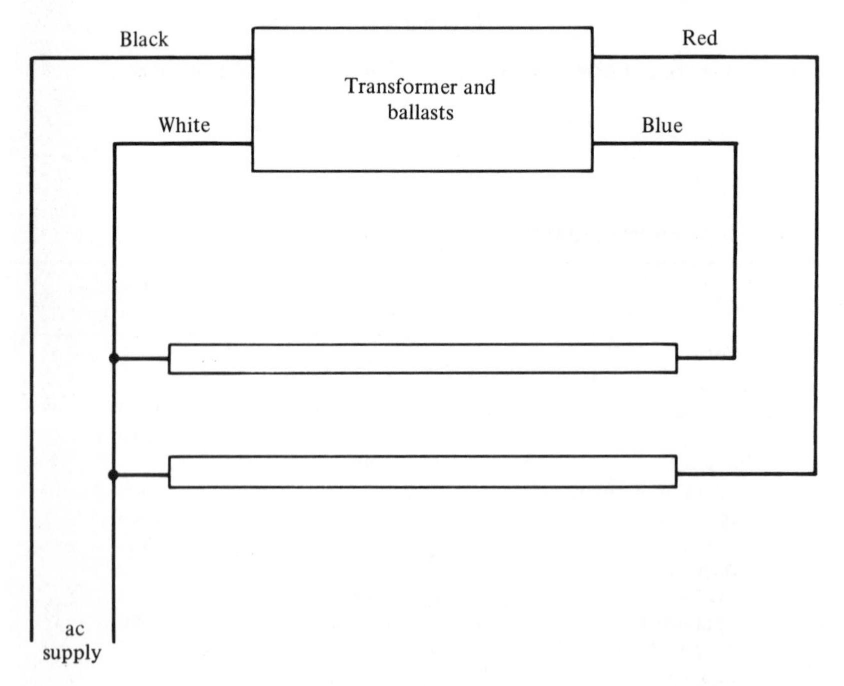

Fig. 12-6. Typical fluorescent luminaire with slimline instant-start lamps.

Table 12-3. White slimline fluorescent lamp data

Size	Length (in.)	Watts	Current (mA)	Initial Lumens after 100 hr
T-8	72	34	120	1770
T-8	72	49	200	2600
T-8	72	56	300	3100
T-12	48	23	200	1600
T-12	48	39	425	2700
T-12	96	45	200	3400
T-12	96	74	425	5700

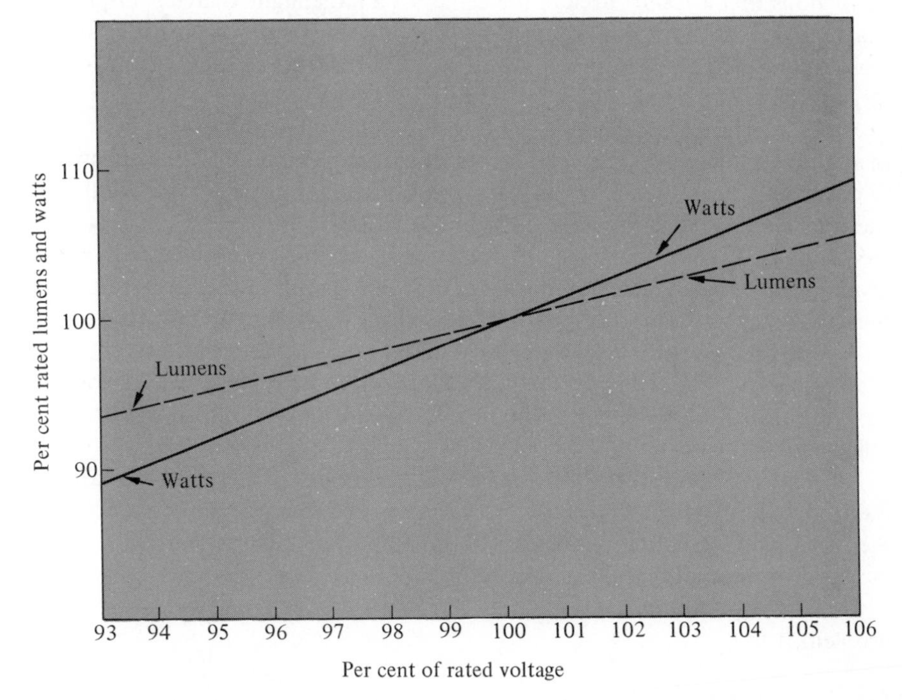

Fig. 12-7. Lumen output and watts input of fluorescent luminaires vs. voltage.

12-4 ***Efficiency of luminaire—selection of CU.*** A luminaire has been described as a lighting unit complete with lamps, all auxiliary equipment, and necessary reflectors or diffusers for a particular application. It is a relatively simple matter to determine the lumen output of a particular luminaire by multiplying the lumens per lamp by the number of lamps. However, all the lumens produced by the lamps are not available at the work surface, even at a distance close to the luminaire. Simply assigning a number to a luminaire and stating that is is 70% efficient does not take into account the many variables that need to be considered when designing a lighting system.

Let us assume that we wish a light intensity of 75 fc at desk-top level in a large office. This would require 75 lm on each square foot of office area. The theoretical approach might be to provide a 75-lm source on each square foot of ceiling area; however, this is impractical because light sources are usually concentrated. The full lumen output of the light source will not be realized since some of the light will be absorbed by walls and by the luminaire itself. A larger area requires less lumens than a small room for the same light intensity because of light absorption by walls in the smaller room. It should be obvious that a 100 ft^2 area in the center of a large, lighted area is brighter than a 100-ft^2 room with the same lumen input per 100 ft^2. The smaller area is surrounded by *light-absorbing* walls, while the 100 ft^2 in the center of the larger area is surrounded by other *light-producing* sources.

The principal problem in lighting design is to determine what fraction or multiplying factor of the total lumen output at the lamps will be effective in producing the desired fc at the work plane. This multiplying factor is called the *coefficient of utilization* and is abbreviated CU.

The CU for a particular luminaire is determined experimentally and given by each manufacturer. Charts are prepared giving consideration to the relative light-absorbing qualities of the ceilings, walls, floors, and the luminaire. Also shown is the relative pattern of light, illustrating what percentage is reflected up to the ceiling or down. Obviously a white or dark ceiling would have little effect on the CU of a luminaire if 100% of its lumen output were reflected downward.

Figure 12-8 gives the CU for various conditions of a typical two-lamp fluorescent luminaire. To include the variables of room size and distance to the work plane, a column entitled "Room Ratio" or "Room Index" is also included in the chart.

The equation for determining the room ratio for a particular lighting problem is

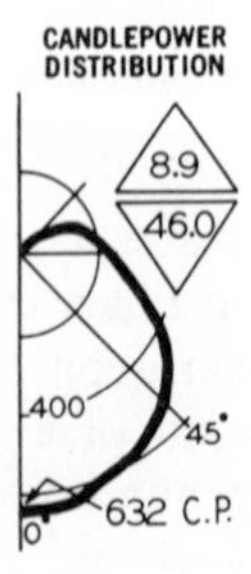

COEFFICIENTS OF UTILIZATION

Floor	30%				10%			
Ceil.	80%		50%		80%		50%	
Walls	50%	30%	50%	30%	50%	30%	50%	30%
Room Index J	.22	.17	.20	.16	.21	.17	.19	.16
I	.27	.23	.25	.21	.26	.22	.24	.20
H	.31	.25	.27	.23	.29	.24	.26	.22
G	.36	.30	.32	.28	.33	.29	.30	.27
F	.39	.34	.35	.30	.36	.32	.33	.29
E	.44	.39	.39	.35	.40	.36	.37	.34
D	.48	.43	.42	.38	.43	.40	.39	.36
C	.50	.46	.44	.41	.45	.42	.41	.38
B	.54	.50	.47	.44	.48	.45	.43	.41
A	.62	.54	.49	.47	.50	.48	.45	.43

Fig. 12-8. Typical chart for a two-light fluorescent luminaire. *Courtesy of Litecontrol Corporation, Watertown, Mass.*

$$\text{Room ratio} = \frac{(W)(L)}{H(W+L)} \tag{12-1}$$

where W and L are the width and length of the room and H is the distance from the luminaire to the work plane, all expressed in the same unit (feet). Usually the work plane is assumed to be $2\frac{1}{2}$ ft from the floor.

Some charts use a letter to designate a specific room ratio. Table 12-4 lists the calculated room ratio and its corresponding letter.

Table 12-4. Room ratios and corresponding letters

Calculated Room Ratio	Corresponding Letter
0.6	J
0.9	I
1.0	H
1.25	G
1.5	F
2.0	E
2.5	D
3.0	C
4.0	B
5.0	A

Example **12-1**

(a) Determine the room ratio for a small office area measuring 30 ft × 40 ft with a 12-ft ceiling. Assume the work plane to be $2\frac{1}{2}$ ft from the floor. Luminaires are secured on the ceiling.
(b) Determine the room ratio if the luminaires are suspended 4 ft from the ceiling.
(c) Select a CU from Fig. 12-8 for (a) and (b) and explain the difference.

Solution

(a)
$$\text{Room ratio} = \frac{(W)(L)}{H(W+L)} \tag{12-1}$$
$$= \frac{(30)(40)}{(9.5)(30+40)} = \frac{1200}{(9.5)(70)}$$
$$= 1.81 \quad \text{or } E \quad \text{(Table 12-4)}$$

(b)
$$\text{Room ratio} = \frac{(30)(40)}{(5.5)(30+40)} = \frac{1200}{(5.5)(70)} \tag{12-1}$$
$$= 3.12 \quad \text{or } C \quad \text{(Table 12-4)}$$

(c) If we should determine that the ceiling and walls have maximum reflective ability, the CU for (a) is 0.44 and for (b) is 0.50. The higher CU for (b) is a result of the shorter distance from the luminaires to the work plane.

12-5 ***Consistency of light sources—selection of MF.*** Figures 12-3 and 12-5 show the decrease in the lumen output of lamps as a function of their age. This deterioration is very slow and difficult to detect without a light meter. However, replacing lamps only after they burn out is not good practice because at this point the lamp is producing only 70 to 80% of its initial lumen output.

Another cause of reduced illumination is poor housekeeping in an environment having high air pollution. Lamps and reflectors accumulate dirt and dust—some luminaires more than others. Frequent maintenance is necessary to assure optimum light output.

Manufacturers suggest a maintenance factor (MF) for each particular luminaire. Some may use the equivalent term *depreciation factor* (DF). The MF combines all factors that cause the lumen output to decrease with age. An MF of 0.8 implies that with average use only 80% of the lumen output should be considered available. It may be necessary to decrease the MF if luminaires are subject to a highly dusty location. If luminaires are regularly cleaned and the lamps replaced before the lumen output decreases, a higher MF may be used.

MF has *no* dependency on CU; however, both must be very carefully considered if the desired fc (lumens per square foot) level of illumination is to be obtained.

Example **12-2.** A group of 10 luminaires, each with two 40-W, cool-white fluorescent lamps, are installed in a classroom. An appropriate CU of 0.65 is selected from the manufacturer's chart, which also suggests an MF of 0.85. Determine the total lumen output of the luminaires.

Solution

From Table 12-2, each lamp has an initial output of 2900 lumens.

Total lumen output of lamps

$$= \left(\frac{\text{lumens}}{\text{lamps}}\right)(\text{number of luminaires})\left(\frac{\text{lamps}}{\text{luminaire}}\right)$$

$$= (2900)(10)(2) = 58{,}000 \text{ lm}$$

$$\text{Total lumen output of luminaires} = (58{,}000)(\text{CU})(\text{MF})$$

$$(58{,}000)(0.65)(0.85) = 32{,}100 \text{ lm}$$

12-6 ***Distribution of lumens.*** Example 12-2 shows that 32,100 lumens are available at the work plane of the original 58,000 lumens produced.

The selection of 0.65 as the CU gave consideration to the distance from the luminaire to the work plane. Before we can determine the intensity of the illumination that can be provided by these 10 luminaires we must consider how much area at the work plane is receiving the 32,100 lumens.

It should be obvious that for a given number of lumens the smaller the area the more lumens can be provided for each square foot of area. As lumens per square foot is foot-candles as indicated on a light meter, we then must divide the 32,100 lumens by the area that the 10 luminaires must serve. If the classroom in Ex. 12-2 measures 30 ft $\times$ 20 ft, then

$$\text{Lumens per square foot} = \text{foot-candles} = \frac{32{,}100 \text{ lm}}{600 \text{ ft}^2}$$

$$= 53.5 \text{ foot-candles}$$

We now can develop a simple equation that will enable us to determine the necessary lumens for any lighting problem:

$$\text{Lumens} = \frac{(\text{foot-candles})\ (\text{area})}{(\text{CU})(\text{MF})} \qquad (12\text{-}2)$$

where lumens is the total output of all lamps; foot-candles is the desired level of illumination at the work plane with the conventional light meter; area is the dimension of the room in feet; CU is the coefficient of utilization, selected from the chart prepared by the manufacturer of the particular luminaire to be used; and MF is the maintenance factor recommended by the manufacturer or amended to suit conditions.

This equation also may be used to determine the foot-candles of illumination that would result from installing a specified number of lamps in a particular type of luminaire. It would then be written

$$\text{Foot-candles} = \frac{(\text{lumens})(\text{CU})(\text{MF})}{\text{area}} \qquad (12\text{-}2\text{a})$$

Figures 12-9, 12-10, and 12-11 list the CUs for seven other types of luminaires.

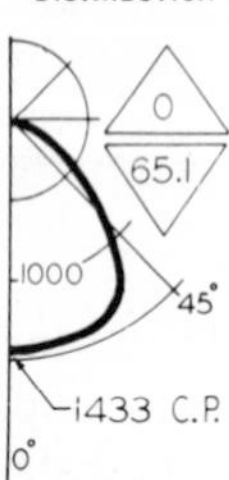

COEFFICIENTS OF UTILIZATION

Floor	30%				10%			
Ceil.	80%		50%		80%		50%	
Walls	50%	30%	50%	30%	50%	30%	50%	30%
Room Index J	.32	.27	.31	.27	.31	.27	.30	.26
I	.41	.35	.39	.35	.39	.35	.38	.34
H	.47	.41	.45	.40	.44	.40	.43	.39
G	.52	.47	.50	.45	.49	.45	.48	.44
F	.57	.51	.53	.49	.53	.48	.51	.47
E	.62	.57	.58	.54	.57	.53	.55	.52
D	.66	.61	.61	.58	.59	.56	.57	.55
C	.69	.64	.63	.60	.61	.58	.59	.57
B	.72	.68	.66	.64	.63	.61	.61	.59
A	.75	.71	.68	.66	.65	.63	.63	.61

Fig. 12-9. Coefficients of utilization for a flush luminaire. *Courtesy of Litecontrol Corporation, Watertown, Mass.*

CANDLEPOWER DISTRIBUTION

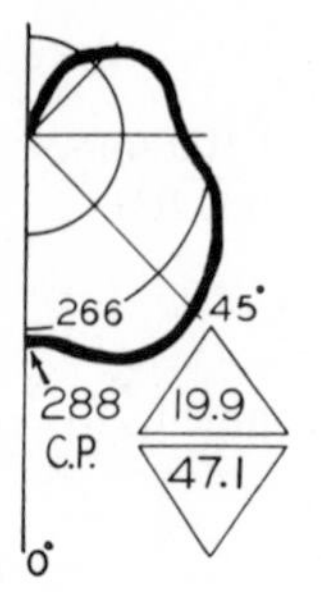

COEFFICIENTS OF UTILIZATION

Floor		30%				10%			
Ceil.		80%		50%		80%		50%	
Walls		50%	30%	50%	30%	50%	30%	50%	30%
Room Index	Room Ratio								
J	0.6	.22	.16	.19	.15	.21	.16	.18	.15
I	0.8	.29	.22	.25	.20	.27	.22	.24	.20
H	1.0	.34	.27	.29	.24	.33	.27	.28	.24
G	1.25	.40	.33	.34	.30	.38	.32	.32	.28
F	1.5	.44	.38	.37	.32	.41	.36	.35	.31
E	2.0	.51	.44	.42	.37	.46	.41	.39	.36
D	2.5	.55	.49	.45	.41	.50	.45	.43	.39
C	3.0	.58	.53	.48	.44	.52	.48	.45	.41
B	4.0	.63	.58	.52	.48	.56	.52	.48	.45
A	5.0	.66	.62	.54	.51	.58	.55	.50	.48

Fig. 12-10. Coefficients of utilization for a corridor luminaire. *Courtesy of Litecontrol Corporation, Watertown, Mass.*

Example 12-3. An office 100 ft × 75 ft with a 9-ft ceiling is to be illuminated to 60 fc using luminaires flush with the ceiling, as described in Fig. 12-9, each with two 40-W, cool-white lamps. Reflective surfaces are floor, 30%; ceiling, 50%; and walls, 30%. Calculate the number of luminaires necessary.

Solution

$$\text{Room index} = \frac{(W)(L)}{H(W + L)} = \frac{(100)(75)}{(6.5)(175)} \qquad (12\text{-}1)$$

$$= 6.6 \quad \text{or A} \quad \text{(Table 12-4)}$$

Select CU of 0.66 from Fig. 12-8.

TYPICAL DISTRIBUTION	Ceiling	80%			70%			50%			30%			
	Walls	50%	30%	10%	50%	30%	10%	50%	30%	10%	50%	30%	10%	
	Floor	10%			10%			10%			10%			
	Room RATIO	COEFFICIENTS OF UTILIZATION *												
FLUX RATIO .300 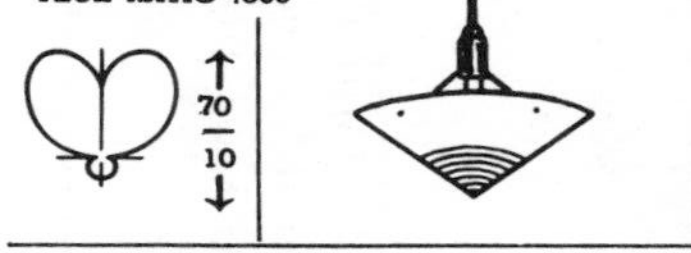 (↑ 70 / 10 ↓)	0.6	.24	.19	.15	.22	.17	.13	.17	.13	.10	.12	.09	.07	Luminous indirect fixture, using inside-frosted lamps — high quality lighting, minimum reflected glare — soft shadows, pleasing appearance. Suggested D F —.65
	0.8	.31	.25	.20	.28	.23	.18	.21	.18	.14	.15	.12	.10	
	1.0	.36	.30	.25	.32	.27	.23	.25	.21	.18	.18	.15	.12	
	1.25	.41	.35	.30	.36	.31	.27	.28	.24	.21	.20	.17	.15	
	1.5	.44	.39	.34	.40	.35	.30	.31	.27	.24	.22	.19	.17	
	2.0	.50	.44	.40	.44	.40	.36	.34	.31	.28	.24	.22	.20	
	2.5	.53	.48	.44	.48	.44	.39	.36	.33	.30	.26	.24	.22	
	3.0	.56	.51	.47	.50	.46	.42	.38	.35	.32	.27	.25	.23	
	4.0	.59	.55	.52	.52	.49	.46	.40	.38	.35	.28	.27	.25	
	5.0	.61	.58	.55	.54	.52	.49	.42	.40	.37	.29	.28	.26	
FLUX RATIO .515 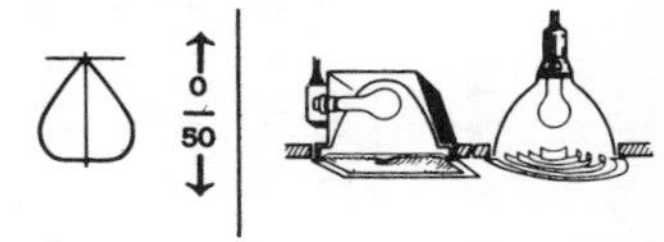 (↑ 0 / 50 ↓)	0.6 (J)	.28	.24	.21	.27	.24	.21	.27	.24	.21	.26	.23	.21	Recessed fixture — in general, good cutoff of luminaire reduces distraction — often used in combination with fluorescent system in stores. Suggested D F —.70
	0.8 (I)	.33	.29	.26	.32	.29	.26	.32	.29	.26	.31	.28	.26	
	1.0 (H)	.36	.33	.30	.36	.33	.30	.36	.32	.30	.35	.32	.30	
	1.25 (G)	.40	.37	.34	.40	.36	.34	.39	.36	.34	.38	.36	.34	
	1.5 (F)	.42	.39	.37	.42	.39	.37	.41	.39	.36	.40	.38	.36	
	2.0 (E)	.45	.43	.40	.44	.42	.40	.44	.42	.40	.43	.41	.40	
	2.5 (D)	.47	.45	.43	.46	.44	.43	.45	.44	.42	.44	.43	.42	
	3.0 (C)	.48	.46	.44	.48	.46	.44	.47	.45	.44	.46	.44	.43	
	4.0 (B)	.50	.48	.46	.49	.48	.46	.48	.47	.46	.47	.46	.45	
	5.0 (A)	.50	.49	.48	.50	.49	.48	.49	.48	.47	.48	.47	.46	
FLUX RATIO .415 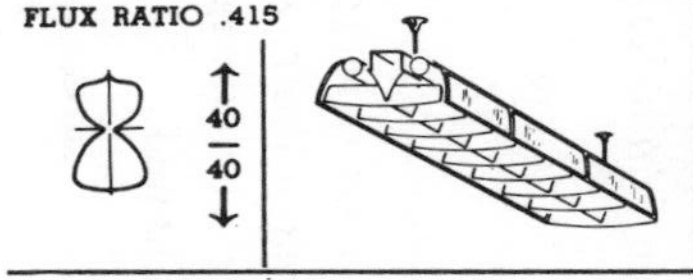 (↑ 40 / 40 ↓)	0.6 (J)	.27	.22	.18	.26	.21	.17	.23	.19	.16	.20	.18	.15	Typical direct-indirect fixture with luminous sides — many variations — best for lengthwise viewing — widely used for general lighting. Suggested D F—.75
	0.8 (I)	.34	.28	.24	.32	.27	.24	.29	.25	.22	.26	.22	.20	
	1.0 (H)	.39	.34	.30	.37	.32	.28	.33	.29	.26	.29	.26	.23	
	1.25 (G)	.44	.39	.35	.42	.37	.33	.38	.34	.30	.33	.30	.27	
	1.5 (F)	.48	.43	.39	.46	.41	.37	.41	.37	.34	.36	.33	.30	
	2.0 (E)	.53	.49	.45	.51	.46	.43	.45	.42	.39	.39	.37	.35	
	2.5 (D)	.57	.53	.49	.54	.50	.47	.48	.45	.42	.42	.39	.37	
	3.0 (C)	.60	.56	.52	.56	.53	.50	.49	.47	.44	.43	.41	.39	
	4.0 (B)	.63	.60	.53	.59	.56	.54	.52	.50	.47	.45	.44	.42	
	5.0 (A)	.65	.62	.59	.61	.58	.56	.53	.52	.50	.47	.45	.44	
FLUX RATIO .350 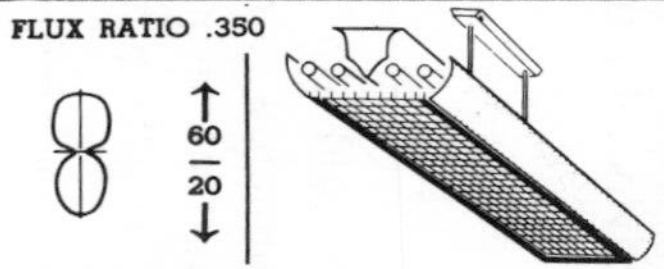 (↑ 60 / 20 ↓)	0.6	.26	.20	.16	.23	.19	.15	.19	.16	.13	.15	.12	.10	Semi-indirect luminaire — used where uniform, low brightness and soft atmosphere desired. Suggested D F—.70
	0.8	.32	.27	.22	.29	.24	.20	.24	.20	.17	.18	.16	.13	
	1.0	.36	.31	.27	.33	.29	.25	.27	.24	.21	.21	.18	.16	
	1.25	.41	.36	.32	.38	.33	.29	.31	.27	.24	.24	.21	.19	
	1.5	.45	.40	.36	.41	.36	.33	.33	.30	.27	.26	.23	.21	
	2.0	.50	.45	.43	.46	.42	.38	.37	.34	.31	.28	.26	.24	
	2.5	.53	.49	.45	.49	.45	.41	.39	.36	.33	.30	.28	.26	
	3.0	.55	.52	.48	.51	.47	.44	.40	.38	.35	.31	.29	.28	
	4.0	.58	.55	.50	.53	.50	.47	.42	.40	.38	.33	.30	.30	
	5.0	.59	.57	.54	.54	.52	.50	.43	.42	.40	.34	.32	.31	
FLUX RATIO .395 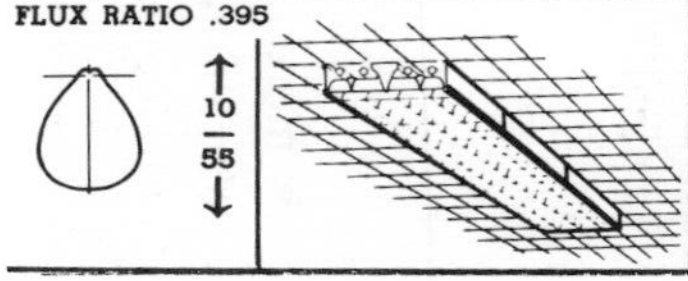 (↑ 10 / 55 ↓)	0.6 (J)	.26	.22	.18	.26	.21	.18	.25	.21	.18	.24	.21	.18	Semi-direct ceiling-mounted fixture used in stores and low-ceiling interiors — suitable for school and office lighting when equipped with low-brightness side panels and louvers or lens enclosure. Suggested D F—.75
	0.8 (I)	.33	.28	.24	.32	.27	.24	.31	.27	.24	.30	.26	.23	
	1.0 (H)	.38	.33	.29	.37	.31	.29	.35	.31	.28	.34	.30	.28	
	1.25 (G)	.42	.38	.34	.42	.37	.34	.40	.36	.33	.38	.35	.32	
	1.5 (F)	.46	.41	.38	.45	.41	.37	.43	.39	.37	.41	.38	.36	
	2.0 (E)	.50	.46	.43	.49	.46	.42	.47	.44	.41	.45	.42	.40	
	2.5 (D)	.53	.49	.46	.52	.48	.46	.49	.47	.44	.47	.45	.43	
	3.0 (C)	.55	.52	.49	.54	.51	.48	.51	.49	.47	.49	.47	.45	
	4.0 (B)	.58	.55	.53	.56	.54	.52	.54	.52	.50	.52	.50	.49	
	5.0 (A)	.60	.57	.55	.58	.56	.54	.56	.54	.53	.53	.52	.51	

Fig. 12-11. Data for five luminaires.

$$\text{Lumens} = \frac{(\text{foot-candles})(\text{area})}{(\text{CU})(\text{MF})} \times \frac{(60)(7500)}{(0.66)(0.85)} \tag{12-2}$$

$$= 800{,}000 \text{ lm necessary from light sources}$$

Each 40-W, cool-white lamp produces 2900 lm.

$$\text{Number of lamps necessary} = \frac{\text{total lumens}}{\text{lumens/lamp}}$$

$$= \frac{800{,}000}{2900} = 275 \text{ lamps}$$

$$\text{Number of luminaires necessary} = \frac{\text{number of lamps}}{\text{lamps/luminaire}}$$

$$= \frac{276}{2} = 138 \text{ luminaires}$$

After the number of luminaires has been determined it is relatively simple to locate them to give best light distribution. If four lamp luminaires were used in Ex. 12-3, then only half as many luminaires would be necessary, but they would be located twice as far apart, resulting in a poorer distribution of light.

Fluorescent luminaires are usually installed in continous rows to facilitate installation. However, for lower levels of illumination it may be necessary to arrange spacing in both directions or the rows will be too far apart.

Allowing 48 in (4 ft) for each luminaire in Ex. 12-2, 24 can be accommodated in each row. The number of rows is determined by the number of luminaires divided by the luminaires per row or

$$\frac{138}{24} = 5.76 \text{ rows}$$

This would require six rows (of 24 units per row) with the addition of six more luminaires to make the rows equal. If light at the ends of the area were unimportant, using 23 units per row, with six rows, the original 138 units could be accommodated.

Example **12-4.** A corridor 75 ft long, 6 ft wide, and with a ceiling 10 ft high is to be illuminated with 4-ft luminaires, described in Fig. 12-10. Use one 40-W, cool-white fluorescent lamp in each luminaire. Assume 30% reflectance from the floor and 50% from the ceiling and walls.

(a) Determine the number and spacing of the luminaires if they are installed on the ceiling if the light intensity at floor level is to be 15 fc.
(b) Determine the foot-candles of illumination at floor level if the luminaires are installed in a continuous row.

Solution

(a)

$$\text{Room ratio} = \frac{(W)(L)}{(H)(W+L)} = \frac{(6)(75)}{(10)(75+6)} = \frac{450}{810} \tag{12-1}$$

$$= 0.55 \quad \text{or J} \quad \text{(Table 12-4)}$$

Select a CU of 0.23 from Fig. 12-9.

$$\text{Lumens} = \frac{\text{(foot-candles) (area)}}{\text{(CU)(MF)}} = \frac{(15)(75)(6)}{(0.23)(0.85)} \tag{12-2}$$

$$= 34{,}500 \text{ lm}$$

$$\text{Number of luminaires needed} = \frac{\text{total lumens}}{\text{lumens/luminaire}}$$

$$= \frac{24{,}500}{2900} = 11.9 \quad \text{or 12 luminaires}$$

In this lighting problem the results show that the 4-ft luminaires require (12)(4) = 48 ft. The corridor is 75 ft long; therefore, 75 − 48 = 27 ft to be divided into 13 sections. Each section is $\frac{27}{13} = 2.08$ ft. Therefore, luminaires should be installed about 2 ft apart with the same spacing at the end of corridor.

(b) If the rows of luminaires are continuous, $\frac{75}{4} = 18.8$ or 18 luminaires can be accommodated.

$$\text{Total lumens} = (18)(2900) = 52{,}200 \text{ lm}$$

Using Eq. (12-2a),

$$\text{Foot-candles} = \frac{\text{(lumens)(CU)(MF)}}{\text{area}}$$

$$= \frac{(52{,}200)(0.23)(0.85)}{(75)(6)}$$

$$= 22.7 \text{ fc}$$

The use of Eq. (12-2a) enables us to determine the level of illumination from a predetermined number of lumens. The selection of room ratio, CU, and MF is the same regardless of the number of luminaires that may be used in a given area.

Figure 12-11 lists the CU, MF, and light distribution charts for two incandescent and three fluorescent luminaires. Examination of these charts shows how a white ceiling can contribute to the efficiency of a luminaire.

Number 1 is an indirect type, as 70% of its lumen output is reflected upward to the ceiling. The CU for the 80% ceiling reflectance is about double that of the 30%. However, luminaire No. 2, a direct type, shows practically no difference in the CU.

The proper determination of ceiling and wall reflectance is relative and is not easy to evaluate without experience.

The spacing of rows of luminaires in large areas must be considered. Widely spaced rows result in a drop in illumination between the rows. A contributing factor is the pattern of light distribution as shown on the chart for each luminaire. A narrow pattern requires closer spacing than one that could distribute light horizontally. For the average luminaire that diffuses rather than focuses its light, a spacing of about $1\frac{1}{2}$ times the height above the work plane is permissible. When using luminaires that have about a 90° beam spread, the spacing should be about equal to the height. Some luminaire charts include in their data the ratio of spacing to height.

12-7 ***Calculation of current and power.*** In this chapter we have considered light sources only with concern for their lumen output as stated in Tables 12-1 to 12-3. From desired levels of illumination we have determined the number of specific light sources necessary.

The calculation of current and power for incandescent lighting loads is simplified, since power factor is unity. The total power is obtained from the addition of all light sources in terms of their power rating. Dividing this power by the system voltage, we can determine the necessary current, as shown in Ex. 12-5.

Example ***12-5.*** Calculations show that an area requires 150,000 lumens; 100-W incandescent lamps are used.

(a) Calculate the power and current if the supply is 115–230 V and single phase.

(b) Calculate the number of 15-A branch circuits necessary.

Solution

(a) Lumens per lamp from Table 12-1 is 1540 lm.

$$\text{Number of lamps} = \frac{\text{lumens}}{\text{lumens/lamp}} = \frac{150{,}000}{1{,}540} = 97.5 \quad \text{or 98 lamps}$$

$$\text{Total power} = (\text{lamps})\left(\frac{\text{watts}}{\text{lamp}}\right) = (98)(100) = 9800\text{ W}$$

$$\text{Current} = \frac{9800\ \text{W}}{230\ \text{V}} = 42.6\ \text{A}$$

The load must be divided equally on each side of a three-wire system.

(b) To operate all lighting simultaneously, branch circuits must be limited to 12 A. 42.6 A/(12 A/branch circuit) = 3.55 or four branch circuits for each side or a total of eight branch circuits. See Fig. 11-5.

Fluorescent lighting requires more consideration when calculating current and power. Current rather than power determines the sizing of distribution equipment. The NEC requires that all fluorescent lighting units be plainly marked with their voltage and current ratings, including ballasts (reactors) and transformers. The frequency also must be included. The stated current rating is lower for a pf *corrected unit* than for one uncorrected having the same power rating. This feature provides a capacitor built into the ballast and is identified in technical data as HPF.

The NEC does not require that the power rating be given. It can be determined by adding the power of the lamps and then adding about 20% to provide for auxiliary equipment losses. A high pf unit would operate at about 0.90; an uncorrected unit, from 0.5 to 0.6.

Example ***12-6.*** A luminaire uses four 40-W lamps. The equipment is designed to operate at 115 V at 60 Hz. Calculate (a) the current rating if the pf is corrected to 0.9 and (b) the current rating if the pf is uncorrected at 0.5.

Solution

(a)

$$\text{Total lamp power} = (\text{lamps})\left(\frac{\text{W}}{\text{lamp}}\right) = (40)(4) = 160\ \text{W}$$

$$20\%\ \text{loss in equipment} = (160)(0.2) = 32\ \text{W}$$

$$\text{Total power of luminaire} = 192\ \text{W}$$

$$I = \frac{P}{(E)(\text{pf})} = \frac{192\ \text{W}}{(115)(0.9)} = 1.86\ \text{A} \qquad (4\text{-}5)$$

(b)

$$I = \frac{P}{(E)(\text{pf})} = \frac{192\ \text{W}}{(115)(0.5)} = 3.34\ \text{A} \qquad (4\text{-}5)$$

From these calculations we note that only three units may be installed on a 15-A branch circuit if uncorrected, but if the pf is corrected, six may be accommodated. This is a strong argument for the relatively inexpensive capacitor-corrected unit.

When designing panel boards and branch circuits for incandescent lighting, it must be considered that larger lamps may be used than originally intended. The NEC does not specifically state the number of lighting outlets permitted on each branch circuit. The watts per square foot criterion is a minimum requirement. When conditions exist requiring that the originally designed branch circuit loads be increased, it is good practice to make provisions to limit the current to less than 80% of circuit rating.

Incandescent lighting units may be less expensive to install; however, they produce less lumens per watt than fluorescent units. This added power from incandescents also increases the air-conditioning load.

PROBLEMS

12-1 Determine the initial lumen output of the following groups of lamps when they are operating at rated voltage: (a) fifty 100-W incandescent lamps, (b) twenty 200-W incandescent lamps, (c) fifty 40-W, cool-white fluorescent lamps, and (d) one hundred T-12, 96-in., 200-mA slimline fluorescent lamps.

12-2 Calculate the lumen output of the following lamps when they have been operating for one-half their rated life: (a) 300-W incandescent lamps, (b) 40-W daylight fluorescent lamps, and (c) 20-W cool-white fluorescent lamps.

12-3 Incandescent lamps rated at 115 V and 100 W are operated at 112 V. Determine (a) the lumen output of each lamp and (b) the life of the lamps.

12-4 An area illuminated with one hundred 200-W incandescent lamps operates continuously, 168 hr/week.
(a) Determine the initial total lumen output.
(b) Calculate the cost per 24-hr day based on the initial values. (The energy cost is 2 cents/kWh.)
(c) Calculate the lumen output after the lamp life has extended to 140% of its rated life.
(d) Calculate the cost per 24-hr day under the conditions of (c).
(e) Compare the relative cost of lumens under the conditions of (b) and (d), and compare them to the cost of relamping at 30 cents each.

12-5 Determine the lumen output of a 40-W, 48-in., cool-white fluorescent lamp after 5000 hr of normal use.

12-6 A luminaire uses four 40-W, cool-white fluorescent lamps. The ballasts are rated at 118 V and require 20% of the rated power of the lamps. Determine the lumen output and power input at (a) the rated voltage, (b) 120 V, and (c) 110 V.

12-7 An area requires 100,000 lumens using 40-W, cool-white fluorescent lamps, operating at a rated voltage of 118 V.
(a) How many lamps are necessary?
(b) How many lamps would be necessary if the voltage were 110 V?

12-8 A two-lamp luminaire uses two T-12, 96-in., 200-mA slimline lamps. Auxiliary equipment is rated at 277 V and requires 20% of the power of the lamps. Assuming the unit operates at 0.9 pf, determine the lumen output (a) at rated voltage, (b) at 265 V, and (c) at 265 V at the switch board, if the feeder is designed to tolerate a 3% voltage drop.

12-9 Calculate the voltage that will permit 115-V incandescent lamps to double their rated life.

12-10 Calculate the voltage that will cause 115-V incandescent lamps to last only half their rated life.

12-11 (a) At what voltage do 115-V incandescent lamps give half their lumen output?
(b) Calculate the power required of 100-W lamps at this voltage.

12-12 If two 100-W, 115-V incandescent lamps are connected in series, calculate (a) the voltage across each lamp, (b) the lumen output of both lamps, (c) the power consumed by both lamps, and (d) the current required by both lamps and (e) explain why such operation is not approved by the NEC in permanent installations.

12-13 Compare the lumens per watt for the following conditions for a 200-W incandescent lamp, using the mean value of lumen output: (a) at rated voltage of 115 V, (b) at 110 V, and (c) at 120 V.

12-14 Calculate the lumen output from a 40-W fluorescent lamp after 3000 hr of normal use.

12-15 An area requires 400,000 lm. Slimline T-12, 96 in, 200-mA lamps are used. Add 20% of the power of the lamps for ballasts and assume the units operate at 0.9 pf. Determine the current necessary for the lighting load if the power supply is (a) 118–236 V, single phase, and three wire, and (b) 277–480 V, three phase, and four wire.

12-16 (a) For photographic use, incandescent lamps are used on a higher voltage to obtain more light. Using 100-W, 115-V lamps, calculate the voltage that will increase their lumen output 50%.
(b) Calculate the power used by the 100-W lamp under these conditions, and compare this with its rating.
(c) Calculate the current required by the 100-W lamp, and compare this with its current at 115 V.
(d) Determine the approximate life of the lamps.

12-17 Calculate the room ratio for a classroom 25 ft × 35 ft with a 10-ft ceiling. Luminaires are suspended 2 ft from the ceiling.

12-18 Calculate the room ratio for a small drafting office 18 ft × 15 ft with an 8-ft ceiling. The luminaires are secured directly on the ceiling.

12-19 Calculate the room ratio for an auditorium 75 ft × 50 ft with a 28-ft ceiling. The lighting units are suspended 5 ft from the ceiling.

12-20 An office area measures 30 ft × 30 ft. It is illuminated with sixty 40-W, cool-white fluorescent lamps.
(a) Calculate the total lumens of the lamps.
(b) If conditions permit luminaires to be 65% efficient, calculate the available lumens.
(c) If the average use and dust accumulation causes only 75% of available lumens to be effective, calculate the useful lumens at the work plane. Include the efficiency of (b).
(d) Determine the available lumens for each square foot of office area.

12-21 A shop 50 ft × 30 ft is illuminated with 28 luminaires, each equipped with a 150-W incandescent lamp.
(a) Calculate the total lumen output of the lamps.
(b) If a CU of 0.68 and an MF of 0.75 are selected from the luminaire chart, how many lumens are available at the work plane?
(c) Calculate the foot-candles of illumination at the work plane.

12-22 Calculate the room ratio for a large store 150 ft × 100 ft with 12-ft ceilings. The luminaires are secured to the ceiling.

12-23 A hobby and shop area in a basement 28 ft × 40 ft with $7\frac{1}{2}$ ft ceiling requires 150 fc of illumination. Use the luminaires of Fig. 12-9. The floor, wall, and ceiling have 30, 80, and 50% reflectability. Design the lighting system.

12-24 An office area 50 ft × 50 ft with 9-ft ceilings has five continuous rows of luminaires. Using No. 4 of Fig. 12-11 with four T-12, 96-in., 200-mA lamps in each luminaire, calculate the resultant illumination in foot-candles at the work plane, $2\frac{1}{2}$ ft from the floor, if (a) the maximum reflectabilities of the surfaces are assumed and (b) the ceiling has 50% and walls 30% reflectability and (c) explain any difference between (a) and (b) and (d) calculate the foot-candles if 425-mA lamps are used, as in (a).

12-25 A supermarket 150 ft × 125 ft × 11 ft high requires 75 fc at the work plane. Install luminaire No. 3 of Fig. 12-11, using T-12, 96-in., 425-mA lamps. Assume 70 and 50% ceiling and wall reflectability, respectively. Design the lighting system, using sufficient panel boards to limit the branch circuit length to 75 ft. The service entrance is located at the center of a 150-ft wall.

12-26 A room 15 ft × 12 ft × 8 ft high has one luminaire (No. 1 of Fig. 12-4) at the center. Assume maximum reflection from the surfaces. Calculate the foot-candles of illumination at the work plane, $2\frac{1}{2}$ ft from the floor, if (a) a 40-W lamp is used, (b) a 100-W lamp is used, and (c) a 300-W lamp is used.

12-27 A drafting room is 20 ft × 40 ft × 8 ft high and has four 40-W, cool-white fluorescent lamps. Use luminaire No. 4 of Fig. 12-11; the ceiling and walls are white and give maximum reflectance. Design the system, including appropriate rows and spacing.

12-28 (a) Install 100 No. 2 luminaires of Fig. 12-11 in a gymnasium 60 ft × 60 ft × 15 ft high. Use 100-W lamps. Calculate the foot-candles of illumination at the normal work plane, $2\frac{1}{2}$ ft from the floor.

(b) Calculate the foot-candles at the end of the lamps' rated life.

(c) Calculate the cost of illumination per 15-hr day if the energy cost is 2 cents/kWh.

Chapter Thirteen

Electric Space Heating

13-1 *Why electric heat?* Space heating may be defined as the process of heating an area in which the structure of a building or a vehicle is to retain the heat. Man's desire to maintain a comfortable environment regardless of climatic conditions and rapid temperature changes has fostered many methods of space heating. The open fire on the floor of a cave; the fireplace; the potbellied stove; and the warm air, steam, and hot water furnaces using wood, coal, gas, or oil have all had their day. The modern approach to space heating is by electric heat.

There are many reasons for this approach. The use of electrical energy has doubled every 10 years. This increased volume plus the development of larger generating units and higher transmission voltages have made it possible to generate and deliver electrical energy at lower cost. The cost of heating oil has tripled since 1940. During this same period, the unit cost of electricity in many areas has decreased to one-third of its 1940 price.

This factor of nine has made electricity more competitive with other fuels for space heating, although per heating unit electricity has a higher cost. Strict adherence to a few details, however, can make electric heating overall costs about the same as those of other fuels.

The lack of any combustion in the building permits a "tighter" construction. Careful attention to insulation reduces the heat loss. Careful design of the heating elements in each area prevents one heating unit from loafing while another cannot keep up with its heating assignment. The occupants of the building must use caution in conserving heat as much as possible.

There are many advantages to heating electrically. Most utilities permit an approximate 35% rate reduction per kilowatt-hour for all residential energy when electric space heating is installed. This means a saving of 35% in the regular electrical bill. A separate thermostat in each room permits the ultimate in zoning. Electric heating is clean, free of any odor, and completely safe from explosion or fire. The entire system requires little maintenance, and the initial installation is less expensive than any other heating system. Increased efficiency in generating electric power will undoubtedly continue the trend toward less expensive electrical energy.

13-2 ***Fundamentals of space heating.*** We shall define and explain some of the nomenclature and units that will be encountered in this chapter:

BRITISH THERMAL UNIT. Btu. A measurement of heat energy. One Btu of heat energy will raise the temperature of 1 lb of water 1°F. One watt of electrical power is the equivalent of 3.41 Btu/hr. Heating equipment in general is heat-power rated in Btu's per hour, but to relate to electric power systems we shall convert to, and use, watts.

DESIGN TEMPERATURE DIFFERENCE. DTD. The difference between inside and outside design temperatures, expressed in degrees Fahrenheit. The inside temperature is usually 70°F but could be higher. The outside temperature is not the lowest on record locally, but the lowest likely to occur several times a season.

DEGREE DAYS. DD. A measure of the work a heating system must provide. It is the number of degrees Fahrenheit that the average temperature over a 24-hr day falls below 65°F. For example, if the average temperature for a particular day is 40°F, that day would contribute 25 DD to the seasonal total. This data is calculated by local weather bureaus each day. Daily, monthly, and seasonal totals are printed in most newspapers.

THERMAL RESISTANCE. R. A measure of a material or a combination of materials to resist the flow of heat. Commercial insulations of various types and thicknesses use R as a measure of their effectiveness. The exact R of any combination of building materials can be determined by adding the individual resistances of each material that the heat must pass through. These are available in heating and architectural handbooks. R is the reciprocal of U.

COEFFICIENT OF HEAT TRANSFER. U. Amount of heat power flow, expressed in Btu's per hour per square foot per degree Fahrenheit DTD. However, we shall convert and express this in electrical units as watts per square foot per degree Fahrenheit DTD.

$$U = \frac{1}{R} \tag{13-1}$$

$$W = \frac{U}{3.41} \tag{13-2a}$$

or

$$W = (0.293)(U) \tag{13-2b}$$

INFILTRATION. Heat loss through cracks around doors and windows and normal use of doors. It is measured by air changes per hour. A rate of one-half or three-quarters of one air change per hour is suggested for average construction. The frequent use of doors in commercial buildings increases the infiltration rate.

An architect once defined a building as a structure designed to keep the people in and the weather out. The problem of heating a building is not as much to keep the weather out as to keep the heat in.

To compete favorably with less expensive fuels, an electrically heated building must be built with enough insulating materials to reduce heat loss. Comparison of a building with no insulation to one well insulated is like comparison of a milk bottle to a vacuum bottle.

Electric heat is 100% efficient. Efficiency of the combustion of other fuels may vary from 50 to 75%. More effective insulation for these fuels, therefore, can save only a portion of the 50 to 75% that is actually effective.

13-3 ***Heat loss through various building materials.*** To properly design and predict the operating cost of an electrical heating system, we must be able to evaluate heat loss accurately. Heat loss depends on the building material and the temperature design difference. Usually the proper insulating material will reduce the heat loss from 10 to 20% of that of an uninsulated building—as the vacuum bottle keeps the coffee hot 5–10 times longer than a milk bottle.

Tables 13-1 through 13-5 have been developed to enable us to select an appropriate multiplier to apply to each square foot of a building to determine the total heat loss. W by definition is the heat loss for each square foot per degree Fahrenheit, while the values listed for various design temperature differences are DTD multiplied by W under different insulation conditions.

Table 13-1 is for masonry buildings in which the insulation may be expanded polystyrene sheets of varied thicknesses.

Table 13-2 is for frame buildings of wood. Insulation may be rock wool or fiber glass.

Table 13-3 lists heat loss through walls and floors below grade. This table assumes a constant ground temperature, below frost, of 50°. Therefore, the DTD will be only 20°F.

Table 13-4 gives heat loss through doors and windows. Table 13-5 lists multipliers for each cubic foot of volume to compensate for infiltration. Fireplaces, even with dampers, contribute to infiltration heat loss. From 5 to 15 W/°F should be included in the heat loss total, depending on the fit of the damper and whether the fireplace is on an outside or inside wall.

Table 13-1. Heat loss through masonry walls in watts per square foot

Construction of Wall	W/No. Insulation	With 2-in. Polystyrene Insulation, $R = 8.3$				With 3-in. Polystyrene Insulation, $R = 12.5$				With 4-in. Polystyrene Insulation, $R = 16.5$			
	Watts per square Foot per °F	Temperature Difference of				Temperature Difference of				Temperature Difference of			
		50°	60°	70°	80°	50°	60°	70°	80°	50°	60°	70°	80°
10- in. concrete, $R = 1.68$	0.175	1.46	1.75	2.05	2.34	1.03	1.24	1.45	1.65	0.8	0.96	1.12	1.28
4-in. brick plus 8-in. concrete block, $R = 2.46$	0.12	1.36	1.63	1.90	2.18	0.98	1.18	1.37	1.57	0.77	0.93	1.08	1.23
12-in. concrete block, $R = 2.93$	0.1	1.3	1.56	1.82	2.08	0.96	1.14	1.33	1.53	0.755	0.905	1.06	1.21
4-in. brick plus 8-in. cinder block $R = 3.1$	0.095	1.28	1.54	1.8	2.05	0.94	1.12	1.31	1.5	0.75	0.9	1.05	1.2

Table 13-2. Heat loss through frame walls and ceilings in watts per square foot

Type of Construction	Watts per Square Foot per °F	Temperature Difference of 50°	60°	70°	80°
Walls of wood shingles, ¾ in. sheathing, lath and plaster, with 3 in. mineral insulation, *R* = 11	0.0215	1.07	1.28	1.5	1.71
Walls of wood shingles, ¾-in. sheathing, lath and plaster, with 3⅝-in mineral insulation, *R* = 13	0.0185	0.925	1.11	1.3	1.48
Ceilings of tile or plaster with 6-in. insulation, *R* = 19	0.0129	0.645	0.775	0.9	1.03
Ceilings of tile or plaster, with 6-in. insulation, *R* = 24	0.0117	0.585	0.7	0.82	0.935

Table 13-3. Heat loss through walls and floors at or below grade level

Type of Construction	Heat Loss (W/ft^2)
Concrete walls below grade, no insulation, ground temperature = 50°F, DTD = 20°F	3.5
Concrete walls below grade, 3-in. insulation, R = 11, ground temperature = 50°F, DTD = 20°F	0.4
Concrete floors below grade, ground temperature = 50°F, DTD = 20°	0.6
Concrete slab floor at grade level with 2-in. perimeter insulation, 24-in. vertically downward or horizontally	Watt per linear foot of exposed slab edge for DTD of

50°	60°	70°	80°
8	9	10	12

Table 13-4. Heat loss through windows and doors in watts per square foot

Type of Window or Door	Temperature Difference of 50°	60°	70°	80°
Single glass	16.5	19.7	23	26.5
Double glass, ½-in. airspace	8.2	9.7	11.3	13
Single wood sash	14.5	17.5	20.5	23.5
Single metal sash	16.5	19.7	23	26.5
Wood sash, wood storm window	7	8.3	9.7	11.1
Wood sash, aluminum storm window	7.9	9.2	10.7	12.2
Aluminum sash and aluminum storm window	8.6	10.3	12	13.8
1⅜-in wood exterior door	7	8.4	10	11.2
1⅜-in. wood door with glass storm door	4.5	5.5	6.4	7.3

Table 13-5. Heat loss through infiltration in watts per cubic foot

Number of Air Changes per Hour	Temperature Difference of 50°	60°	70°	80°
One-half	0.135	0.16	0.185	0.21
Three-quarters	0.2	0.24	0.28	0.315
One	0.27	0.32	0.37	0.42
Two	0.54	0.64	0.74	0.84

Example 13-1. The wall of a frame building has the following materials with the given resistances:

Material	R
Wood shingles	0.87
5/8-in. plywood	0.78
Building paper	0.06
Gypsum lath and sand plaster	0.41
Airspace in wall	0.95

(a) Calculate the total resistance, U, and W for the uninsulated wall.
(b) Calculate the resistance, U, and W, if insulation with $R = 13$ is used.
(c) Determine the multiplier to apply to each square foot of wall in (a) and (b) for a DTD of 80°F.

Solution

(a)

$$R = 0.87 + 0.78 + 0.06 + 0.41 + 0.95$$

$$= 3.07 \text{ units}$$

$$U = \frac{1}{R} \tag{13-1}$$

$$= \frac{1}{3.07} = 0.326 \text{ Btu/hr}$$

$$W = \frac{U}{3.41} \tag{13-2a}$$

$$= \frac{0.326}{3.41} = 0.096 \text{ W heat loss/ft}^2/^\circ\text{F}$$

(b)

$$R = 0.87 + 0.78 + 0.06 + 0.40 + 13 = 15.11 \text{ units}$$

$$U = \frac{1}{R} = \frac{1}{15.11} = 0.0662 \text{ Btu/hr}$$

$$W = \frac{U}{3.41} = \frac{0.0662}{3.41} = 0.0194 \text{ W/ft}^2/^\circ\text{F}$$

(c) Uninsulated:

$$\text{Multiplier} = (0.096)(80) = 7.65 \text{ W/ft}^2$$

Insulated:

$$\text{Multiplier} = (0.0194)(80) = 1.55 \text{ W/ft}^2$$

The heat loss through the uninsulated wall is approximately five times the loss through the insulated.

Tables 13-1 through 13-5 list enough information to calculate the heat loss of most construction methods. However, Ex. 13-1 gives the procedure for calculating the proper multiplier for any construction if the resistances are known. It should be noted that the insulation used is the principal obstruction to heat passing through the wall. The type of building material actually makes very little difference.

13-4 ***Determination of total heat loss.*** When calculating heat loss it is only necessary to consider walls, ceilings, or floors that are exposed to unheated areas. A five-story building has a heat loss through only one ceiling, the fifth floor. We then plan our heating system room by room to replace the heat loss.

If heating units were installed of the same capacity as the heat loss, theoretically they would operate continuously when the outside temperature was at the design temperature. Tight construction, in particular well-fitted doors and windows, and more insulation than required are factors that decrease heat loss below design levels. To enable the system to respond fast to sudden changes in temperature and to prevent continual operation, a good practice is to install heating units 10–25% larger than heat loss calculations. Once again this becomes a specification detail; however, before we exceed a particular design value we must base our initial figures on sound calculations rather than on "guesstimates."

Example ***13-2.*** Calculate the heat loss for a room of frame construction, 15 ft × 15 ft, with an 8-ft ceiling. The floor is over a heated basement, two walls and the ceiling are exposed to the outside. There are two windows with wood sashes and storm windows. The ceiling insulation has $R = 24$; the walls, $R = 13$. The lowest expected temperature is 0°F. The DTD is 70°. For infiltration, assume three-quarters of an air change per hour.

Solution

Correct multipliers, watts per square foot for 70° DTD, are selected from Tables 13-2 and 13-4. The infiltration multiplier is (0.37)(0.75) = 0.28, from Table 13-5.

	Multiplier	Area	Heat Loss (W)
Gross wall area, ft^2		240	
Window area, ft^2	9.7	12	116
Net wall area, ft^2	1.30	228	296
Ceiling area, ft^2	0.82	225	185
Room volume, ft^3	0.28	1800	500
			1097

A 1250-W heating unit would properly heat this room.

*Example **13-3.*** Calculate the heat loss and select the appropriate heating units for each room and basement of a single-family home.

Solution

Details are shown on the plans in Fig. 13-1. Design details and the work sheet are shown in Fig. 13-2.

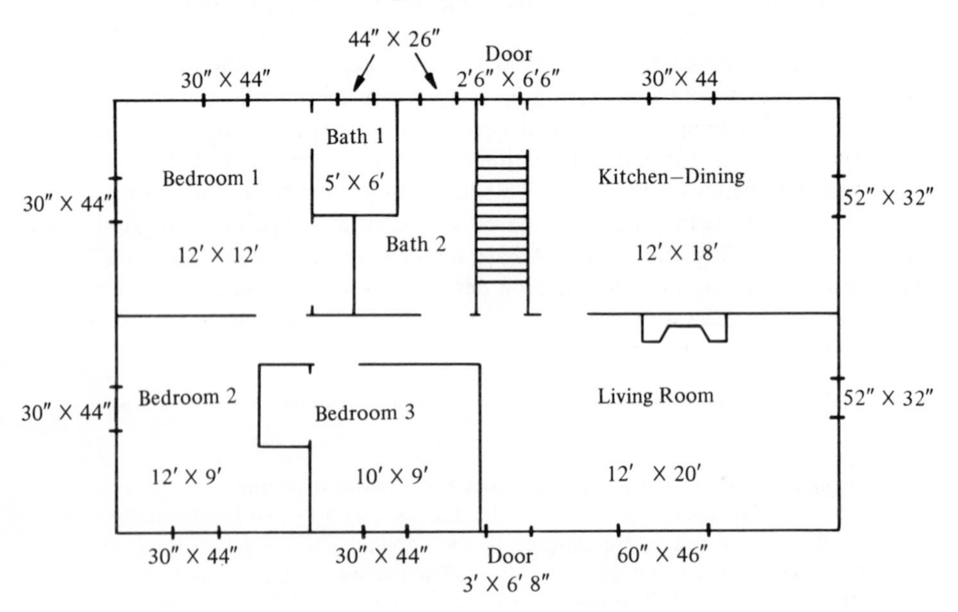

Fig. 13-1. House plan for Ex. 13-3.

The following assumptions have been made: Frame construction over heated basement, aluminum storm windows over wooden sashes. Ceiling insulation R = 19, walls R = 13. DTD is 70°. Infiltration, three quarters of an air change per hour. Ceiling height = 8 ft.

House	Multiplier	Bedroom 1	Bedroom 2	Bedroom 3	Bath 1	Bath 2	Kitchen-Dining	Living
Gross Wall Area ft^2	–	192	168	72	48	24	240	256
Window Area ft^2	10.7	18.3	18.3	9.15	7.94	7.94	20.65	30.7
Door Area ft^2	6.4	–	–	–	–	–	16.25	20
Net Wall Area ft^2	1.30	173.7	149.7	62.85	40.06	16.06	203.1	205.3
Room Volume ft^3	0.21	1056	864	720	240	288	1728	1920
Ceiling Area ft^2	0.90	132	108	90	30	36	216	240
Total Heat Loss (W)	–	772.5	668.7	411.7	214.6	198.7	1145	1342
Unit Selected (W)	–	1000	1000	750	500	500	1500	1750

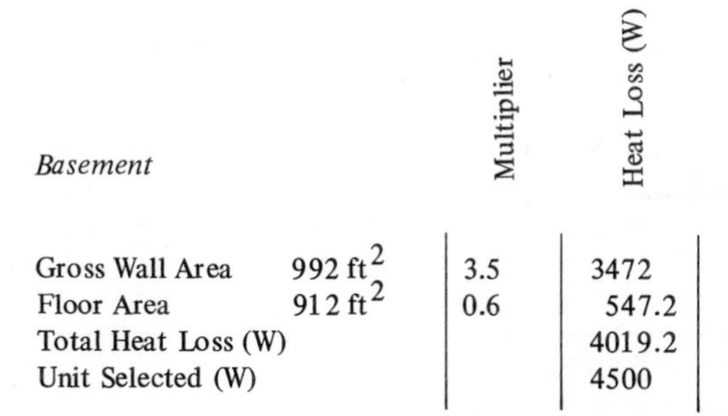

Basement		Multiplier	Heat Loss (W)
Gross Wall Area	992 ft^2	3.5	3472
Floor Area	912 ft^2	0.6	547.2
Total Heat Loss (W)			4019.2
Unit Selected (W)			4500

Fig. 13-2. Solution for Ex. 13-3.

When selecting the heating equipment in Ex. 13-3, about 15% has been added to the heat loss. With a combination of various lengths of baseboard units, increments of 250 W are possible. Note that the bathroom heaters are much larger than necessary. This enables bathrooms to be heated above normal very quickly.

Example **13-4.** Calculate the heat loss for an auditorium built of face brick and 8-in. concrete blocks with 3-in. polystyrene insulation. The dimensions are 50 ft × 90 ft × 18 ft high. The ceiling is very heavily insulated; $R = 30$.

There are 20 windows with wood sashes and aluminum storm windows, each measuring 10 ft × 4 ft. The back end of the auditorium is next to another heated area and, therefore, has no heat loss. The inside temperature is to be 70°F; the minimum outside temperature is 0°F. Allow one air change per hour.

Solution

The multiplier for the walls, from Table 13-1, is 1.37 W/ft². To calculate the multiplier for the ceiling, with $R = 30$,

$$U = \frac{1}{R} \qquad (13\text{-}1)$$

$$= \frac{1}{30} = 0.033 \text{ Btu/hr/°F}$$

$$W = \frac{U}{3.41} \qquad (13\text{-}2a)$$

$$= \frac{0.0333}{3.41} = 0.00975 \text{ W/ft}^2\text{/°F}$$

For DTD at 70°, (0.00975)(70) = 0.685 W/ft² (the multiplier for the ceiling). The multiplier for the windows, from Table 13-4, is 10.7 W/ft². The multiplier for one-half air change per hour, from Table 13-5, is 0.185 W/ft³.

		Multiplier	Heat Loss (W)
Gross wall area	4,140 ft²		
Window area	800 ft²	10.7	8,550
Net wall area	3,340 ft²	1.37	4,570
Ceiling	4,500 ft²	0.685	3,080
Infiltration	81,000 ft³	0.185	15,000
	Total heat loss		31,200

13-5 ***Estimating heating costs.*** Where electric heat is to be used, much attention must be directed to assure tight construction and insulation requirements. This will make heat loss calculations more accurate and assure properly sized heating elements. Also of great importance is the ability to predict heating costs accurately, especially if they must be compared with other methods of space heating.

In addition to the building characteristics, the factors affecting heating costs are the calculated heat loss, the number of degree days, the cost of energy, and the design temperature difference. These factors together with a constant give the following equation:

$$\text{Kilowatt-hours} = \frac{(\text{kilowatt loss})(\text{DD})(18.5)}{\text{DTD}} \tag{13-3}$$

where kilowatt-hours is the total energy required for a complete heating season; DD is the number of degree days, determined locally; 18.5 is a constant; and DTD is the difference between inside and outside design temperature.

The constant is given here as 18.5. It is a conservative figure, which will make most estimates realistic. It would have to be increased if the occupants wished a temperature higher than 70°F or were not careful in conserving heat or if the building were subject to consistently high winds. Conversely, this constant could be decreased if the building could benefit from solar heat or if heat were reduced in unused rooms. This constant might vary between 15 and 20, and only experience can actually dictate its correct value.

Once the yearly consumption has been determined the cost will be

$$\text{Cost} = (\text{kilowatt-hours})(\text{electric rate per kilowatt-hour}) \tag{13-4}$$

Although most electric utilities may have step rates for the first few hundred kilowatt-hours each month, we use the lowest rate in Eq. (13-4). This is justifiable because electrical space heating is an additional load and the higher initial rates are applied to the nonheating load. Of course, with electrical heating, the rate is at least 30% lower.

Example ***13-5.*** Calculate the seasonal heating cost for the house in Ex. 13-3 if the house is located (a) in New York City, number of degree days is 5000 and the rate is 1.4 cents/kWh; (b) in Lexington, Massachsetts, 5900 degree days, and rate 1.6 cents/kWh; and (c) in Louisville, Kentucky, 4350 degree days, and rate 1.2 cents/kWh. (All three locations have approximately the same DTD, 70°F.)

Solution

(a)
$$\text{Kilowatt-hours} = \frac{(\text{kilowatt-hours loss})(\text{DD})(18.5)}{\text{DTD}} \qquad (13\text{-}3)$$
$$= \frac{(9104)(5000)(18.5)}{70}$$
$$= 12{,}000 \text{ kWh}$$
$$\text{Cost} = (12{,}000)(0.014) \qquad (13\text{-}4)$$
$$= \$168$$

(b)
$$\text{Kilowatt-hours} = \frac{(9104)(5900)(18.5)}{70} \qquad (13\text{-}3)$$
$$= 14{,}200 \text{ kWh}$$
$$\text{Cost} = (14{,}200)(0.016) \qquad (13\text{-}4)$$
$$= \$228$$

(c)
$$\text{Kilowatt-hours} = \frac{(9104)(4350)(18.5)}{70} \qquad (13\text{-}3)$$
$$= 10{,}450 \text{ kWh}$$
$$\text{Cost} = (10{,}450)(0.012) \qquad (13\text{-}4)$$
$$= \$125$$

13-6 ***Distribution of electrical load.*** Example 13-3 shows that the largest heaters for individual areas are 2250 W for a living room and 4000 W for a basement. For individual thermostat control only one branch circuit should be provided for each area. To reduce the current to each heater, 230-V heaters are most extensively used. This would require a 15- or 20-A heater for the living room and a 30-A heater for the basement.

The heating load in residences is supplied from panel boards equipped with double-pole circuit breakers. If fuses are used, two fuses must be provided for each 230-V branch circuit. A 20-A, 230-V branch circuit is usually provided for each room. When multiple units are used in one room they are connected in parallel.

No demand factor less than 100% can be applied to any heating load. Although it is seldom necessary to operate entire heating loads simultaneously, it is sometimes required. Therefore, the equipment must be able to provide the full rating of installed heaters.

Example 13-6. Calculate the current necessary for the heating load of Ex. 13-3, and list the branch circuits necessary. Calculate the minimum feeder size to the panel board if only heating is supplied.

Solution

	W	A
Living Room	2,250	9.8
Kitchen-dining area	1,750	7.6
Bath 1	750	3.26
Bath 2	750	3.26
Bedroom 1	1,250	5.45
Bedroom 2	1,250	5.45
Bedroom 3	750	3.26
Basement	4,000	17.4
	12,750	55.4

The minimum feeder size is No. 6 AWG RH copper conductors, selected from Table 9-4. A panel board with a minimum of 16 circuits is necessary for this heating load: two 30-A circuits, and fourteen 15- or 20-A branch circuits. The branch circuits must be rated at least 125% of the heating load.

13-7 ***Heating equipment and control.*** The most commonly used heating equipment for domestic or commercial installations is baseboard units or forced-air wall heaters. It makes little difference how the required wattage is radiated into the room. However, a more comfortable heating effect is obtained if the heating elements are dispersed. Two extremes might be heaters of equal kilowatt rating: one a very small but red-hot heater in the center of the room, and the other heater raising the temperature of the entire surface of the walls and ceiling to slightly over 70°. The best compromise between these extremes is to use low-density heating elements—but to use many of them because their temperatures are lower.

Baseboard heaters are manufactured with densities from 175 to 250 W per linear foot. Sometimes space limitations necessitate using higher densities. Figure 13-3 illustrates a typical baseboard unit.

Built-in units with fans to force air around a heating element also are useful when space is limited. They may be used when fast recovery is desired, such as in basements where continual heating is not necessary. They are

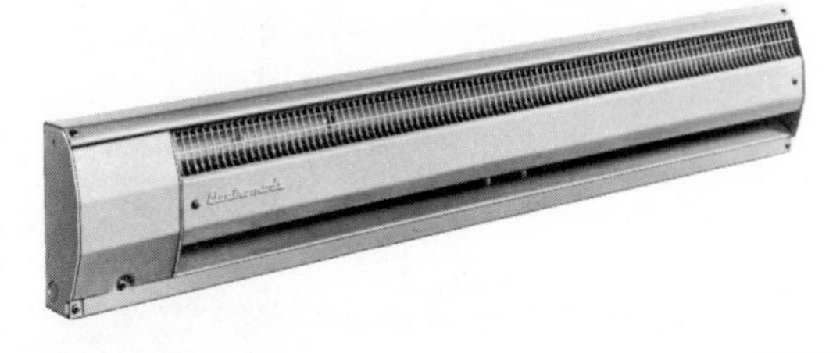

Fig. 13-3. Baseboard heater. *Courtesy of Singer Co., Electromode Division.*

Fig. 13-4. A forced-air heater. *Courtesy of Singer Co., Electromode Division.*

available from 500 W to as high as 5 kW. Figure 13-4 illustrates a typical forced-air heater equipped with its individual thermostat.

Figure 13-5 illustrates a compact fan-type heater, equipped with plugs that can readily change its power rating to suit conditions. The unit illustrated may be adjusted for 1000, 1500, or 2000 W.

A very comfortable heating system is possible by heating the entire ceiling to about 80 to 90°. This is done with heating cables directly imbedded in the plaster. About 360 ft of cable is necessary for each 1000 W of heating load.

Fig. 13-5. Fan-type heater. *Courtesy of Singer Co., Electromode Division.*

One of the major advantages of electric heat is that separate control for each room or area is possible. Thermostats designed to operate as a switch in series with the heating load are seldom rated over 5 kW. Large heating leads in auditoriums also must be controlled by a single thermostat. A magnetic contactor switches the main line. Figure 13-6 shows a 15-kW load controlled by a line or high-voltage thermostat.

A low-voltage control system utilizes more sensitive, quiet thermostats. Some manufacturers provide a relay-and-transformer device that may be installed directly in the wiring space of the heater or may be grouped in an enclosure at a central location. Figure 13-7 shows a diagram of a typical low-voltage control circuit.

Another method of controlling a large load with one thermostat uses a series of relays that permit a delay of 30 sec to 1 min before each subsequent heater is energized. The relays use a low-voltage source provided by a

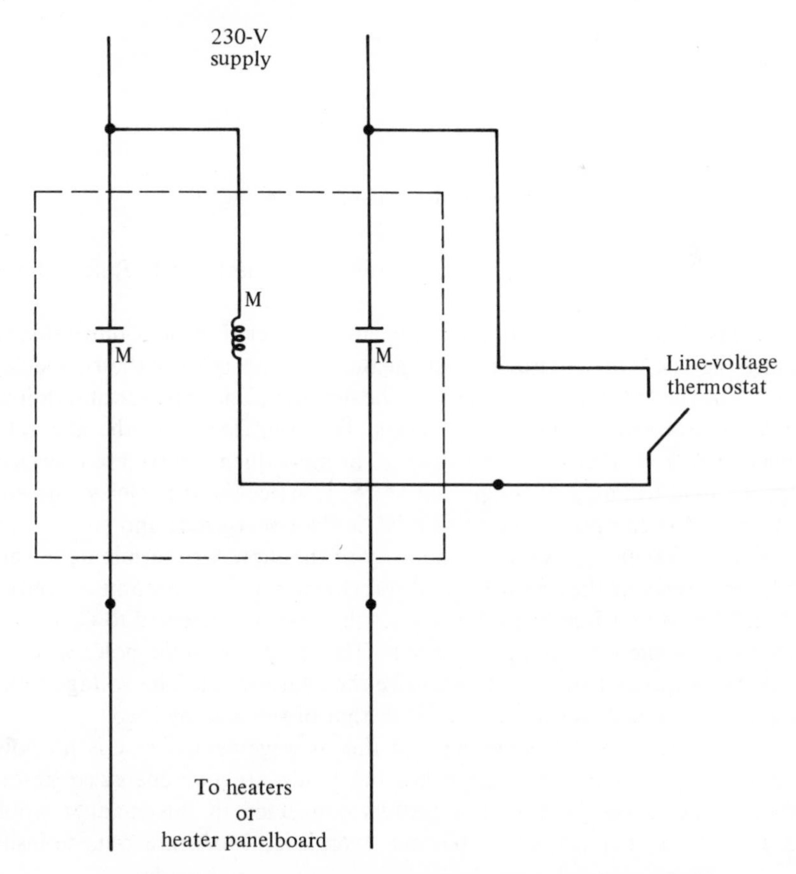

Fig. 13-6. Magnetic contactor controlling a large heating load with a line-voltage thermostat.

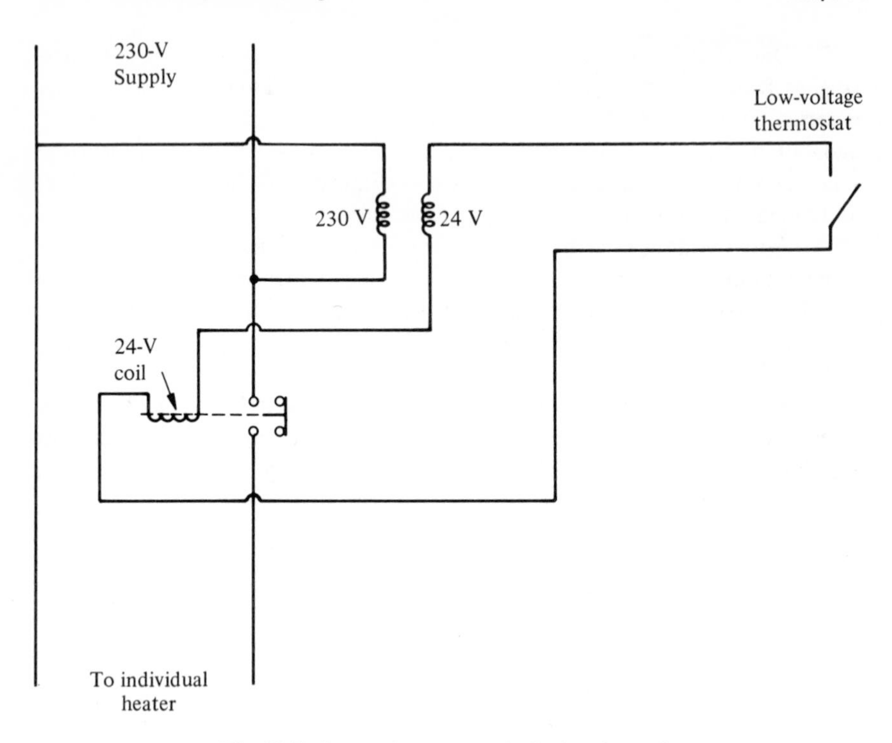

Fig. 13-7. Low-voltage control of a heating unit.

built-in transformer for the control circuit and thermostat. Refer to Fig. 13-8.

The control circuit allows a very small heater to bend a bimetal strip, closing the line voltage circuit to the heater. This constitutes the time delay.

The thermostat operates the first heater only. The thermostat terminals are short-circuited on subsequent relays. The transformer of the first relay is energized from the line. Transformers of subsequent relays are energized from the load terminal of the preceding relay. After the time delay, the next heating load is energized; the third relay is then energized; and so on. Caution must be taken to reverse the polarity of the input line terminals to each subsequent relay or the control circuit transformer will not become energized. In Fig. 13-8, *A* and *B* refer to the ac line; the load is connected to *C*; *T* is the thermostat or the low-voltage terminals. The relay is a single-pole switching device but requires two poles to energize the transformer. The voltage rating of the relay transformer must agree with that of the heating load.

There are several advantages to this arrangement. There is no noise from magnetic contactors. Large heating loads are not energized at one instant, causing line surges. Ten heaters connected in this manner would require 7 or 8 min for all to become energized. Moderate outside-inside

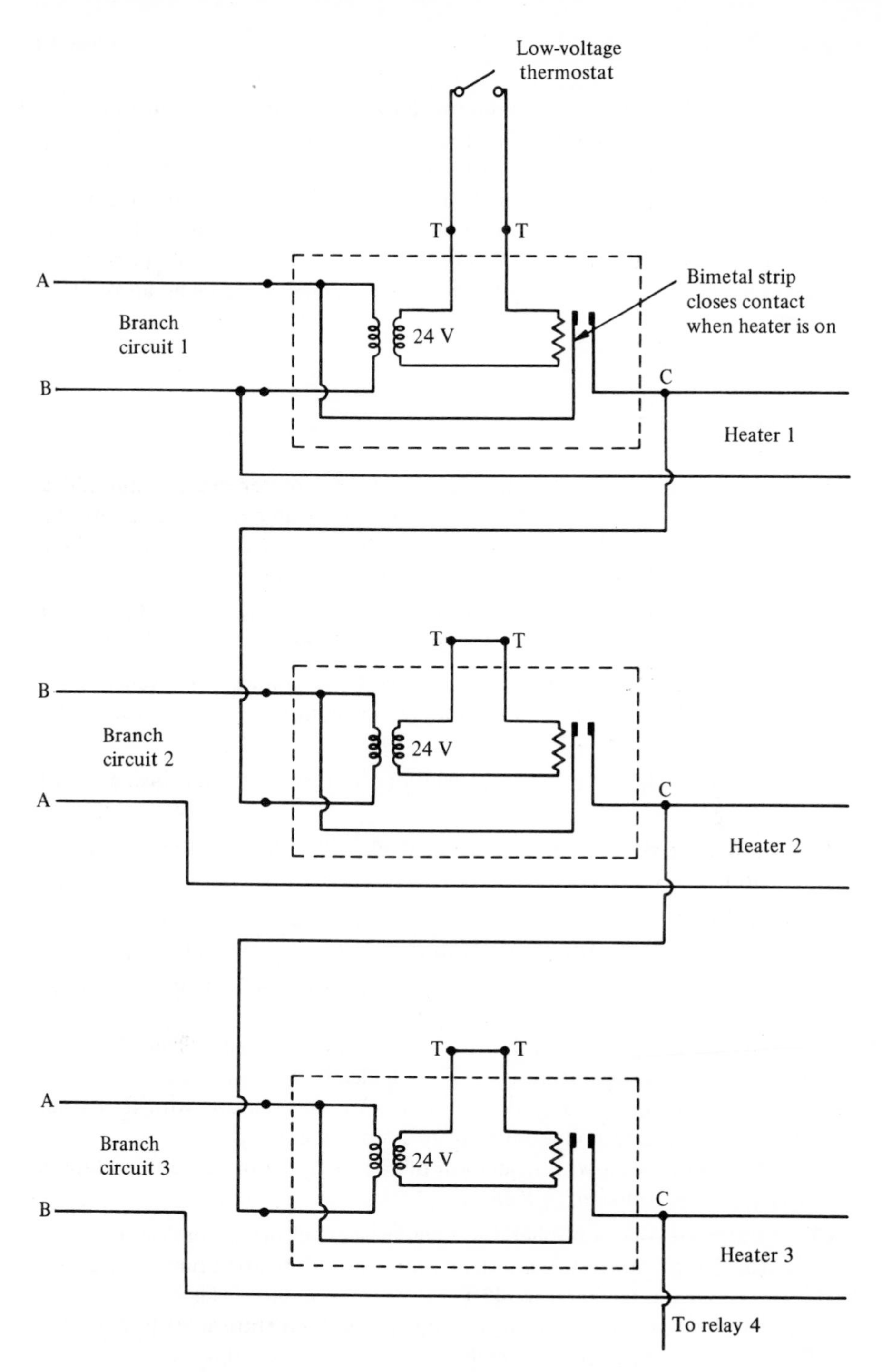

Fig. 13-8. Low-voltage control of a large heating load.

temperature differences might require only a few minutes of heating-load operation to maintain thermostat differential. It must be noted that all heating loads are designed to heat the building at a very low outdoor temperature. This low temperature occurs very infrequently. For outdoor temperatures of 40–50°F, the heating load may only function a few times per hour. Hot water heating systems sometimes use this principle to modulate the temperature of the circulating water according to the outside temperature using an outdoor thermostat.

PROBLEMS

13-1 Determine W (watts heat loss per square foot per degree Fahrenhiet) for the following types of construction: (a) frame walls with an insulation value of $R = 13$, (b) ceilings with $R = 24$, (c) walls of 12-in. concrete block, and (d) walls of 10-in. concrete.

13-2 Calculate W for a frame wall with the same resistance as Prob. 13-1 when insulation with (a) $R = 6$ is added and (b) $R = 10$ is added

13-3 Calculate W for a frame wall with $R = 3$ and a brick veneer with $R = 0.45$ with (a) no insulation, (b) insulation $R = 5$, and (c) insulation $R = 8$.

13-4 Calculate W for a basement wall of 10-in. concrete and insulation of $R = 8$.

13-5 Calculate W for a 12-in. concrete block wall with insulation with $R = 5$ added.

13-6 Calculate the heat loss through a double-glass window measuring 40 in. $\times$ 70 in. The inside temperature is 70°F; outside, −10°F.

13-7 Determine the correct multiplier to apply to each square foot of exposed building area for each of the following:

(a) 10-in. concrete wall with insulation value and $R = 12.5$. The inside temperature is 70°F; outside, 10°F.

(b) 8-in. concrete block with 4-in. brick and insulation with $R = 16.5$. The inside temperature is 70°F; outside, 0°F.

(c) Frame wall with insulation of $R = 11$. The inside temperature is 70°F; outside, −10°F.

13-8 Calculate the multiplier for each of the following conditions:

(a) 10-in. concrete, no insulation, inside temperature 70°F, and outside temperature 40°F.

(b) 10-in. concrete, no insulation, inside temperature 70°F, and outside temperature −20°F.

(c) 12-in. concrete block wall, insulation value $R = 10$, inside temperature 70°F, and outside temperature 10°F.

(d) Wood frame wall, insulation value $R = 13$, inside temperature 70°F, and outside temperature 40°F.

13-9 Calculate the heat loss through a double-glass window measuring 46 in. × 70 in. The inside temperature is 70°F; outside, −10°F.

13-10 Calculate the heat loss through a concrete floor 30 ft × 36 ft, below grade. The inside temperature is 70°F; ground temperature, 50°F.

13-11 Calculate heat loss through the 10-in. concrete walls of a basement 30 ft × 36 ft and 7 ft high (half of the basement is below frost; the ground temperature is 50°F; half is subject to 0°F, the outside temperature; the inside temperature is 70°F) (a) if no insulation is used and (b) if insulation with $R = 11$ is used, and (c) determine the cost of heating the basement in (a) and (b) if the season has 6000 DD and the energy cost is $0.014/kWh.

13-12 Calculate the heat loss through the walls of a building 30 ft × 40 ft with a ceiling height of 8 ft. Ten per cent of the total wall area is double glass. The walls are of 8-in. concrete block and 4-in. brick, with 3-in. polystyrene insulation with $R = 12.5$; the inside temperature is 70°F; outside, 0°F.

13-13 Calculate the heat loss through the ceiling of a building 30 ft × 40 ft. The ceiling is tile with insulation value $R = 24$. The inside temperature is 70°F; outside, 0°F.

13-14 Calculate the heat loss through the concrete floor at the grade level of a building 30 ft × 40 ft. Two-inch perimeter insulation is installed. The inside temperature is 70°F; outside, 0°F.

13-15 (a) Calculate the heat loss due to infiltration in a building 30 ft × 40 ft, allowing one air change per hour. The temperature difference is 70°F; the ceiling height is 8 ft.

(b) Calculate the total heat loss of the building in Probs. 13-12 through 13-15(a).

(c) Calculate the heating cost per year for 6000 DD. The energy cost is 1.5 cents/kWh.

13-16 Calculate the heat loss for a small store measuring 30 ft × 40 ft with a ceiling height of 9 ft. The floor and ceiling have no heat loss as they are adjacent to heated areas. The construction is 4-in. brick and 8-in. concrete block with 2-in. polystyrene insulation. The only windows are two single-plate glass display windows measuring 6 ft × 12 ft each. The front and rear glass doors measure 3.5 ft × 7 ft. The design temperature difference is 60°F. A double-door entry allows a minimum of two air changes per hour.

13-17 Calculate the heat loss through the ceiling of a room 12 ft × 18 ft, (the design temperature difference is 70°F; the ceiling resistance is three units) when insulation is (a) $R = 19$, (b) $R = 24$, and (c) $R = 30$.

13-18 Calculate the heat loss through a 12-in. concrete block wall 30 ft × 8 ft with an 8-ft ceiling, using 70°F as the design temperature difference if (a) no insulation is used, (b) 3 in. of expanded polystyrene is used, and (c) if the insulation cost is 35 cents/ft^2, how long will it take to pay for itself? The heating season is 6000 DD. The energy cost is 0.015 per kWh.

13-19 Calculate the heat loss through a single-plate glass store window 6 ft × 11 ft. The inside temperature is 70°F; outside, 0°F.

13-20 A small house has eight windows measuring 28 in × 46 in, each of wood sash, and two measuring 24 in. × 38 in. each. Calculate (a) the total heat loss through the windows if the temperature difference is 70°F, (b) the total heat loss through the windows if wooden storm windows are installed, and (c) the savings in heating costs between (a) and (b) for a heating season of 6000 degree days. The energy cost is 1.4 cents/kWh.

13-21 Calculations for a six-room house show a total heat loss of 12,600 W based on an inside temperature of 70°F and an outside temperature of 0°F. Calculate the seasonal heating costs at the following locations (all have approximately the same minimum temperature, 0°F):
(a) Salt Lake City, Utah, where the heating season consists of 5875 degree days and the energy cost is 1.3 cents/kWh, (b) Boston, Massachusetts, 5800 DD, 1.6 cents/kWh, (c) Newark, New Jersey, 5250 DD, 1.3 cents/kWh, (d) Nashville, Tennessee, 3500 DD, 0.75 cents/kWh, and (e) Amarillo, Texas, 4350 DD, 1.35 cents/kWh.

13-22 A small shack for a parking lot is to be heated with a portable electric heater. The outside temperature is −10°F, inside, 40°F. It measures 5 ft × 5 ft × 7 ft high and is built of plywood that has a resistance (R) of 1.00. The 2 ft × 6 ft door is half glass (single). Assume five air changes per hour. What size portable heater is necessary?

13-23 The basement of a house measures 30 ft × 38 ft with a 7 ft ceiling high. Four single-glass windows measure 16 in × 24 in. each. Half the wall is below frost with a ground temperature at 50°F. Above this point use an outside temperature of 0°F. The inside temperature is 70°F.

(a) Calculate the heat loss with no wall insulation.
(b) Calculate the heat loss with insulation of $R = 11$ and windows equipped with wood storm sashes.
(c) Determine the saving in heating costs for a season of 5000 DD. The energy cost is 1.35 cents/kWh.

Chapter Fourteen

Transformers in Interior Wiring Systems

14-1 ***Selection of voltages for lighting and power.*** The basic problem of determining the proper voltage at which an electrical system should operate has many variables to consider. Nearly all electrical equipment must be physically sized according to the current that the system must carry. To simply say "The higher the voltage, the lower the current" makes good sense, but obviously there must be an optimum voltage choice.

Practially all electrical loads fall into one of three categories: lighting, heating, or power. Reference to power implies electrical-to-mechanical conversion or the use of electric motors. For power, a three-phase supply is most desirable.

The most practical voltage for lighting in residential, commercial, or industrial applications is 115 V. The three-wire, 115–230-V distribution system doubles the permissible load of the two-wire, 115-V system. Three-wire distribution is a good investment: 100% more load with an increase of only 33% conductors. The single-phase, three-wire power supply is explained in Sec. 4-13.

The development of lighting units designed to operate at 265 V saves a considerable amount of copper. Voltages of this magnitude are allowed only in commercial or industrial installations. This system is supplied from a three-phase, four-wire, 265–460-V source. The availability of a three-phase, 460-V supply makes this an ideal system to provide both lighting and power.

The correct voltage selection for power has several contributing factors. Single-phase motors are impractical for anything other than the fractional horsepower sizes for domestic applications or other establishments where three-phase power is unavailable. These motors are designed to operate on 115 V or, by a combination of winding connections, on 115 V or 230 V.

Most three-phase motors are designed for operation on one of three commercial voltages: 230, 460, or 575 V. These are approximate voltages and are all multiples of 115 V. The power supplies could just as well be identified as 240, 480, or 600 V since these would be multiples of 120 V. Supply voltage can vary as much as 10% from specific motor nameplate voltage rating without materially affecting performance. Voltage a little above nameplate rating might be preferred to a lesser value since less current in the same proportion of increased voltage would be required for a particular load.

The ensuing articles explore each of these system voltages and discuss in detail the factors that help make a decision. The transformer's function in providing these various supplies is investigated and comparisons are made with alternate methods, including appropriate examples.

14-2 ***Lighting systems at* 115-230 V.** Residential electrical systems are always 115–230 V. Lighting and small appliances up to about 1300 W are operated at 115 V. Appliances rated over 1300 W are usually designed for 230 V. The neutral conductor is grounded. To assure the grounding of all screw shells of incandescent lamps, 115 V must be used for residential lighting. This same system is used for general lighting in stores, schools, office buildings, and industrial plants. Commercial lighting installations involve large areas that are to be illuminated continuously and, therefore, require detailed attention to design.

Industrial plants that have large power requirements require a three-phase service at the appropriate voltage, 230, 460, or 575 V. None of these voltages would be suitable for lighting; therefore, a 115–230-V supply also must be provided. Electric utility companies provide both these voltages by means of separate service entrance conductors and metering equipment. Small plants whose average load is in the order of 50 kW find this a practical arrangement. Where power requirements are larger, the utility company can furnish power at a lower rate per kilowatt-hour if only one type of service through one meter is provided. This supply must be three phase because of power requirements. This would necessitate that the plant furnish its own transformers to supply 115–230 V for the lighting load.

When a building is large in area, in particular the one- or two-floor modern manufacturing establishments, the distribution of 115–230-V power for lighting becomes a problem, not only because of the large loads but because of the distances involved.

This problem is greatly simplified by the proper use of distribution transformers, which under many conditions can save considerable cost of conductors, raceways, and protective equipment.

The use of transformers in interior wiring systems can be classified into two basic techniques: One installation can have all feeders originate in the switch room where either the utility company provides services for both lighting and power or the customer can maintain his own transformers in the service entrance area of the building. Such an arrangement is shown in Fig. 14-1. In large buildings this can result in long feeders and prohibitive voltage drops in the 115–230-V system. In smaller buildings, however, this arrangement is acceptable.

The other general method of power distribution involves the installation of higher voltage three-phase or single-phase feeders throughout the building. The 115-V loads such as lighting then can be supplied by trans-

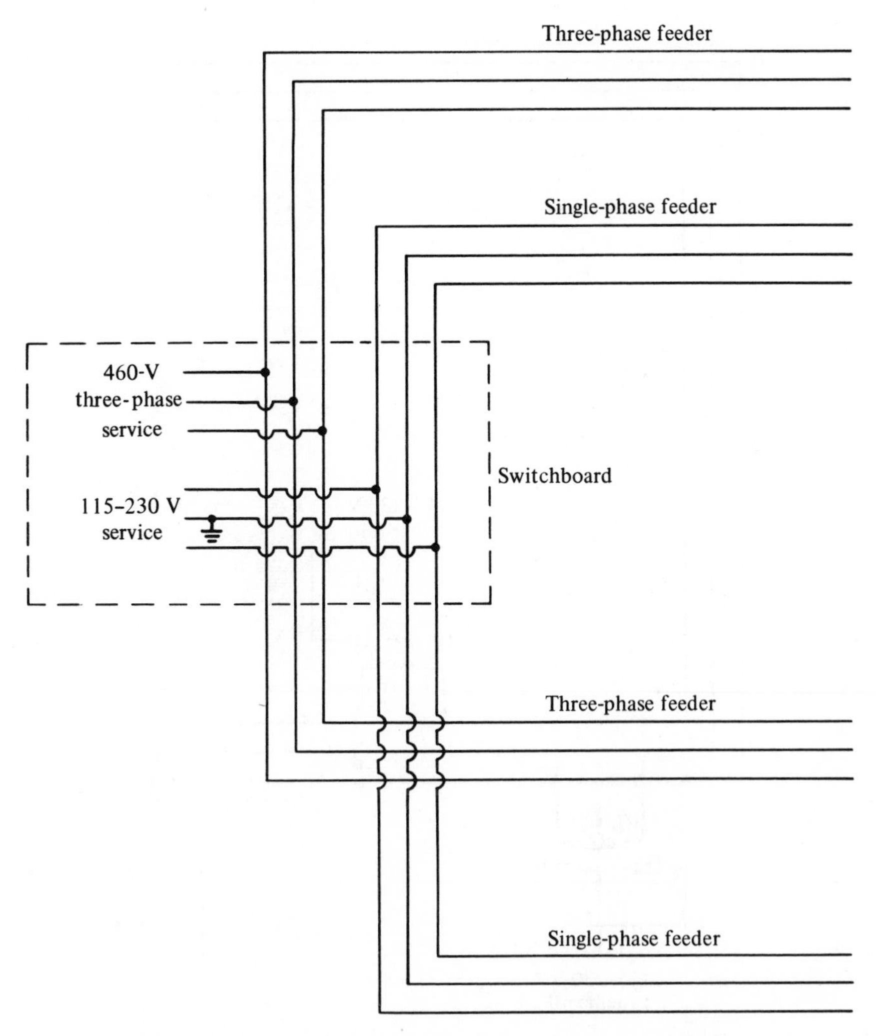

Fig. 14-1. Arrangement of separate feeders for power and lighting.

formers located near the load. By this method the load is conducted the longer distances by smaller conductors carrying smaller currents. An illustration of this is shown in Fig. 14-2. Some companies provide a transformer and panel-board assembly complete with connecting ducts that can be secured to columns in industrial buildings.

Figure 14-2 shows three-phase feeders with a transformer and panel board supplied from each phase. A separate single-phase feeder (two wires) could be provided from the main switchboard. However, this might prove to be uneconomical. Chapter 5 explained that a three-phase feeder can

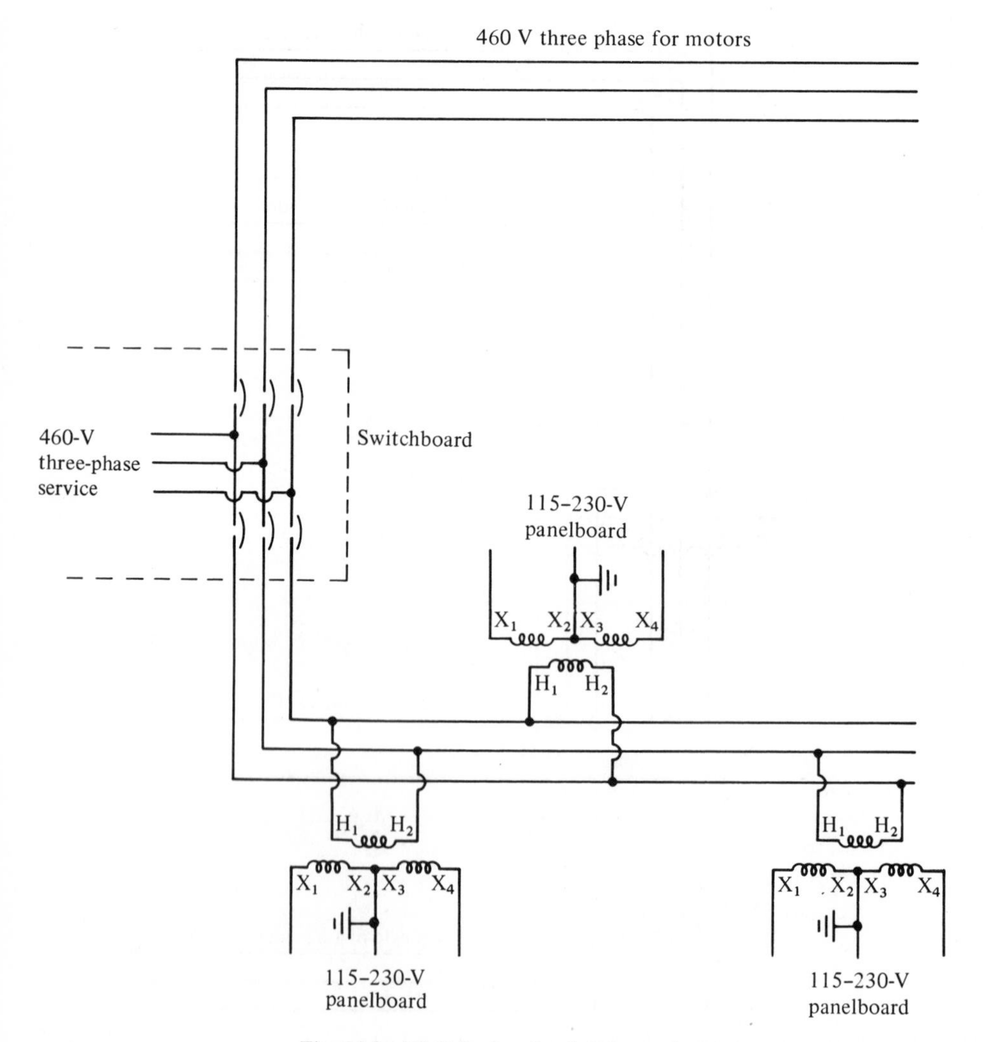

Fig. 14-2. 460-V feeders for lighting and power.

provide 1.732 times the power of the single-phase one at the same current and voltage. When following the layout of Fig. 14-2 the same size conductors must be used to the last transformer—unless another overcurrent device is used. The NEC requires that all conductors be protected according to their ampacities.

Another advantage of the three-phase feeder system is the availability of the three-phase system for motors at any point. The NEC permits tapping conductors of at least one-third the ampacity of the larger if the smaller conductors do not extend more than 25 ft. They must be suitably protected against mechanical injury and terminate in a single set of overcurrent devices that will limit the current to their ampacity. Note that this does not permit the installation of panel boards for branch circuits unless the sum of all branch circuit currents is equal to the ampacity of the smaller feeder.

Example ***14-1.*** A store 150 ft × 100 ft is to be provided with 3 W/ft^2. Two panel boards equally divide the 115–230-V lighting load. The copper feeders are 50 and 125 ft, respectively. Because of short distances the voltage drop need not be considered. Use TW conductors.

(a) Calculate the size of each feeder if a 115–230-V supply is used. Assume a unity power factor.

(b) Calculate the size of each feeder if a 460-V, single-phase supply is used to supply a transformer at each panel board location.

Solution

(a) Total power = (150)(100)(3) = 45,000 W

$$\text{Current in each feeder} = \frac{22{,}500}{(230)(1)} = 98 \text{ A} \qquad (4\text{-}6)$$

From Table 9-4 select No. 2 AWG RHW conductors. Three conductors require $1\frac{1}{4}$ in. conduit or EMT.

(b) $\text{Current in each feeder} = \dfrac{22{,}500}{460} = 49 \text{ A}$

From Table 9-4 select No. 6 AWG RHW conductors. Two conductors require 1-in. conduit.

Example 14-1 shows that for short distances the 460-V feeder does not offer much saving in equipment when compared to the 115–230-V feeder. Figure 14-3 shows the wiring arrangement of Ex. 14-1.

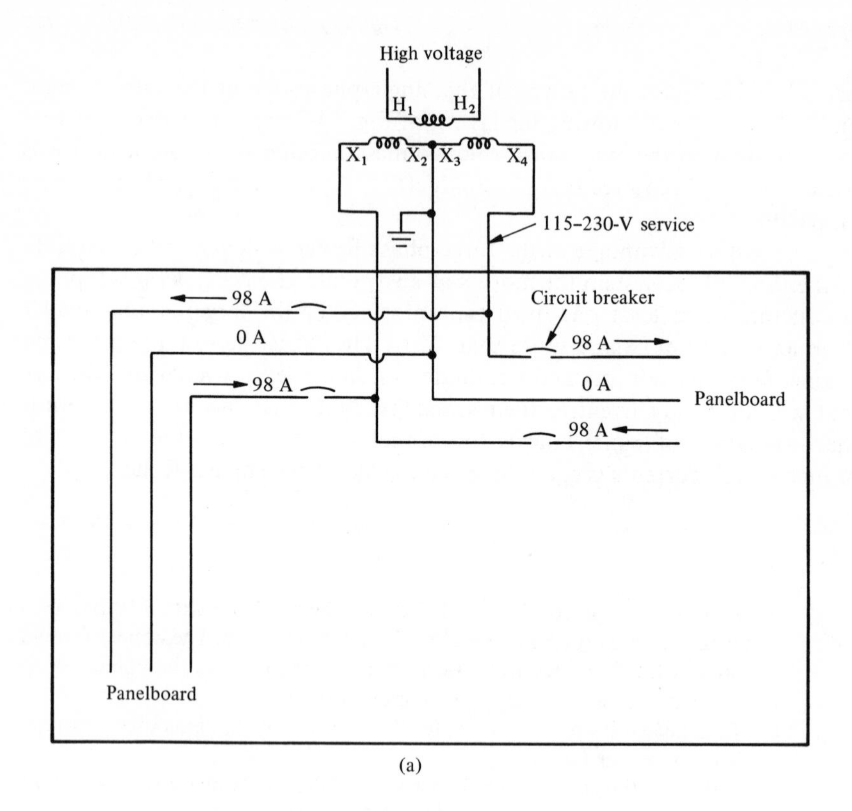

(a)

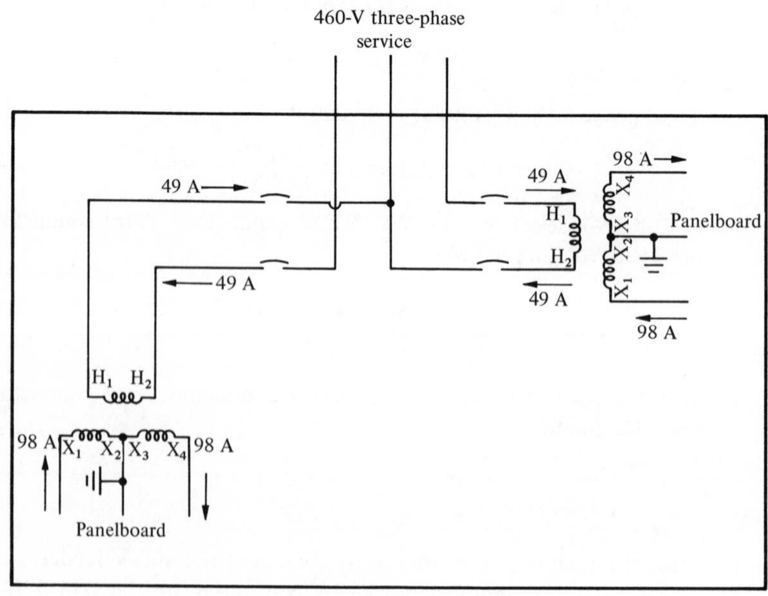

(b)

Fig. 14-3. Diagram and solution of Ex. 14-1.

Example **14-2.** A panel board is supplied with copper TW feeders 350 ft long and distributes 22.5 kVA at 115 V.

(a) Calculate the feeder size if the feeder is 115–230 V. Assume unity pf and a 2% voltage drop.

(b) Calculate the feeder size if 460 V and single phase is to supply a transformer at the location of the panel board.

Solution

(a) The size feeder at 115–230 V (three wires) is

$$I = \frac{22{,}500}{230} = 98 \text{ A} \qquad (3\text{-}4)$$

$$\text{cmil} = \frac{(\rho ID)(2)}{V_{\text{drop}}} = \frac{(10.4)(98)(350)(2)}{(0.02)(230)} \qquad (9\text{-}1)$$

$$= 155{,}000 \text{ cmil}$$

The next largest standard size is No. 000 AWG. The ampacity is 195 A. Three conductors require 2-in. conduit or EMT.

(b) The size feeder at 460 V (two wires) is

$$I = \frac{22{,}500}{460} = 49 \text{ A}$$

$$\text{cmil} = \frac{(10.4)(49)(350)(2)}{(0.02)(460)} = 38{,}800 \text{ cmil} \qquad (9\text{-}1)$$

The next largest standard size is No. 4 AWG. The ampacity is 70 A. Two conductors require 1-in. conduit or EMT.

From the results of Exs. 14-1 and 14-2 we note that long feeders (over 300 ft) can be installed with smaller equipment when the transformers are located adjacent to the panel boards. There are no firm guidelines to follow in this matter. When the electric utility company provides a 460-V, three-phase power supply, the transformers for 115-V lighting must be in the building somewhere. Therefore it is good design practice to locate them as close to the lighting load as possible. Figure 14-4 shows the wiring diagram of installations of Ex. 14-2. Example 14-2 shows a definite advantage when using the 460-V feeders.

14-3 ***Operating* 120-V *lighting from three-phase feeders.*** Chapter 5 revealed that a three-phase system can provide 1.732 times the power of a single-phase system with the same current and voltage. This principle is worthy of consideration when designing feeders for lighting systems.

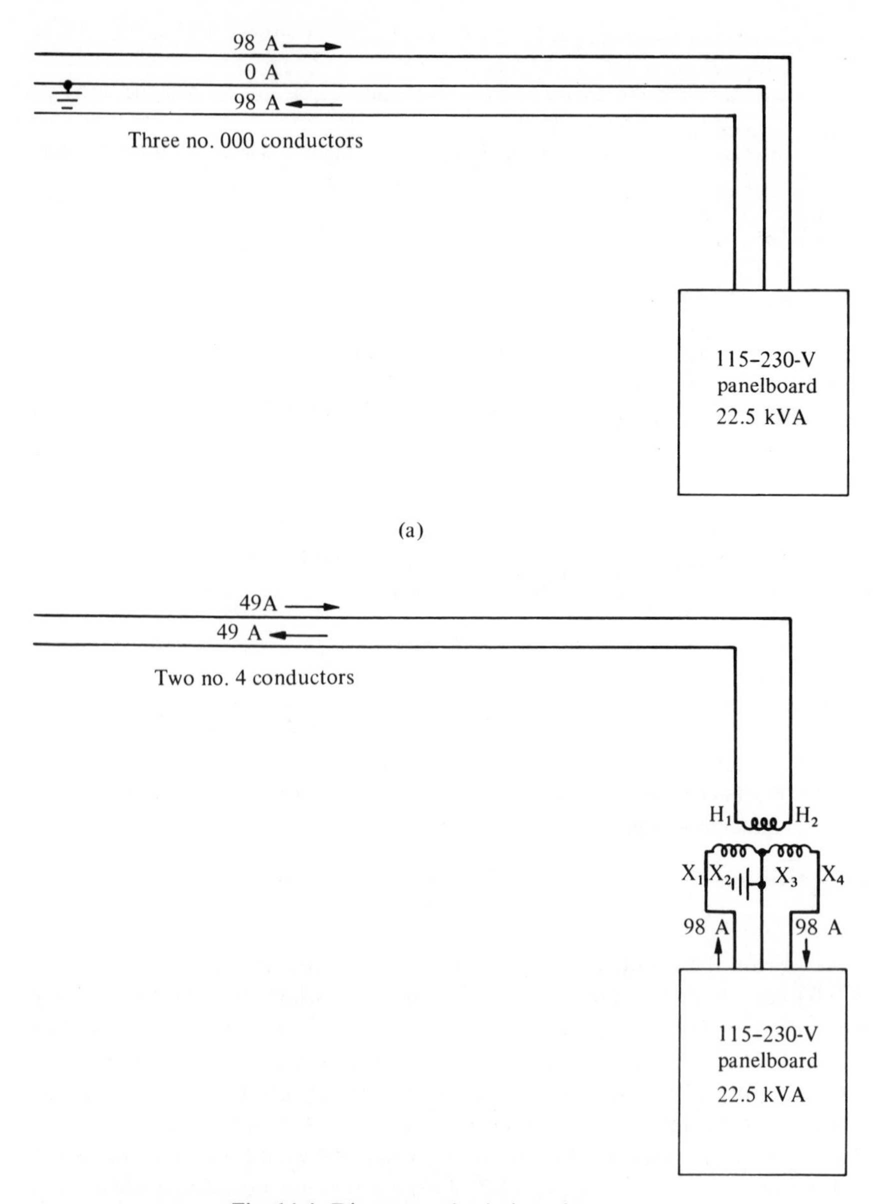

Fig. 14-4. Diagram and solution of Ex. 14-2.

This requires a 120-208-V, three-phase, four-wire power supply, as shown in Fig. 5-11. If the transformer secondaries were each rated at 115 V, it would be a 115–200-V system. Figure 11-5 shows the internal connections of a panel board using a three-phase feeder as compared to a single phase.

The loads on each phase should be equal. The feeder must include the neutral conductor to assure the grounding requirement of the NEC. The neutral is also necessary to carry unbalanced currents when part of the lighting load is in operation and is unbalanced.

In large buildings with many panel board locations the use of transformers adjacent to the panel boards to obtain 120–208 V may not be practical. Each panel board would require three transformers. Section 11-5 proved the importance of limiting the length of the branch circuits. This imposes a limit on the area to be served by each panel board.

The conditions that require the panel boards to be dispersed, thereby limiting the load that each must allocate, also require the transformers to be dispersed. When high-voltage feeders are used they may be three phase or single phase (two wires) tapped from a three-phase feeder.

The merits of single-phase and three-phase feeders may be compared in Ex. 14-3. Note that (b) results in a 115–200-V, four-wire system.

Example ***14-3.*** A 30-kVA, 115-V lighting load is supplied from a panel board, 225 ft from the service entrance location. The supply is 460 V and three phase.

(a) Calculate the size of copper TW feeders if a 30-kVA transformer is located adjacent to panel board. Limit the voltage drop to 2%.

(b) Calculate the size of copper TW feeders if three 10-kVA transformers connected in delta wye are located adjacent to the panel board.

Solution

(a) The current if the feeder is single phase is

$$I = \frac{VA}{E} = \frac{30{,}000}{460} = 65.2 \text{ A}$$

$$\text{cmil} = \frac{(\rho ID)(2)}{V_{\text{drop}}} = \frac{(10.4)(65.2)(225)(2)}{9.2} \tag{9-1}$$

$$= 33{,}200 \text{ cmil}$$

The next largest standard size is No. 4 AWG. The ampacity of No. 4 copper TW is 70 A. Two No. 4 conductors require $1\frac{1}{4}$-in. conductor EMT

(b) The current if the feeder is three phase is

$$I = \frac{VA}{E\sqrt{3}} = \frac{30{,}000}{(460)(1.732)} = 37.6 \text{ A}$$

$$\text{cmil} = \frac{(10.4)(37.6)(225)(\sqrt{3})}{9.6} = 15{,}800 \text{ cmil} \tag{9-1}$$

The next largest standard size is No. 8 AWG. The ampacity of No. 8 copper TW is 40 A. Three No. 8 conductors require $\frac{3}{4}$-in. conduit or EMT.

The savings in the conductors and raceway of Ex. 14-3(b) must be weighed against the cost of three 10-kVA transformers and one rated at 30 kVA and relative installation. Figure 14-5 shows the wiring diagram in each case.

The service equipment for many buildings can be smaller if a 120–208-V, three-phase supply is installed instead of a 120–240-V, single-phase one. If this is the only service connection to the building, the utility company provides the three-phase transformation necessary.

The 208 V would not be as good a supply for three-phase motors since fewer motors are available rated for 208 V compared to 220–240 V. With 120 V available for lighting, an efficient distribution of heating and lighting loads is possible. Apartment buildings can use this three-phase supply. However, 208-V appliances, ranges, dryers, and air conditioners would be necessary, which sometimes presents a problem in availability.

When the three-phase, four-wire system is used in apartment buildings, three phases are not necessary for each apartment. Three conductors, only two lines, and the neutral need supply each apartment through individual meters. This permits meter connections as in Fig. 11-12a. This is not the same power supply as 120–240 V single phase; the voltage between the two line wires is 208 V. Therefore large appliances must be rated for 208 V to assure proper operation.

---

*Example **14-4**.* An office building requires a 120-V lighting load of 15 kW at unity pf. A space-heating load of 55 kW may use 208- or 240-V equipment. Calculate the size of the service equipment if the power supply is (a) 120–240 V, single phase and (b) 120–208 V, three phase.

Solution

(a)
$$I = \frac{70{,}000}{230} = 304 \text{ A}$$

Service equipment must include three 350 MCM copper THW conductors in 3-in. conduit and a 400-A main switch with 300-A fuses or a 300-A circuit breaker.

(b)
$$I = \frac{70{,}000}{(208)(\sqrt{3})} = 194 \text{ A}$$

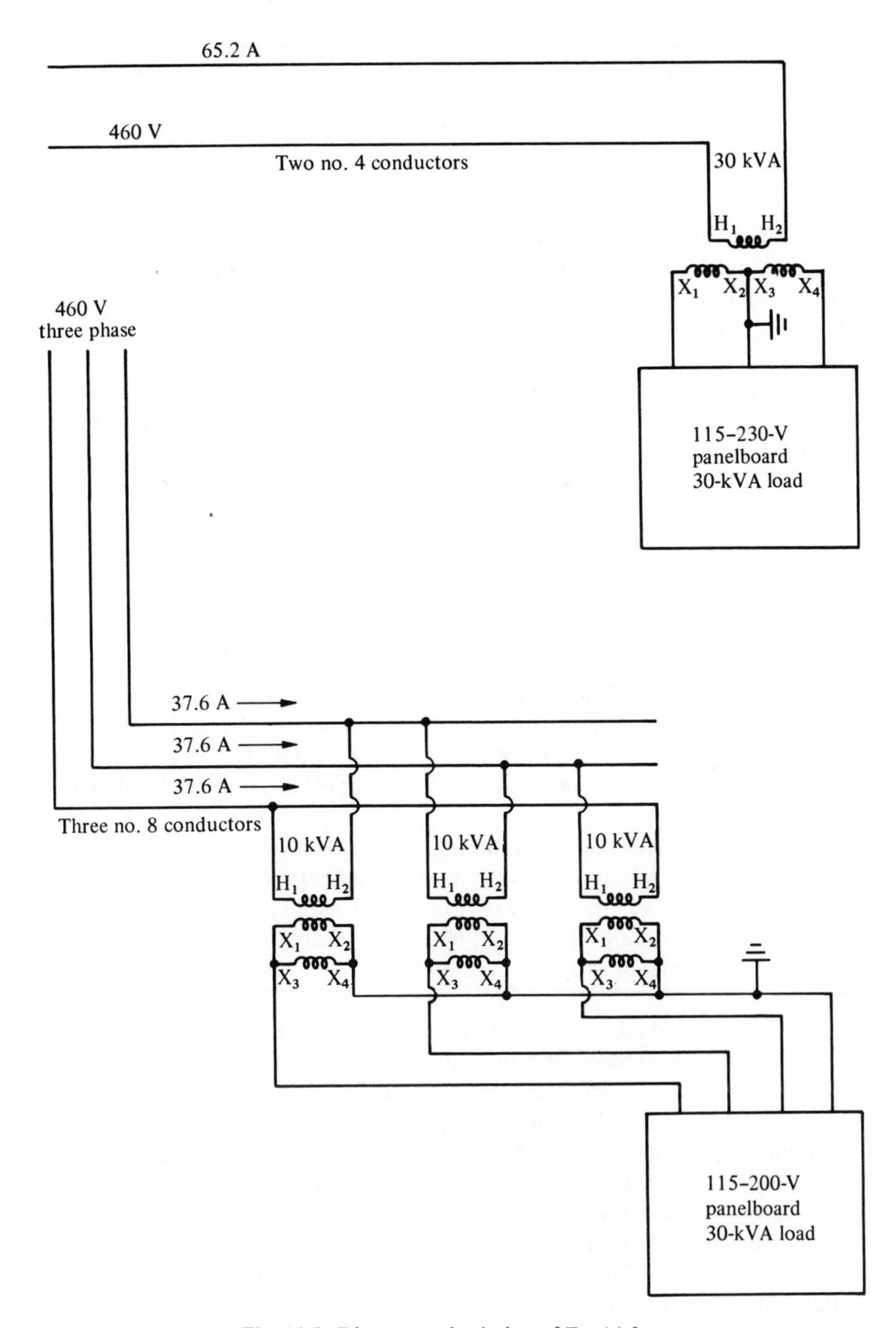

Fig. 14-5. Diagram and solution of Ex. 14-3.

Service equipment must include four 000 AWG copper THW conductors in $2\frac{1}{2}$-in. conduit and a 200-A main switch with 200-A fuses or a 200-A circuit breaker.

The obvious economy in service equipment shown in Ex. 14-4 continues when feeders and distribution equipment are designed. The 120–208-V, three-phase system is worthy of consideration for lighting or heating loads; however, when extended long distances the voltage drop must be carefully considered.

The neutral conductor of 115–230 V or 120–208 V always must be grounded as close as possible to the transformer bank, regardless of its location within the building. This is now the point of entrance referred to in Sec. 10-2.

14-4 ***Lighting loads at* 265 V.** Section 11-5 illustrates the limited distance that 115-V circuits should extend. The use of 265-V lighting units permits a branch circuit $2\frac{1}{2}$ times longer with the same percentage voltage drop. This is a very important advantage in large buildings or those with high ceilings such as factories or arenas.

The 265–460-V supply is obtained from any high-voltage source by connecting the secondaries of a transformer bank in wye as in Fig. 5-11, except that each phase or transformer secondary would be a 265-V winding. This results in a line voltage of 460 V. This supply is sometimes identified as 277–480 V.

These transformer connections could be the utility company's service conductors to the building. This makes a convenient service for both lighting and power, as 460 V, three phase is available for motors. There invariably will be some equipment requiring 115 V, which can be obtained either by separate service conductors or the use of a 460/115–230-V transformer inside the building.

Panel boards designed for three-phase feeders, as shown in Fig. 11-5b, require four-wire feeders. We now should consider the merits of a three-phase, four-wire, 265–460-V feeder supplying 265-V lighting against those of Ex. 14-3(a), where a three-phase, three-wire, 460-V feeder supplies 115-V lighting. It appears to be a similar problem when designing the feeders except that this situation has one more conductor. The factor that determines the best decision is the requirements dictated by length of branch circuits. If they must extend *over* 100 ft, the three-phase, four-wire feeder is the *best* solution, using 265-V lighting units. Example 14-5 compares the feeders with the single phase of Ex. 14-2(b).

Example **14-5.** A four-wire panel board is supplied by means of a 265–460-V copper TW feeder, 350 ft long. The lighting load is 22.5 kVA operating at 265 V at unity pf. Allow a 20% voltage drop. Calculate the size of the feeder necessary.

Solution

$$I = \frac{22{,}500}{(460)(\sqrt{3})(1)} = 28.2 \text{ A} \tag{5-4}$$

$$\text{cmil} = \frac{(10.4)(25.2)(350)(\sqrt{3})}{(0.02)(460)} = 19{,}300 \text{ cmil} \tag{9-1}$$

The next largest standard size is No. 6 AWG. The ampacity is 55 A. Four No. 6 AWG conductors require a 1-in. conduit or EMT. Figure 14-6 shows this solution.

The results of Ex. 14-5 compared to Ex. 14-2(b) show no advantage in feeder design; however, the longer branch circuits offered by the 265-V lighting system of Ex. 14-5 make a much more efficient system.

The load of 22.5 kVA might be considered small. In Ex. 14-6 we shall use a larger load and make the same comparisons.

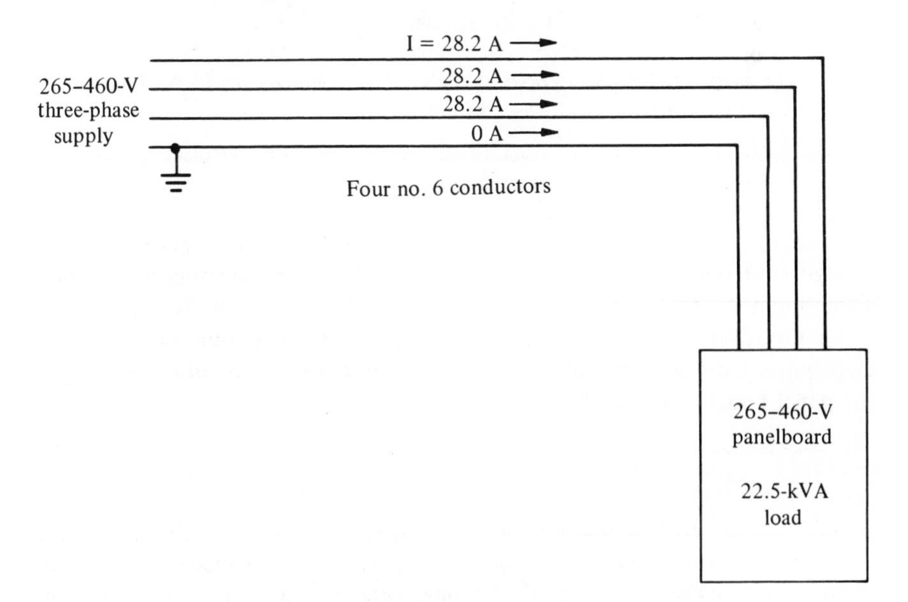

Fig. 14-6. Diagram and solution of Ex. 14-5.

Example 14-6. A 50-kVA load is distributed by a panel board 350 ft from the 460-V switchboard. In each case permit a 2% voltage drop; use copper TW conductors. Calculate the size and make comparisons if (a) a four-wire, three-phase, 265–460-V feeder supplies 265-V lighting units from a four-wire panel board and (b) a two-wire, single-phase, 460-V feeder supplies a 50-kVA, 460/115–230-V transformer adjacent to a three-wire panel board supplying 115-V lighting.

Solution

(a) The current if a three-phase feeder is used is

$$I = \frac{50{,}000}{(460)(\sqrt{3})} = 62.7 \text{ A}$$

$$\text{cmil} = \frac{(10.4)(62.7)(350)(\sqrt{3})}{(0.02)(460)} = 43{,}000 \text{ cmil} \qquad (9\text{-}1)$$

The next largest standard size is No. 3 AWG. The ampacity is 80 A. Four No. 3 AWG conductors require $1\frac{1}{4}$ conduit or EMT.

(b) The current if a single-phase feeder is used is

$$I = \frac{50{,}000}{460} = 109 \text{ A}$$

$$\text{cmil} = \frac{(10.4)(109)(350)(2)}{(0.02)(460)} = 86{,}500 \text{ cmil} \qquad (9\text{-}1)$$

The next standard size is No. 0 AWG. The ampacity is 125 A. Two No. 0 AWG conductors require $1\frac{1}{2}$ conduit or EMT.

Example 14-6 does not offer much choice in feeder equipment. Although four conductors are necessary in (a), a smaller raceway is possible. Therefore the 265-V lighting system again will be the best overall design. It is doubtful if a 115-V lighting load as large as 50 kVA could be effectively distributed from a single panel board. Example 14-7 illustrates the size of the panel boards in each situation of Example 14-6.

Example 14-7. A 50-kVA lighting load is to be distributed from a single panel board using 15-A branch circuits. Determine the minimum number of circuits necessary if (a) the lighting operates at 265 V from a three-phase, four-wire panel board, as in Ex. 14-6(a), and (b) the lighting operates at 115 V from a single-phase, three-wire panel board, as in Ex. 14-6(b).

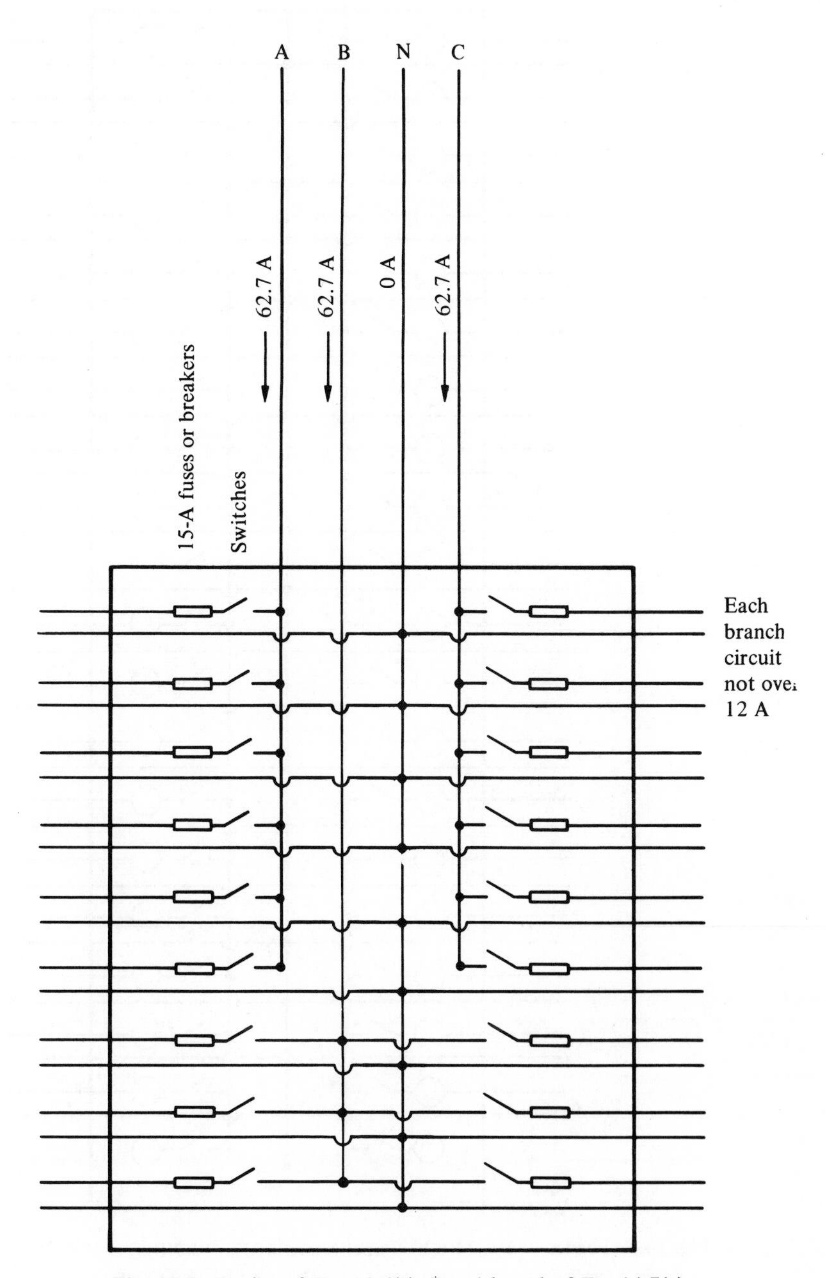

Fig. 14-7. Feeder of Ex. 14-6(a). Panel board of Ex. 14-7(a).

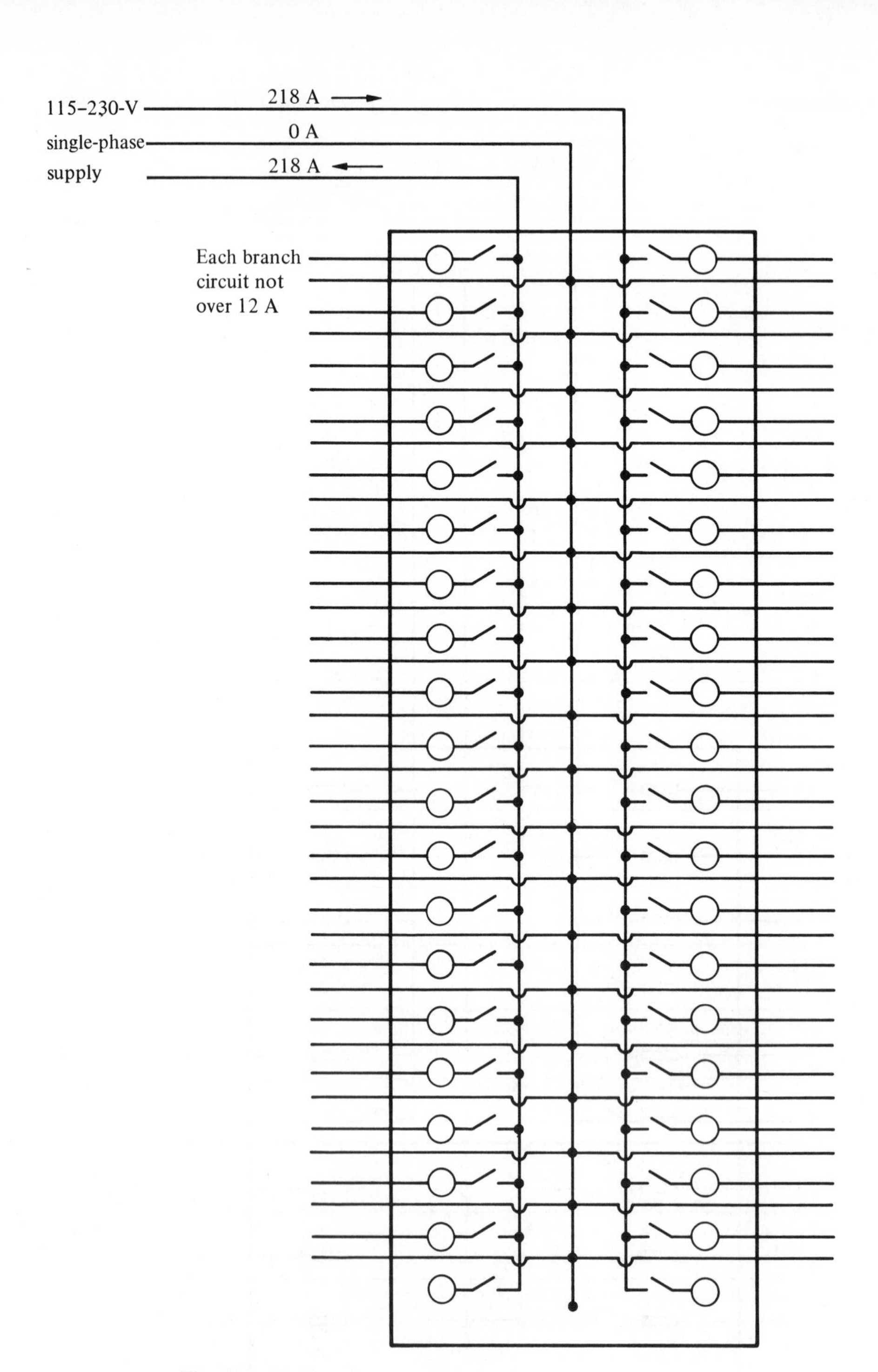

Fig. 14-8. Feeder of Ex. 14-6(b). Panel board of Ex. 14-7(b).

Solution

(a)
$$I = \frac{50{,}000}{(460)\sqrt{3}} = 62.7 \text{ A in each line}$$

The maximum current per 15-A circuit is 80% of 15 A = 12 A. The number of circuits for each line is 62.7/12 = 5.2 or 6 branch circuits. There are three line wires; the total branch circuits'= (6)(3) = an 18-circuit panel board (minimum).

(b)
$$I = \frac{50{,}000}{230} = 218 \text{ A}$$

The number of circuits for each line = $\frac{218}{12}$ = 18.2 or 19 branch circuits. There are two line wires; the total branch circuits = (19)(2) = a 38-circuit panel board (minimum).

Example 14-7 illustrates the further advantage of 265-V lighting with regard to cost and simplicity of distribution equipment. Figure 14-7 illustrates the solution of Exs. 14-6(a) and 14-7(a) using four-wire, 460-V feeders for 265° lighting. Figure 14-8 illustrates the solution of Ex. 14-6(b) and 14-7(b) using three-wire, 230-V feeders for 115-V lighting. A 20-circuit panel board is shown.

We note from these examples that to operate 265-V lighting from a 460-V, three-phase feeder the neutral conductor must be included. The three-wire feeder intended to supply a motor load could not be used. The grounding of the neutral, resulting in 265 V to ground throughout the entire secondary system, is recommended. This depends on the local option.

14-5 ***Autotransformer applications.*** Practically any problem of providing proper voltage for a particular use can be solved by the manipulation of proper transformer ratios. In this respect the use of autotransformers can be of considerable value. As explained in Sec. 6-11, the rating of a transformer when used in this manner can supply a much larger load than a transformer with a conventional primary and secondary. A few of these techniques are illustrated here, but applications are virtually unlimited.

A 230-V, three-wire, three-phase feeder system is very practical for moderate motor loads. Lighting loads must be supplied by a separate system of 115–230-V, single-phase feeders. Considerable duplication of feeders can be eliminated by autotransformers connected to provide a 115-V, two-wire supply. The problem is not to transform power from one voltage to another but simply to provide some means of tapping half of one phase of the 230-V supply.

However, the NEC prohibits the use of autotransformers to supply lighting and appliance branch circuits unless a grounded conductor is common to both the primary and secondary circuits. Three-phase circuits often are not grounded, except neutrals of wye-connected secondaries. This imposes limitations on this technique. The decision of when or when not to ground a system is thoroughly discussed in Chapter 10.

Figure 14-9 shows how a 115-V load can be connected to one phase of a 230-V supply with a one-to-one ratio transformer. The 230-V supply must be grounded; connections cannot be made to phase *AB*.

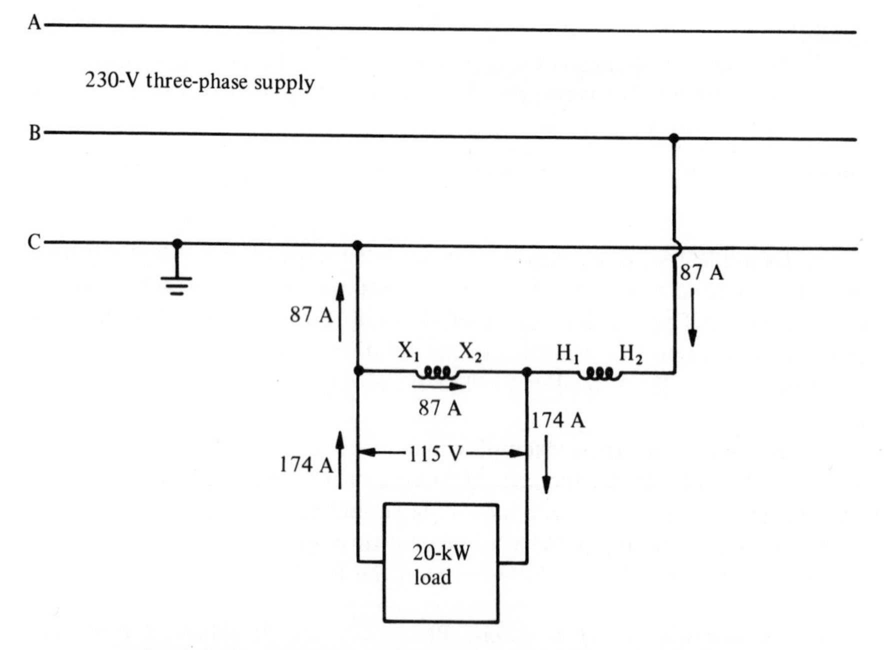

Fig. 14-9. A one-to-one ratio transformer used as an autotransformer to operate a 115-V load from a 230-V grounded supply.

Example 14-8

(a) What must be the rating of the transformer of Fig. 14-9 if a lighting load of 20 kW at unity pf is connected to a 115-V circuit?

(b) What rating is required if the pf is 0.9?

Solution

(a) The current to a 115-V load is

$$I = \frac{(\text{kilowatts})(1000)}{E} = \frac{(20)(1000)}{115} = 174 \text{ A}$$

the current from a 230-V line is

$$I = \frac{(\text{kilowatts})(1000)}{E} = \frac{(20)(1000)}{230} = 87 \text{ A}$$

These currents and relative directions are shown in Fig. 14-9. The transformer rating, using the winding voltage and current above, is

$$\text{Kilovolt-amperes} = \frac{I_1 E_1}{1000} = \frac{(87)(115)}{1000} = 10 \text{ kVA}$$

Transformer rating

$$\text{kilovolt-amperes} = \left(\frac{E_L}{E_H + E_L}\right)(20) = \left(\frac{115}{115 + 115}\right)(20) = 10 \text{ kVA} \qquad (6\text{-}5)$$

(b) If the pf is 0.9, the current from a 230-V line is

$$I = \frac{(\text{kilowatts})(1000)}{(E)(\text{pf})} = \frac{(20)(1000)}{(230)(0.9)} \qquad (4\text{-}5)$$

$$= 96.5 \text{ A}$$

The transformer rating is

$$\text{Kilovolt-amperes} = \frac{I_1 E_1}{1000} = \frac{(96.5)(115)}{1000} = 11.1 \text{ kVA}$$

or

$$\text{Transformer kilovolt-amperes} = \left(\frac{E_L}{E_H + E_L}\right)(\text{load kVA}) \qquad (6\text{-}5)$$

$$\text{Load kilovolt-amperes} = \frac{P}{\text{pf}} = \frac{20}{0.9} = 22.2 \text{ kVA}$$

$$\text{Transformer kilovolt-amperes} = \left(\frac{115}{230}\right)(22.2) = 11.1 \text{ kVA}$$

Because of the lighting panel construction and the larger current required, long two-wire, 115-V feeders are seldom used. The three-wire, 115–230-V system is desired.

Many conditions may find this technique a practical solution. Figure 14-9 shows a 115-V lighting load supplied from a grounded phase of a 230-V, three-phase feeder. This requires a 115-to-115-V transformer or one 230-V winding, center tapped.

Autotransformers can be used to advantage in three-phase circuits.

Both 460- and 575-V, three-phase supplies are extensively used for motors. They are not compatible because motors must be designed for one or the other. Many motors are wound for either 230 or 460 V, making these voltages more versatile than 575 V.

When it becomes necessary to operate 575-V motors on 460-V, three-phase systems, the proper selection of three transformers using the auto-transformer technique provides an economical solution.

Refer to Fig. 14-10a. We first connect the three primaries rated at 460 V each in delta to the 460-V source, as illustrated in Fig. 6-8. With the rated voltage on each primary winding, the proper excitation is assured. The desired output voltage from A_2 to B_2 is 575 V or 115 more than from A_1 to B_1. The voltage in each X_1X_2 winding of Fig. 14-10a are 60° apart, as shown in the vectors of Fig. 14-10b. The output line voltage of 575 V must be the vector sum of E_{ac}, E_{AB}, and E_{ab}. The vector diagram shows that the sum of E_{ac} and E_{ab} is 30° from the input line voltage E_{AB} and must be equal to 128 V. Graphically or mathematically it can be shown that 128 V added at 30° to 460 V is equal to 575 V, our desired output. The voltage E_{ac} and E_{ab} are 60° apart; therefore each must equal $128/\sqrt{3}$ or 74 V. This is the necessary secondary voltage required for each transformer, as shown in Fig. 14-10c.

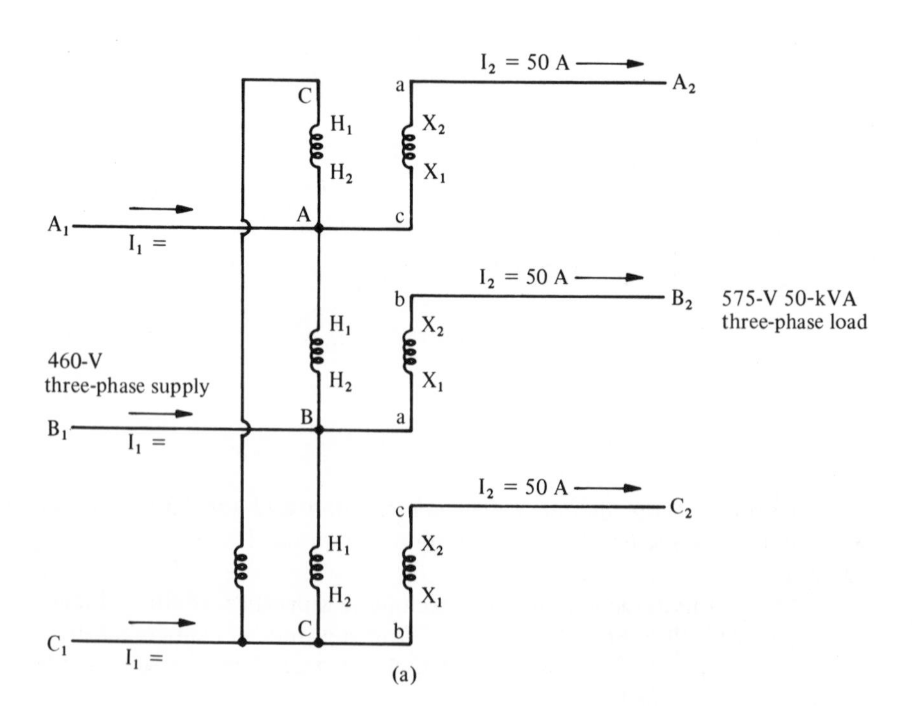

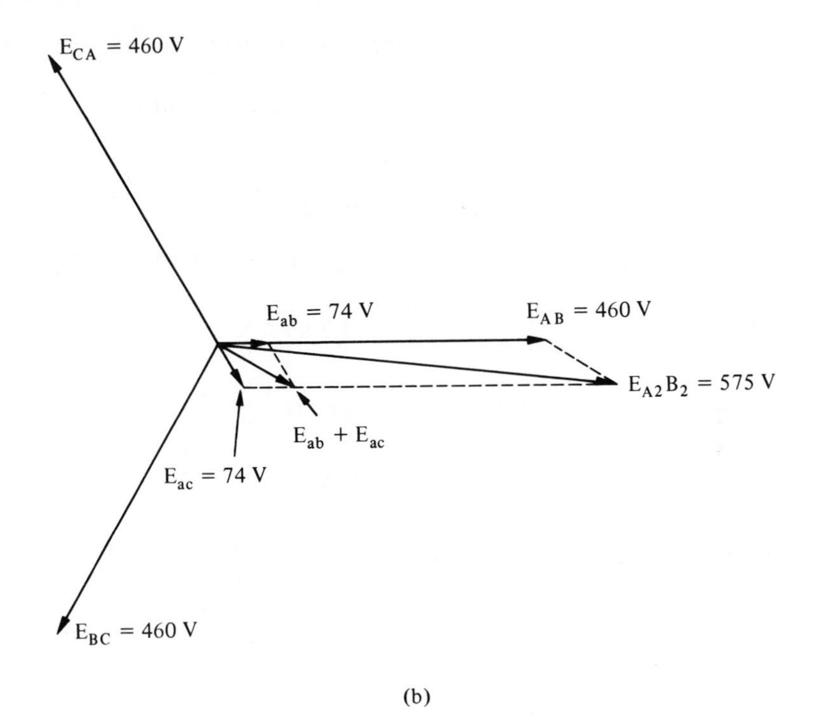

(b)

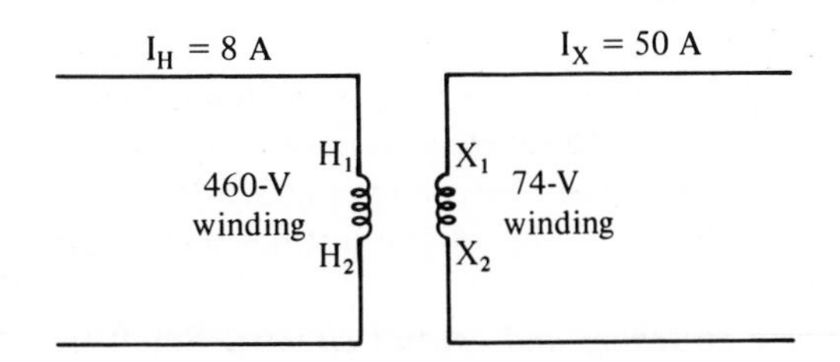

Each transformer rated at 3.7 kVA
to provide 575 V three phase for
50-kVA load

(c)

Fig. 14-10. (a) Autotransformer connections to obtain 575 V from a 460-V three-phase supply. (b) Vector additions of voltages of Fig. 14-10(a). (c) Necessary windings of transformers to obtain 575 V from a 460-V three-phase supply.

The same procedure will result in 575 V between B_2 and C_2 and also between C_2 and A_2.

For a 575-V, 50-kVA load, the line current to the load is

$$I_2 = \frac{VA}{(E)(\sqrt{3})} = \frac{50,000}{(575)(\sqrt{3})} = 50 \text{ A}$$

The required rating of each transformer for the 50-kV load is

$$\frac{E_2 I_2}{1000} = \frac{(74)(50)}{1000} = 3.7 \text{ kVA}$$

The primary current in the transformer winding is

$$I = \frac{3700}{460} = 8.0 \text{ A}$$

The line current from the 460-V supply is

$$I_1 = \frac{50,000}{460\sqrt{3}} = 62.7 \text{ A}$$

According to the technique of autotransformers, Fig. 14-10 shows 50 A flowing from the supply directly to the load. Only 8 A at 460 V is the primary loading, and 50 A at 74 V is transformed. From these values the size of each transformer is obtained.

We note from the transformer rating that the combined rating for three is approximately equivalent to 20% of the load. The calculations in this explanation are accurate; however, a variation up to 15% in the booster winding voltage will not materially change the 575-V output. Close examination of the vector addition of Fig. 14-10b justifies this observation.

14-6 ***Low-voltage switching and control circuits.*** Small transformers rated from 10 to 100 VA are very economical power supplies for low-voltage devices operating at 6, 12, or 24 V. These transformers are considered low-energy sources; when short-circuited, no equipment or wiring is subject to overcurrent. These circuits do not require conventional wiring methods; wiring may consist of any type of cord or open wiring.

Several companies manufacture equipment consisting of relays and momentary contact switches for controlling individual or groups of lights from many locations. The wiring of these switches is simplified as only a three-wire cord need be extended to each location. This eliminates voltage drop in switching circuits: The relay performs the switch's function.

Figure 14-11 shows three remote switches controlling a light. The

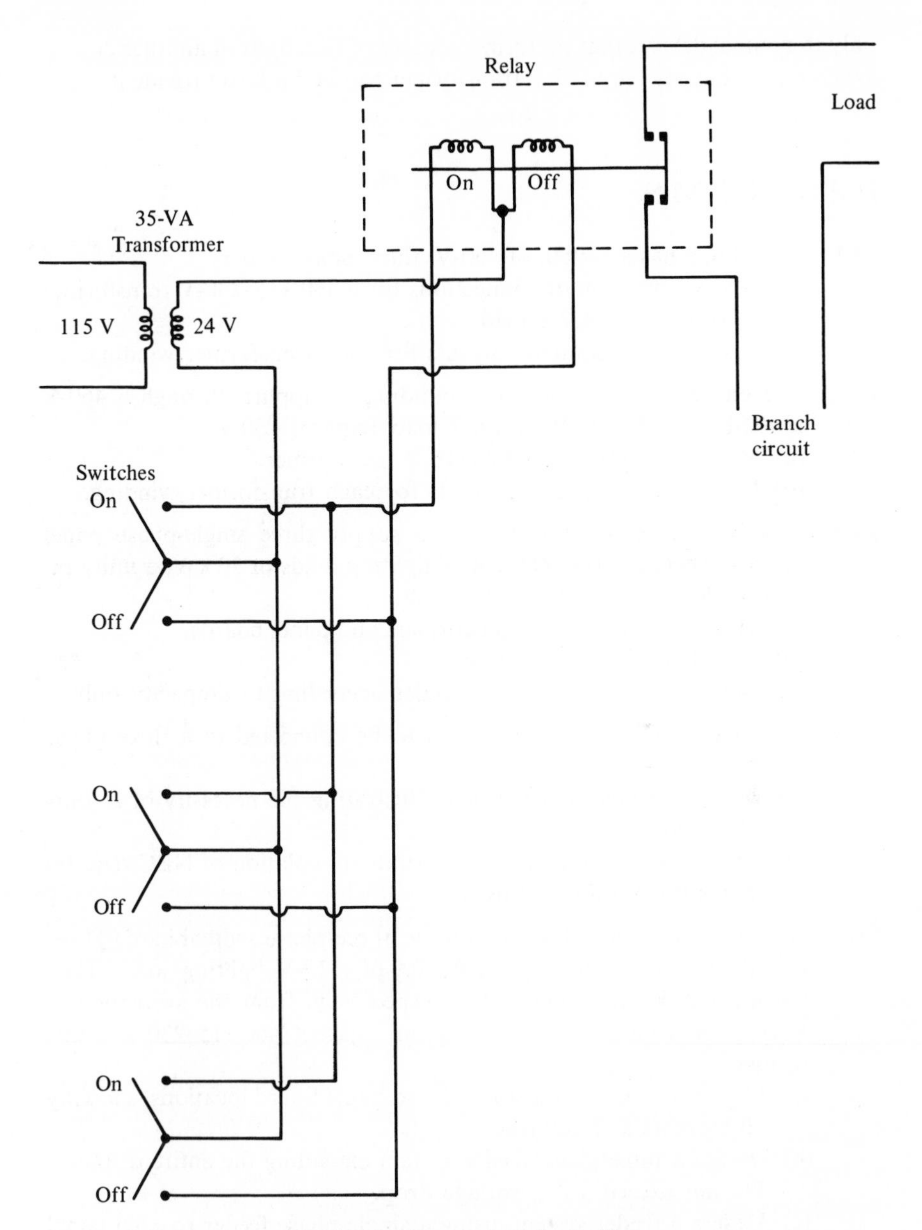

Fig. 14-11. Remote control switching.

relay is held in the ON or OFF position with a spring. The low-voltage circuit provides current only during the switching process; all switches are normally open.

Transformers can provide the appropriate voltage for circuits designed only to control devices that perform a variety of functions in another circuit. When any voltage is required, a transformer can be built to provide it.

PROBLEMS

14-1 A building has a system of 460-V, three-phase feeders.

(a) Make a diagram of connections for a 460/115–230-V transformer to supply a 20-kVA load.

(b) Indicate the required currents for each transformer winding.

14-2 The entire electrical load of a building is supplied through a 460-V switchboard. A 30-kW heating load requires 230 V.

(a) Make a diagram of a 460/115-V transformer.

(b) Indicate the required currents for each transformer winding.

14-3 A 460-V, three-phase feeder is to supply three single-phase panel boards. There are two 115–230-V lighting loads of 30 kW, a unity pf, and a 230-V heating load of 25 kW.

(a) Make a sketch of the transformers and panel boards.

(b) Calculate the feeder current.

(c) Determine the size of the feeder according to ampacity only.

14-4 A 30-kVA, 265-V lighting load is to be connected to a three-phase feeder.

(a) Make a sketch of the system illustrating the necessity of a four-wire feeder.

(b) Show how the system might operate in violation of NEC requirements if neutral is not used.

14-5 A building is supplied from a 460-V, three-phase switchboard. Three panelboards distribute a 54-kW, 0.9 pf, 115-V lighting load. They are located 50, 150, and 250 ft, respectively, from the switchboard. Each panel board is supplied by an adjacent 460/115–230-V transformer.

(a) Sketch one of the transformer and panel board locations, showing all currents and voltages.

(b) Design a three-phase feeder system extending the entire distance. Do not exceed a 2% voltage drop.

(c) Design a feeder system, using a single-phase feeder to each panel board. Do not exceed a 2% voltage drop.

14-6 A delta-to-delta 1100/115–230-V transformer bank is supplying a 230-V, three-phase, 60-kW, unity pf heating load.

(a) Make a sketch of the transformers and panel board. (Each heater is 1500 W and is connected to a separate branch circuit.)

(b) Calculate the rating of each transformer

(c) If one transformer should become defective, make a sketch of the secondaries and calculate the heating load that can be used without exceeding the kilovolt-ampere rating of transformers. (This is an open delta operation.)

14-7 It is necessary to operate a 25-kVA, 265-V lighting load from one phase of an ungrounded 230-V, three-phase supply with an autotransformer connected to obtain 265 V.

(a) Sketch the transformer connections, showing necessary voltages.

(b) Calculate the line and load currents.

(c) Calculate the kilovolt-ampere rating of the transformer.

14-8 A 25-kVA, 265-V lighting load is to be supplied from a 230-V, three-phase supply using three autotransformers.

(a) Sketch the transformer connections.

(b) Calculate the secondary voltage necessary to obtain 265 V.

(c) Sketch the load and panel board connections.

(d) Calculate the kilovolt-ampere rating of each transformer.

14-9 A factory has retained some 240-V, *two-phase* machinery that must be supplied from a 460-V, *three-phase* switchboard. The output of all two-phase motors is 85 hp. Assume that all the motors must operate simultaneously and are 80% efficient.

(a) Draw a diagram of the Scott transformer connections necessary.

(b) Calculate the kilovolt-ampere rating of the transformers.

(c) Calculate the line current in the three-phase feeder.

14-10 A factory has a 460-V, four-wire switchboard. The load is allocated as follows: motors, 75 kW at 0.75 pf, 460 V; fluorescent lighting, 60 kW at 0.9 pf, 265 V; heating equipment, 50 kW at 1.0 pf, 230 V; and incidental lighting and equipment, 15 kW at 1.0 pf, 115 V.

(a) Calculate the total power of the load.

(b) Design the feeders for the fluorescent-lighting load showing all currents. (Consider conductor ampacity only.)

(c) Design the transformers for the heating load using a delta-to-delta configuration. Show all currents, voltages and kilovolt-ampere ratings.

(d) Design the transformer for the 115-V load.

(e) Calculate the total kilovolt-ampere load.

(f) Calculate the total current at 460 V.

(g) Repeat (a), (b), (e), and (f) if 265-V heating equipment is used. (Combine with the feeders of (b).)

(h) Discuss the best arrangement.

14-11 A decision must be made regarding a 100-kVA fluorescent lighting load. Should 265- or 115-V units be installed? For convenience of operation, three panel boards are necessary, located in tandem 125 ft apart; the first is 125 ft from the 460-V switchboard. This will permit the branch circuits to be under 100 ft in length.

(a) Sketch the 460/115–230-V transformers and necessary feeders if they are located near the switchboard. Include the currents, voltages, and kilovolt-ampere ratings. Allow a 2% voltage drop.

(d) Repeat (a) but locate one 460/115–230-V transformer at each panel board location.

(c) If 265 V units are installed, sketch the feeders to the panel boards, using the four-wire, 265–460-V switchboard supply.

(d) If 265-V units are installed, repeat (b).

(e) Which is the best installation? Why?

14-12 A group of motors require 575 V and three phase. Only 460 V is available. The combined motor nameplate rating is 100 hp. Design the transformer bank; include currents, voltages, and kilovolt-ampere ratings. Assume a pf of 0.8; the efficiency of the motors is 85%.

Index